T0251277

Rivers

Originally published in 1982, this book presents a detailed review of alluvial river form and process and integrates the distinct but related approaches of geomorphologists, geologists and engineers to the subject. It outlines the environmental catchment factors that control the development of channel equilibrium and provides a detailed account of the sediment transport processes that represent the physical mechanisms by which channel adjustment occurs. Where possible it evaluates theoretical analyses in the context of the empirical evidence. *Rivers* is a valuable textbook for geomorphology students on advanced undergraduate courses on river behaviour and will also be of interest to students of hydraulics and sedimentology and to those concerned with civil and environmental engineering, river management and channel design, maintenance and management in the water industry.

Rivers

Form and Process in Alluvial Channels

Keith Richards

Routledge
Taylor & Francis Group

First published in 1982 by Methuen & Co. Ltd

This edition first published in 2024 by Routledge
4 Park Square, Milton Park, Abingdon, Oxon, OX14 4RN
and by Routledge
605 Third Avenue, New York, NY 10158.

Routledge is an imprint of the Taylor & Francis Group, an informa business

© 1982 Keith Richards.

The right of Keith Richards to be identified as the author of this work has been asserted by him in accordance with sections 77 and 78 of the Copyright, Designs and Patents Act 1988.

All rights reserved. No part of this book may be reprinted or reproduced or utilised in any form or by any electronic, mechanical, or other means, now known or hereafter invented, including photocopying and recording, or in any information storage or retrieval system, without permission in writing from the publishers.

ISBN 13: 978-1-032-73758-4 (hbk)
ISBN 13: 978-1-003-46579-9 (ebk)
ISBN 13: 978-1-032-73762-1 (pbk)
Book DOI 10.4324/9781003465799

RIVERS

*Form and process
in alluvial channels*

Keith Richards

METHUEN
London and New York

First published in 1982 by
Methuen & Co. Ltd
11 New Fetter Lane, London EC4P 4EE
Reprinted with revisions 1985

Published in the USA by
Methuen & Co.
in association with Methuen, Inc.
29 West 35th Street, New York, NY 10001

© 1982 Keith Richards

Typeset by Keyset Composition, Colchester, Essex
Printed in Great Britain at
the University Press, Cambridge

All rights reserved. No part of this book may be
reprinted or reproduced or utilized in any form or by any
electronic, mechanical or other means, now known or hereafter
invented, including photocopying and recording, or in any
information storage or retrieval system, without
permission in writing from the publishers.

British Library Cataloguing in Publication Data

Richards, Keith
Rivers.
1. Rivers
I. Title
551.48'3   GB561

ISBN 0-416-74900-3
ISBN 0-416-74910-0 Pbk

Library of Congress Cataloging in Publication Data

Richards, Keith
Rivers, form and process in alluvial channels

Bibliography: p.
Includes index
1. Rivers. 2. Channels (Hydraulic engineering)
I. Title
GV1205. R5   1982   551.4'4   82-8133
ISBN 0-416-74900-3   AACR2
ISBN 0-416-74910-0 (pbk.)

Contents

Acknowledgements

If the aesthetics of river channels appeal to the artist, their ability to create their own regular geometric forms fascinates the scientist. To write a book on the subject demands a somewhat obsessive fascination, which in my case can be at least partially attributed to the stimulating teaching of Dick Chorley, Bruce Sparks, David Stoddart and Barbara Kennedy, who nevertheless deserve no blame for the faults and errors I may have perpetrated. When I first wrote this in 1982, my colleagues were wrestling with government-inspired financial cuts, one of whose inevitable consequences will be to minimize the possibilities of maintaining the kind of active geomorphological research school of which I was fortunate to be a member. My fellow members – Rob Ferguson, Malcolm Anderson, Richard Jarvis, Alan Werritty, Brian Whalley and Bob Bennett – provided a fertile ground for testing and developing ideas. Fortunately, however, I continued to learn new aspects and techniques of geomorphology while lecturing at Hull University from 1978 to 1984, for which I thank Roger Arnett, Steve Ellis, John Pethick and Arthur Fraser, as well as Professor Harry Wilkinson for making available departmental facilities during the writing of this book. In production terms, I am endebted to Keith Scurr and Brian Fisher for the diagrams and Stella Rhind for the manuscript. Finally, my wife Sue's contributions have been enormous; she has bullied me into carrying on with repetitious fieldwork when I became bored with it (usually on the second day), provided 'market research' by reading each chapter (usually on completion, at about 11.00 pm), prepared the bibliography and even volunteered to compile the index.

To all these, my grateful thanks.

The publishers and I would like to thank the following publishers, organizations and individuals for permission to reproduce, quote, or modify copyright material. The figure numbers refer to the present publication. Full citations can be found by consulting captions and the reference list beginning on p. 306. Every effort has been made to identify original sources of illustrations and tables, but if there have been any accidental errors or omissions, we apologize to those concerned.

Addison-Wesley for Figure 6.3(c); Allen & Unwin for Figures 1.2(b) and (d), 1.3(a), 1.6(b) and (c), 2.7(c), 4.7(c), 7.6(c), 7.10(d), 9.1(b), 9.3(e), 10.1(a), 10.5(d), (e) and (f); the American Geophysical Union for Figures 2.2(c) and (e), 3.7(b), 4.1(c), 4.5(b), 5.3(c) and (d), 5.5(e) and (f), 6.7(d), 7.7(a), 8.2(f), 8.3(a), 8.7(b) and (e), 9.3(c) and 9.4(e); the *American Journal of Science* for Table 1.1 and Figures 1.7(d), 2.1(b), 3.7(a), 5.6(d), 6.1(c) and (d), 7.9(c), 7.10(c) and 8.8(a); the American Society of Civil Engineers for Figures 3.4(d), 6.2(d), 6.5(e), 8.7(a), 10.4(a), (b), (c) and (d), and 10.5(c); J. H. Appleton for Plate 10; Blackwell Scientific Publications for Figure 2.2(g); Cambridge University Press for Figures 2.1(c) and 7.12(e); Columbia University Press for Figure 10.5(a); Edward Arnold for Figures 3.2(b) and 4.3(d); Elsevier Scientific Publishing Company and the respective authors for Figures 1.7(b) (J. A. H. Brown), 4.1(d) (D. E. Walling and I. D. L. Foster), 4.1 (f) (I. D. L. Foster), 5.6(b) (G. Pickup and W. F. Warner), 5.6(c) (G. Pickup), 6.1(d) (C. C. Park), 7.10(b) (J. H. McGowen and L. E. Garner), and Plates 1(a) and 5 (M. D. Picard and L. R. High Jr); Exeter University for Figure 2.5(b); W. H. Freeman for Figures 1.2(a), 2.3(a) and 7.8(c) from *Fluvial Processes in Geomorphology* by L. B. Leopold, M. G. Wolman and J. P. Miller © 1964; Gebrüder Borntrager Verlags Buchhandlung for Figures 2.2(d), 2.3(b), 5.1(b) and 8.7(c); Geo Abstracts for Figures 9.4(a) and (e), and Plate 8(b); the Geologists' Association for Figure 9.5(e); the Geological Society of America and the respective authors for Figures 2.7(e) (S. A. Schumm), 6.6(d) (D. N. Wilcock), 7.1(c) (S. A. Schumm), 7.2(b) (T. Lisle), 7.8(e) (J. Lewin), 7.12(b) and (c) (S. A. Schumm), 7.13(b) and (c) (N. D. Smith), 8.2(d) (R. J. Chorley and M. A. Morgan), and 9.5(d) from the *Bulletin of the Geological Society of America*; Gower Publishing Company for Figure 5.5(d); the Institute of Australian Geographers for Figure 2.3(b); Sir C. C. Inglis for Figure 10.3(c); the Institute of British Geographers for Figures 2.7(d), 4.3(b), 7.4(d) and 9.3(a); the Institution of Civil Engineers for Figure 2.5(d); the International Association for Hydraulics Research for Figure 4.4(a); the International Association of Hydrological Sciences for Figures 1.1(b), 2.1(a), 2.4(d) and (e), 2.6(b), 5.5(c) and 5.6(a); E. W. Lane for Figure 10.2(g); J. D. Mollard and Associates for Plate 6; John Wiley and Sons for Figures 1.7(a), 4.2(a), 4.4(d), 6.5(a), 7.7(c), 8.4(f), 8.7(d), 9.2(b), 9.5(a), and 10.3(a) and (b); Kendall/Hunt Publishing Company, Iowa, D. D. Rhodes, Bray and Kellerhals, and Thorne and Lewin for Figures 1.2(c) and 6.5(b); Methuen for Figures 1.3(b), 7.1(a) and 7.8(d); the Office of Naval Research, Geography Branch, Washington, D.C. for Figures 2.3(c) and 8.2(e); the Purdue University Water Resources Centre for Table 10.1; the Royal Geographical Society and J. B. Thornes for Figure 6.4(a); the Royal

Society and respective authors for Figures 4.2(b) (J. E. Abbott) and 4.5(b) (R. A. Bagnold); the Royal Society of Edinburgh and B. J. Bluck for Figures 7.5(a) and (b); Scottish Academic Press and B. J. Bluck for Figure 7.10(a); H. W. Shen for Figures 4.5(d), 9.4(d), 10.2(a) and (g); the Society of Economic Palaeontologists and Mineralogists for Figures 1.5(c), 7.5(d), 8.4(a), (b), (c), (d) and (e); the Soil Conservation Society of America for Figure 4.2(e); Spence Air Photos for Plate 9; the Swedish Society for Anthropology and Geography for Figures 2.6(a), 3.9(d) and 9.3(d); the United States Department of Agriculture for Figures 3.8(a) and 4.6(d); the United States Geological Survey for Tables 6.2 and 7.2, and Figures 1.1(a), 2.6(d), 2.7(a), 3.3(b), (c) and (d), 3.5(a), 3.8(d), 3.9(a), 4.1(b) and (e), 4.2(c), 4.3(a), 4.5(a) and (c), 6.1(a), 6.3(b), 6.6(a), (c) and (e), 7.2(d), 7.6(d) and (e), 7.9(a), 7.11(a) and (b), 7.12(a), 8.1(a), 8.2(a), (b) and (c), 8.8(b) and (c), 9.1(a), 9.3(b), 9.4(b), 10.1(b) and 10.5(b); United States Weather Bureau for Figure 2.5(a); University of Chicago Press for Figures 7.13(a) and 8.5(b); and V. A. Vanoni for Figure 4.5(d); Waikato Geological Society for Figure 5.5(a).

Keith Richards
University of Cambridge
July 1985

1

Alluvial river channels: their nature and significance

Alluvial river channels are 'self-formed'. Their morphology results from the entrainment, transportation and deposition of the unconsolidated sedimentary materials of the valley fill and floodplain deposits across which they flow. Alluvial channel forms are dependent on the environmental controls of hydrology and sedimentology, and while these remain constant in a particular drainage basin, the river morphology remains stable even though the channel may not be static. This stability is reflected in numerous empirical generalizations which demonstrate both adjustment of river form

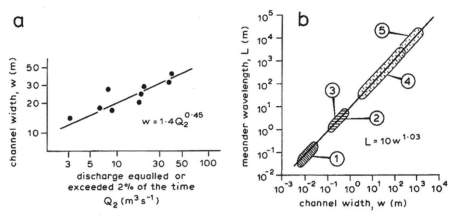

Figure 1.1 (a) Downstream adjustment of channel width to increasing discharge: principal gauging stations on the Brandywine Creek, Pennsylvania (after Wolman, 1955). (b) Relationship between meander wavelength and channel width (after Zeller, 1967). 1 = limestone furrow meanders; 2 = supraglacial stream meanders; 3 = laboratory channel meanders in sand; 4 = Swiss river meanders; 5 = alluvial river meanders (Leopold and Wolman, 1960).

to process (Figure 1.1a) and interrelationships between various aspects of channel form (Figure 1.1b). Equilibrium channel forms result from an interaction between the two sets of variables which measure the 'forces' applied by the flowing water and the 'resistance to erosion' of the sediments. All types of river channel can be referred to a continuum of force-resistance relationships, but those formed in unconsolidated alluvial sediments, with which the flow can readily interact, are distinguished by the adjustable nature of their morphology. Bedrock channels occasionally obey alluvial channel regularities (Figure 1.1b), but their forms are usually governed by lithological and structural influences. Even underfit alluvial streams (Dury, 1964a) with immobile residual bed material inherited from the coarse bedload of more powerful palaeochannels exhibit systematic adjustment of channel form to this sedimentological control (Wilcock, 1967).

The significance of alluvial river channels

In any catchment, areal slope erosion and linear river erosion yield sediment which the rivers transport from the various source areas to the ultimate sediment 'sinks' of the ocean basins. The products of catchment erosion are often only transported short distances to slope bases or floodplain surfaces, but rivers carry an average of 97 t km^{-2} yr^{-1} of suspended sediment and 37 t km^{-2} yr^{-1} of solutes to the sea (Holeman, 1968), much of which represents net denudation of the landscape. Chapter 2 briefly considers the enormous spatial variability of sediment yield, reflecting as it does the influences of climate, vegetation, relief, geology and man on the erosional processes which represent external controls of river channel behaviour. However, most of this book is concerned not with the denudational effect of river transport, but with the physical processes responsible for the observed regularity of alluvial channel behaviour. An understanding of these process–form relationships is of practical significance in watershed management, river management and design, and palaeohydraulic reconstruction. Although Chapters 9 and 10 consider these problems in detail, examples are given here to illustrate the importance of such understanding.

Watershed management

Optimal water resource development must often balance water and power supply with flood alleviation, navigation, recreation and conservation, which together necessitate planning at the catchment scale. Some objectives are met by controlling the river flow by physical structures (dams) which directly influence sediment transport processes. Alternatively, catchment land-use changes may alter hydrological response and sediment yield to affect river processes indirectly and possibly prejudice investment in capital projects.

Catchment management is particularly difficult in sensitive semi-arid environments. For example, gullying in the south-west USA in the early twentieth century reflected a web of causal influences including increased runoff caused by overgrazing and subtle climatic changes (Cooke, 1974). The Rio Puerco (Bryan, 1928) widened from 30 to 85 m and deepened from 1 to 8.5 m between 1845 and 1928, yielding 1200 m^3 yr^{-1} of sediment which aggraded the Rio Grande main stream and caused a 16% loss of storage capacity in Elephant Butte reservoir from 1915 to 1940 (Happ, 1948). Reservoir sedimentation, accelerated by injudicious land-use changes upstream, reduces storage capacity and hence the safe water yield. It thus has a major economic effect on reservoir operation by reducing revenue, shortening the project lifespan, and necessitating expenditure on sediment clearance when traps are installed. For example, even a small Pennine reservoir in Britain entailed a budget of £800 per annum in the 1960s for clearance of 250 m^3 yr^{-1} of bedload trapped on the main inflow stream (Morgan, 1980).

When the bedload is deposited in a reservoir, the clear released water may degrade the stream bed downstream until it is protected by coarse lag sediments. These represent that fraction of the bed material which is rendered immobile because of the reduction of peak discharges caused by the reservoir water storage. Average degradation below ten dams in the American Midwest over varying periods from 1910 to 1960 (Leopold *et al*, 1964, p. 454) was 0.4 m at an average rate of 0.04 m yr^{-1}. Figure 1.2(a) shows marked degradation closest to the dam site, possibly endangering the structure itself (Komura and Simons, 1967). Downstream channel instability may reflect reservoir management policy; deliberate flow regulation could increase the frequency of exceedance of the threshold stress for bed material transport while reducing the frequency of extreme events (Hey, 1976). However, this will depend on regional climatic conditions. In British Columbia increased winter flow in the regulated Peace River resulted in massive spring ice jams which caused flooding to unprecedented levels (Bray and Kellerhals, 1979).

River management and channel design

Widening, deepening or straightening of natural streams (channelization) have often been undertaken to improve navigability or accelerate the passage of flood peaks. Artificial channels are also constructed for irrigation and navigation. In both cases, channel width, depth and gradient are selected to pass the required discharge at a velocity sufficient to maintain transport of sediment without silting, but not so excessive that bed and banks are eroded. Channel design is most successful when it mimics nature because: (i) the resulting stability minimizes maintenance costs necessitated by silting or erosion; (ii) the aesthetics of the channel are enhanced (Leopold, 1969); and (iii) the aquatic ecology can be preserved by maintaining the range of

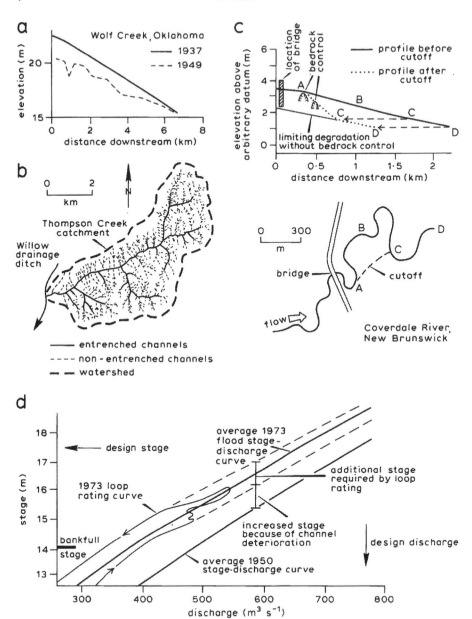

Figure 1.2 (a) River bed degradation downstream from Fort Supply Dam, Wolf Creek, Oklahoma (after Leopold *et al*, 1964). (b) Gullying and network expansion triggered by channelization, Willow Drainage Ditch, Iowa (after Ruhe, 1971). (c) An example of meander bend cut-off which could have endangered bridge foundations had there not been a bedrock control at the upstream end (after Bray and Kellerhalls, 1979). (d) Mississippi River stage–discharge curve showing channel deterioration and 'loop' effects on projected stage at design discharge for levee height (after Noble, 1976).

ecological niches found in natural streams with spatially varying depth, velocity and bed material (Keller, 1976).

Local channelization often ignores the upstream and downstream links of a reach which is part of a wider system. Straightening the Willow drainage ditch (Daniels, 1960) increased the rates of stream energy expenditure because the gradient was steepened, and incision of the channel bed rejuvenated tributaries and initiated gully extension of the drainage network (Figure 1.2b). When the Blackwater River, Missouri, was locally steepened from a gradient of 0.0017 to 0.0031 by straightening (Emerson, 1971), increased flooding occurred downstream, where aggradation of 2 m in 50–60 years resulted from erosion in the channelized reach. In addition, increased turbidity and loss of habitat caused a decline from 600 to 100 kg ha^{-1} in the fish population. The consequences of channelization for existing engineering structures are equally serious (Figure 1.2c). A cut-off on the East Prairie River, Alberta (Parker and Andres, 1976) locally doubled the gradient, and a 3 m high nickpoint created inadvertently at the upstream end receded at 1.6 km yr^{-1} to undermine a bridge pier. Consequently construction of a replacement and of protective stonework on the banks ('rip-rap') was required. Natural readjustment of the channel after 'improvement' also causes engineering problems, such as those of levée design highlighted by the 1973 Mississippi floods (Noble, 1976). A 1950 stage–discharge curve predicted a stage of 17.5 m at the design flood, and levée heights were designed accordingly. However, channel capacity in 1950 had been enlarged by channelization, and by 1973 deterioration resulted in 1–1.5 m higher stages at given discharges. In addition, the looped stage–discharge relation for the 1973 flood indicates a higher water surface elevation as the bed filled on the falling stage than during the rising stage scour. Thus a range of projected stages occurs at the design flood which exceeds the freeboard on the levées (Figure 1.2d).

Palaeohydraulic and palaeohydrological reconstruction

Actively migrating rivers create sedimentary structures which, when preserved in the sedimentary record by burial during aggradation, allow an identification to be made of the type of channel which created the deposits. Meandering streams have a wide range of particle sizes organized in fining-upwards cycles, from basal cobbles through sands to silts (pp. 206–11). Braided streams have coarse sediments (gravels) and structures dominated by lens-shaped channel fills caused by abandonment of braid distributaries (Allen, 1965). Recognition of these structures permits qualitative reconstruction of hydraulic and hydrological conditions; braided streams, for example, are associated with high stream power and unstable sediments, often unvegetated. Smaller-scale sedimentary structures allow more quantitative reconstruction; Figure 1.3(a) illustrates the bedforms associated with different stream power conditions. For a known sand size, this enables an estimation to be made of the range of stream power expenditure in a river characterized by certain bedform structures, with dune cross-bedding implying greater power than the smaller-scale cross-laminae formed by ripple migration.

Schumm's (1968a) study of palaeochannels on the Murrumbidgee riverine plain is a classic illustration of hydraulic and hydrological reconstruction. The two palaeochannels (Figure 1.3b) exhibit different forms and sediments. The older channel is wide, and sections show it to have been shallow; the channel fill is sand and gravel. Application of the Manning

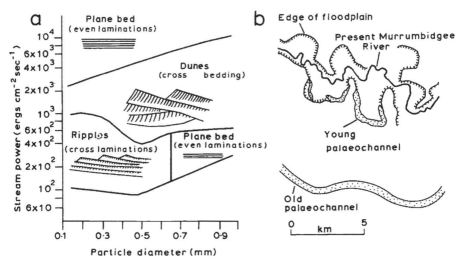

Figure 1.3 (a) The relationship between bedforms, associated bedding characteristics, and stream power and particle size (after Allen, 1970a). 10^7 erg = 1 joule (see Appendix). (b) Part of the riverine plain of the Murrumbidgee River, NSW, sketched from an air photograph (after Schumm, 1969a).

equation (p. 63) allows the velocity of flow to be estimated from the depth, slope and roughness, and multiplication of the velocity by the cross-section area yields an estimated bankfull discharge of 650 m^3 s^{-1}. The fluvial sediments correlate with saline palaeosols, and the channel is interpreted as a bedload-dominated stream of a drier Quaternary climate characterized by rapid runoff and sediment yield. The younger palaeochannel has a smaller width:depth ratio and channel fill deposits of silt and clay; in this case the estimated bankfull discharge is 1450 m^3 s^{-1}. The implied palaeohydrology involves a wetter climate than at present, but comparable vegetation and weathering regime (and alluvium). These palaeochannels preserved on alluvial surfaces are easy to investigate because of their observable continuity; in older, buried sediments and sedimentary rocks, there may only be isolated, discontinuous exposures. Nevertheless, Ferguson (1977a) has shown that the variance of the direction of a river channel is directly related to sinuosity, so that sinuosity can be estimated from the directions of a set of random channel exposures. In conclusion, we may note that the practical importance of palaeohydraulic reconstruction is exemplified by Schumm's (1977) predictions of the locations of heavy mineral placer deposits in floodplain and valley fill sediments.

The alluvial system

'Functional' and 'realist' theory dominate contemporary fluvial geomorphology. The former views reality as 'instances of repeatable and predictable regularities in which form and function can be assumed to be related' (Chorley, 1978, p. 2). Thus Figure 1.1(a) is considered an acceptable basis for estimating peak discharge from measured widths in ungauged catchments (Dunne and Leopold, 1978, p.643). 'Realist' theory eschews this black-box approach, and attempts 'the identification of detailed causal mechanisms and the underlying structures of which the external forms are the artefacts' (Chorley, 1978, p. 2). It is more concerned with explanation than with mere identification of relationships. Functionalism tends to be empirical and inductive in scientific method, whereas realism is theoretical and deductive, and is often focused at the smaller scale of specific processes (Chorley, 1978). Theoretical considerations of sediment transport and flow processes in hydraulics and fluid mechanics (Chapter 3) strengthen a largely empirical fluvial geomorphology by explaining some functional relationships identified empirically. However, realist models of the behaviour of the alluvial channel in its entirety are hampered by the multiplicity of variables, the complexity of feedback between them, and the indeterminacy of process–form relationships at this scale (pp. 24–8).

The predominance of functionalism has encouraged the use of systems analysis. Chorley and Kennedy (1971) define a system as a structured set of objects and/or attributes, which in the present case is the variable set defining alluvial channel form and process. Although criticized for adding nothing fundamentally new to our understanding (Smalley and Vita-Finzi, 1969), systems analysis has two important advantages. First, it emphasizes the measurement process, in which qualitative concepts are translated into quantitative variables, with clear operational definitions which minimize operator variance (Chorley, 1958). Second, it encourages systematic identification of all relevant variables in a given system, as well as their status as dependent or independent variables, and the hypothesized direction of causal links between them. Two types of particularly important systems are the 'morphological' and 'process-response' systems. In a morphological system, variables describing equilibrium channel form and associated sediment size and structure properties are interrelated, and perhaps correlated with the morphological properties of the catchment. The process–response system includes hydrological and sediment transport variables which characterize external, environmental process controls of the alluvial channel; potentially, such a system allows a prediction of the adjustment of the channel system to altered environmental conditions (Chapter 9). The significance of any variable can be defined only in the context of the system of which it is a component; hence, the whole system is greater than the sum of its separate constituents, particularly in terms of our understanding of its operation. Nevertheless, there are limitations to systems analysis, notably in its reliance on linear statistical methods to identify the direction and strength of relationships. Nonlinear hydraulic processes, such as the

Table 1.1 The varying status of river channel variables over different time scales (after Schumm and Lichty, 1965).

River variables	Status of variables during designated time spans		
	Geologic $(10^3 + \text{yr})$	Modern $(10^1 \text{ to } 10^2 + \text{yr})$	Present (1 to 10 yr)
Time	independent	irrelevant	irrelevant
Geology	independent	independent	independent
Climate	independent	independent	independent
Vegetation (type and density)	dependent	independent	independent
Relief	dependent	independent	independent
Palaeohydrology (long-term discharge of water and sediment)	dependent	independent	independent
Valley dimensions (width, depth, slope)	dependent	independent	independent
Mean discharge of water and sediment	indeterminate	independent	independent
Channel morphology (width, depth, slope, shape, pattern)	indeterminate	dependent	independent
Observed discharge of water and sediment	indeterminate	indeterminate	dependent
Observed flow characteristics (depth, velocity, turbulence, etc.)	indeterminate	indeterminate	dependent

variation of entrainment velocity with sediment size (Hjulström, 1935; pp. 80–1), may invalidate systems approaches, or at least restrict their range of application.

In defining the components of an alluvial system, the relevant time scale must be established, since the status of variables and the nature of their interrelationships vary according to the time scale adopted (Schumm and Lichty, 1965). Table 1.1 distinguishes geological, modern and present time scales. This division is arbitrary and qualitative; no absolute time span exists for each division, since this depends on local environmental conditions and the relaxation time (Howard, 1965) of the system. Measurable morphological change may occur in a sensitive semi-arid environment over periods of time when a stable humid landscape appears static. However, in general, alluvial channel forms are indeterminate over the long-term 'geological' time scale of W. M. Davis (1899) and the 'cycle of erosion', which is concerned with a hypothetical landscape evolution under assumed conditions of constant climate, base level and tectonic stability. On the short-term scale, channel forms are determined and independent, and day-to-day variations in streamflow result in varying depths, velocities and sediment concentrations as the water prism adjusts within the section. Over the intermediate, modern, time scale (10^1 to 10^2 + years), sometimes called the 'graded' time scale (Schumm and Lichty, 1965, Kirkby, 1977), equilibrium channel forms develop that are dependent both on the mean discharge of sediment and water from the catchment upstream and on the valley characteristics inherited from the longer time span, particularly the gradient and sedimentology of valley fill deposits. This is the time scale with which this book is particularly concerned.

Morphological variables

Channel geometry is three-dimensional, and the cross-section, plan-form and long-profile properties constituting the complete morphology are closely interrelated; to emphasize one dimension alone (e.g. the long profile) is to achieve only a partial understanding. Wide, shallow cross-sections are associated with braided patterns, and meandering streams have lower gradients than straight channels between the same two end points. To simplify the systematic analysis of channel form, Chapters 6, 7 and 8 deal with the three dimensions in turn, while providing constant reference to their interrelationships.

Figure 1.4 and Table 1.2 identify the major morphological variables. The variables describing the cross-section form relate to its scale and shape. A particular problem here is the operational definition of the section, which is normally defined up to the level at which overbank flow occurs – the 'bankfull' section. However, this must be considered in relation to the magnitude–frequency properties of flows capable of filling the channel to this level (or 'stage'), which are defined by morphological, vegetational and sedimentological criteria (Chapter 5). Some channel properties duplicate information: for example, $p \approx w$ and $R \approx \bar{d}$. However, wetted perimeter and hydraulic radius have hydraulic significance as measures both of the extent of solid–fluid contact and of the channel efficiency (a large cross-section area relative to perimeter length implying less frictional retardation of the flow). Plan form is quantified according to the single-thread:multi-thread dichotomy. The scale and shape measures for single-thread channels in Figure 1.4(b) are most appropriate for smooth, regular meanders; other measures are described in Chapter 7, particularly those based on series analysis (Ferguson, 1975). Figure 1.4(c) indicates a method for obtaining a

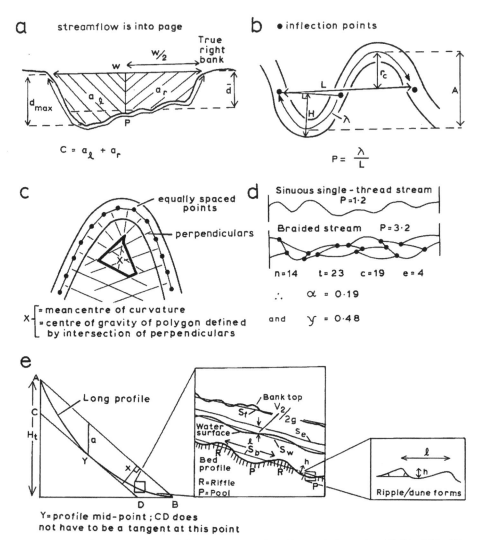

Figure 1.4 (a) Variables describing cross-section size and shape (see Table 1.2); (b) Variables describing meander bend geometry (see Table 1.2); inflection points marked ● (c) Definition of mean centre of curvature in a non-circular bend. (d) Sinuosity and topological indices of braided stream morphology (see Table 1.2). (e) Long-profile geometry at the three scales of the complete profile, the individual reach, and the riffle–pool and ripple–dune bedforms.

mean centre of bend curvature, necessitated by the non-circular nature of many bends. The only index applied to both single-thread and multi-thread channels is 'sinuosity' (Figure 1.4d; Le Ba Hong and Davies, 1979). Finally, long-profile geometry is strongly scale-dependent (Figure 1.4e). Some indices refer to the overall profile form and are useful as generalized

Table 1.2 A summary of the main morphological variables describing river channels (see Figure 1.4).

Variable	Symbol	Definition	Diagram	Comments
Cross-section:				
Width	w		1.4(a)	scale (channel size) variable
Mean depth	$\bar{d}$	$= C/w$	1.4(a)	scale
Maximum depth	d_{max}		1.4(a)	scale; ratio $\bar{d}/d_{max}$ measures shape
Channel capacity	C	$= w\bar{d}$	1.4(a)	scale; cross-section area
Wetted perimeter	p		1.4(a)	scale; $\approx$ width in wide, shallow sections
Hydraulic radius	R	$= C/p$	—	scale; channel efficiency; $\approx$depth in wide, shallow sections
Form ratio	F	$= w/\bar{d}$ or w/d_{max}	1.4(a)	shape; Schumm (1960a) employed d_{max}
Section asymmetry	A_s	$= a_1/a_r$	1.4(a)	shape; ratio of larger to smaller sub-areas (Milne, 1979)
Section roughness	C_R	$= \sigma_{di}/\bar{d}$	—	shape; coefficient of variation of measured depths
Plan form:				
Axial wavelength	L		1.4(b)	bend scale variable
Arc wavelength	λ		1.4(b)	scale
Bend amplitude	A		1.4(b)	scale (lateral development of bend)
Arc height	H		1.4(b)	scale (lateral development, especially of asymmetric bends)
Radius of curvature	r_c		1.4(b,c)	scale; mean or minimum used for non-circular bends
Radius:width ratio	r_c/w		—	shape; dimensionless measure of bend tightness
Sinuosity	P	$= \lambda/L$ for bend	1.4(b,d)	shape; normally channel length divided by reach length
α-index	α	$= \dfrac{t - (n+e) + 1}{2(n+e) - 5}$	1.4(d)	shape; braid intensity; ratio of observed no. of islands to maximum possible for given no. of nodes (n)

Table 1.2 *continued*

Variable	Symbol	Definition	Diagram	Comments
γ-index	γ	$= \dfrac{t}{3(n + e - 2)}$	1.4(d)	as above, but for no. of segments (t); $t = e + c$, e = no. of bisected segments, c = no. of complete segments (Howard *et al.* 1970)
Braid intensity	B_1	$= \dfrac{2 \, (\text{total bar length})}{\text{reach length}}$	–	Brice (1960)
Long profile:				
Profile gradient	S_{1085}		–	Benson (1959), NERC (1975); gradient from 10% to 85% of length upstream from basin mouth
Taylor–Schwartz slope	S_{TS}	$\dfrac{1}{\sqrt{S_{TS}}} = \dfrac{1}{n} \sum \dfrac{1}{\sqrt{S_i}}$	–	NERC (1975); gradient giving same travel time as original profile
Profile concavity	C_L	$= 2a/H_t$	1.4(e)	Langbein (1964b) AB connects headwater and mouth
Profile concavity	C_w	$= x/H_t$	1.4(e)	Wheeler (1979) CD parallel AB, touching profile
Bed slope	s_h		1.4(e)	scale-dependent; needs generalization over bedforms
Water surface slope	s_w		1.4(e)	scale-dependent; defined over reach or locally over bedform
Energy gradient	s_c		1.4(e)	scale-dependent, energy head $= v^2/2g$ at each section
Floodplain slope	s_f		1.4(e)	difficult to define if irregular bank top heights
Bedform wavelength	l		1.4(e)	scale-dependent; use for riffle–pool or ripple–dune forms
Bedform amplitude	h		1.4(e)	scale-dependent; riffle–pool or ripple–dune forms

measures, for example, for flood prediction (NERC, 1975). These may be measured from cartographic or air photograph sources. However, at the scale of individual reach gradients or bedform properties, such as the riffle–pool wavelengths, field survey by levelling and sounding (perhaps by sonic means, Richardson *et al*, 1961) is essential.

Independent 'control' variables

These morphological variables adjust in response to the interaction between variables which reflect independent hydrological and sedimentological aspects of the catchment. In turn they define the fluid force exerted by the streamflow and the resistance to morphological change of the materials in which the channel is formed. However, some variables are semi-independent – adjustable within imposed limits – and these particularly include channel roughness characteristics.

Peak flow events which are competent to transport sediment reflect catchment hydrology; the distribution of their magnitudes is usually characterized by an index of 'dominant' discharge (pp. 135–45) which is commonly treated as a major independent variable controlling channel form, particularly cross-section and bend scale. In fact, the quantity of water imposed on the channel is less important than its capacity to do work. Work done in raising water above a datum gives it potential energy, P_e, which is the product of the weight of water (mass m multiplied by the gravitational acceleration g) and height above the datum (h), that is

$$P_e = mgh. \tag{1.1}$$

As runoff occurs this potential energy is released and converted to kinetic energy, K_e, where

$$K_e = \frac{1}{2}mv^2 \tag{1.2}$$

(v is the velocity), and thence to work involved in transporting sediment, overcoming frictional resistance and generating heat. The 'bucket of water' example in Figure 1.5(a) illustrates loss of P_e over a vertical distance AC; here the mass of water is its density, ρ_w, multiplied by the volume V of the bucket. Embleton and Whalley (1979, p.14) estimate that 37,000 km³ of average annual global runoff from an average height of 800 m provides a total energy of 3×10^{20} J yr^{-1}. In Figure 1.5(a), a discharge of Q m³ s^{-1} occurs over a unit incremental reach AB. At A, P_e is continually replenished by inflow from upstream at the rate $\rho_w gQh_1$, which is measured in J s^{-1} (see Appendix). Between A and B, loss of P_e is

$$\rho_w gQh_1 - \rho_w gQh_2 = \rho_w gQ(h_1 - h_2) = \rho_w gQs \tag{1.3}$$

which is the rate of potential energy expenditure per unit length of channel (the rate of doing work, or stream power). Table 1.3 defines some

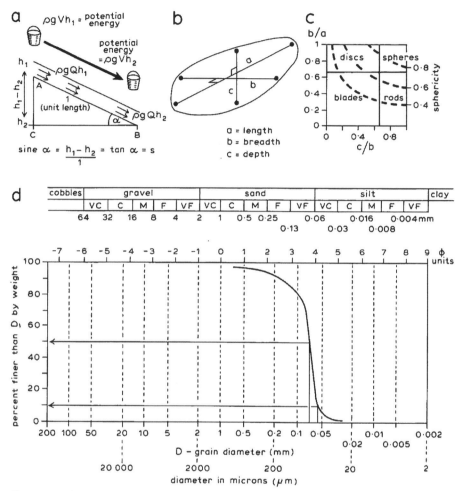

Figure 1.5 (a) Definition diagram illustrating loss of potential energy of streamflow as water travels over an incremental reach of unit length inclined at an angle $\alpha°$. (b) Orthogonal long (a), intermediate (b) and short (c) axes of a pebble, measurable by calipers for gravel coarser than about 8 mm (-3ϕ units). (c) Zingg classification of three dimensional shape (Zingg, 1935), with Krumbein's (1941a) sphericity measure (see Table 1.3). (d) Cumulative grain size distribution; equivalent quantitative sizes for clay, silt, sand and gravel in mm, μm and ϕ units. The example is a well sorted, very fine sand, median diameter (D_{50}) = 0.08 mm (3.8 ϕ units), with 10% silt and clay. VC = very coarse, C = coarse, M = medium, F = fine, VF = very fine.

alternative power criteria, such as power per unit bed area (Bagnold, 1966) which is related to bedload transport rates (pp. 114–17; see Figure 1.3a).

Sediment properties interact with stream power to determine the rates of sediment transport and the stability of the channel perimeter. The direct influence of geology on an alluvial stream is limited, but lithology, structure and weathering regime combine to control the nature of alluvium (Hack,

Table 1.3 A summary of the main independent and semi-independent variables, reflecting stream power and sediment properties (see Figure 1.5).

Variable	Symbol	Definition	Diagram	Comments
Stream power:				
Power per unit length (maximum)	Ωp	$= \rho g Q s_v$	1.5(a)	s_v = valley slope, or steepest possible gradient
Power per unit channel length	Ω	$= \rho g Q s_e$	1.5(a)	s_e = channel slope (energy gradient); channel bed gradient (s_b) is usually an acceptable approximation; τ = mean bed shear stress
Power per unit bed area ω		$= \rho g Q s_e / w =$ $\rho g w d v s_e / w = \tau v$		
Power per unit weight of water	ω'	$= \rho g Q s_e / w d = \rho g v s_e$	—	
Sediment size:				
Long axis	a		1.5(b)	for gravel particles sampled individually; i.e. normally >8 mm (3ϕ units)
Intermediate axis	b		1.5(b)	
Short axis	c		1.5(b)	
Median diameter	D_{50} or ϕ_{50}		1.5(d)	obtained from cumulative grain size distribution
Additional percentile measures	D_{84} or $\phi_{84\,(eg)}$		1.5(d)	
Percent silt and clay	B		1.5(d)	
Sediment distribution characteristics:				
Sorting index	S_0	$= \dfrac{\phi_{90} + \phi_{80} + \phi_{70} - \phi_{30} - \phi_{20} - \phi_{10}}{5.3}$	—	measures variability; if well sorted, $S_0 < 0.5$; if poorly sorted, $S_0 > 1$
Skewness index	S_k	$= \dfrac{\phi_{84} - \phi_{50}}{\phi_{84} - \phi_{16}} - \dfrac{\phi_{50} - \phi_{10}}{\phi_{90} - \phi_{10}}$	—	measures asymmetry of distribution; positive skewness implies a coarse tail to the distribution, negative a fine tail to the distribution, if size distribution is based on % *finer* than stated sizes; symmetry gives $S_k = 0$; range is -1 to $+1$

Table 1.3 continued

Variable	Symbol	Definition	Diagram	Comments
Kurtosis index	K_g	$= \dfrac{\phi_{90} - \phi_{10}}{1.9(\phi_{75} - \phi_{25})}$	—	measures peakedness of distribution; a poorly sorted sediment has a flat (platykurtic) distribution ($K_g < 0.9$); well-sorted sediment is leptokurtic ($K_g > 1.1$)

Particle shape of gravel: (see Figure 1.5c for shape classification based on axis length ratios)

Sphericity	ψ	$= \sqrt[3]{\left(\dfrac{bc}{a^2}\right)}$	1.5(c)	Krumbein (1941a); ranges from 0 to 1, a sphere having a value of 1
Roundness	R	$= \dfrac{2r}{a}$	—	Cailleux and Tricart (1963); $r =$ radius of curvature of sharpest corner in the plane of maximum projection; a highly 'spherical' particle may have low 'roundness' (e.g. a cube), which measures degree of smoothing by attrition
Flatness	F	$= \dfrac{a+b}{2c}$	—	Cailleux (1945); minimum value (1) for equi-dimensional particle, increasing as particles are more disc-like

1957), which may however be sorted by the stream. Sediments influence two forms of resistance, namely, resistance to flow and resistance to entrainment and erosion. Resistance to flow ('channel roughness') affects friction losses and reduces the energy available for transport. 'Skin resistance' is dependent on channel perimeter sediment sizes, sand grains being intrinsically less 'rough' than cobbles. For larger particles (coarser than gravel) shape is also important: discoid pebbles lying flat and overlapping one another project their short axes into the flow, so that although their *b* axes define their size, their *c* axes control their roughness (Figure 1.5b). However, 'form resistance' arises when particles are organized into bedforms (morphological variables in Figure 1.4e, Table 1.2). These increase flow resistance and illustrate the ability of the flow to alter its energy losses by modifying the spatial arrangement of the sediment; hence the concept of 'semi-independent' variables.

Resistance to transport and erosion is also dependent on sediment properties. Chapters 3 and 4 show that bedload and suspended sediment transport rates depend on the sediment size distribution, particle shape, and the packing and arrangement of bed material. Bank erosion reflects grain size distributions when particle detachment occurs, and bulk mechanical properties (pp. 164–7) if large-scale collapse occurs. Since the latter are difficult to measure, size distribution properties are again used as simple approximations (e.g. the percentage of silt and clay; Table 1.3).

To represent the grain size distribution is problematic. Sampling considerations involve sample quantity and site. Quantities required range from about 100 g of well sorted sand to several kilogrammes of bulk sample including pebbles. Wolman (1954) recommends a systematic-random grid-based sampling design to define sampling locations, particularly for gravel. Measurement techniques must reflect the size fraction involved, for example, axis measurement (Figure 1.5b) for gravel, sieving for sand, and the hydrometer or pipette methods based on sedimentation in a still-water column (British Standards Institution, 1975, Akroyd, 1964) for silts and clays. Gravel measurement is usually based on size distribution by numbers of particles, whereas the sand, silt and clay fractions are based on weight distribution. Although Leopold (1970) and Kellerhals and Bray (1971) have suggested conversion factors to enable comparison, these factors assume spherical particles and are of limited use. The application of Stokes' Law (pp. 77–8) in sedimentation also assumes standard particle shapes. Thus it is debatable if a complete size distribution curve (Figure 1.5d) based on three different techniques can be justified. Normally, gravel is treated separately from the fine fraction, with distinct parameters describing the gravel and sand–silt–clay sediments. Size distributions are often log-normal (having a coarse tail) and so sizes are represented by the ϕ unit log-transformation

$$\phi = -\log_2 D, \qquad (1.4)$$

where D is the diameter in millimetres (Figure 1.5d). A range of one ϕ unit at the coarse end of the scale implies a larger absolute size range than at the fine end; this emphasizes the greater physical significance of a given absolute size difference (e.g. 1 mm) among fine sediments rather than among coarse. Size distribution parameters measuring central tendency (median), sorting, skewness and kurtosis (all based on ϕ unit percentiles) are given in Table 1.3. Also listed are some shape measures (sphericity, roundness and flatness) which are particularly applicable to gravel and coarser particles. Figure 1.5(c) illustrates a classification of particle shape based on axis length ratios, and its relationship to the continuous sphericity index.

Equilibrium of alluvial river channels

Alluvial river channels carry a sediment load and erode their bed and banks locally in space and time. However, as they migrate across their valley surfaces, they may maintain stable, or equilibrium, average forms unless during migration they encounter different bank materials or have a new discharge regime imposed on them. The nature of this equilibrium state, defined variously as 'dynamic(al) equilibrium', 'quasi-equilibrium', 'grade', 'regime' and 'steady state' in an extensive literature (Dury, 1966a), can be simply defined with reference to temporal scales of channel adjustment.

Time scales and channel change

Figure 1.6(a) illustrates schematically the possible temporal variations of a single index of channel morphology (width). In the short term, cross-section width varies continually. For example, bank collapse during and after peak flows causes temporary widening. This may be balanced by slower depositional processes, especially in asymmetrical sections on active meanders, where the outer bank retreat is balanced by scroll-bar deposition on the inner bank (Hickin, 1974). In the long term (pp. 8–9), these random variations occur about a progressive trend in mean width caused by climatic change, tectonic influence, or long-term changes in valley gradient reflecting the long, slow, process of landscape degradation. This may be defined as 'dynamic equilibrium' because the stream continually adjusts to maintain equilibrium with its environment. Schumm (1975) suggests a discontinuous long-term equilibrium – dynamic, metastable equilibrium – influenced by thresholds in the fluvial system (Figure 1.6a). For example, sediment output from a rejuvenated laboratory basin is observed to oscillate as a complex response to the initial stimulus of base level reduction (Figure 1.6b). Main-stream deposits are flushed out first, rejuvenating tributaries which then increase sediment yield to the main stream, where a period of sedimentation results. Eventually a threshold main-stream gradient is reached and a new phase of degradation begins; a damped cyclic variation of sediment output thus occurs. In the long-term evolution of a drainage basin, temporal variation of valley floor elevation might be as shown in Figure 1.6(c). Intermittent isostatic uplift compensates for valley floor lowering. This uplift then triggers a complex response, in which bursts of erosion of the main valley floor are separated by depositional phases in the main valley when tributary erosion occurs. These phases are ended, and renewed main valley incision begins, when the deposits are built to a threshold slope (Schumm, 1975).

Over an intermediate, 'modern' time scale environmental controls and

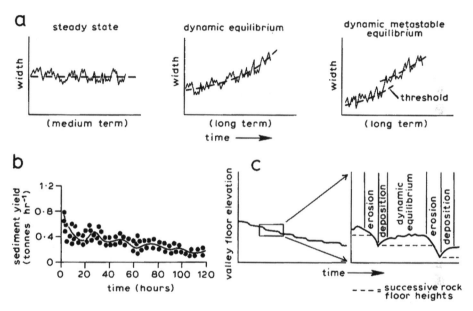

Figure 1.6 (a) Schematic illustration of types of equilibrium in relation to time scales. (b) Sediment yield from experimental basin after rejuvenation (after Schumm, 1975). (c) Suggested episodic erosion of valley floor, with detail of pulsed nature of erosion reflecting coupling of main and tributary valleys (after Schumm, 1975).

resultant channel forms remain roughly constant on average, although short-term random events predominate and even make the distinction between a time trend and stable morphology difficult to define. This condition of stable equilibrium, but not static forms, may be termed 'steady state' (Figure 1.6a). The maintenance of steady state is by internal system adjustments known as 'negative feedback'. For example, Guglielmini (1697) noted that scour of a stream bed reduces the force of the water because it increases cross-section area and reduces velocity. King (1970) recognized that this would initiate renewed fill to maintain the equilibrium form. If Schumm's (1975) model is appropriate, however, disturbance of steady states may occur even if environmental conditions are fixed, because of the complex internal response of the fluvial system to external stimuli. Nevertheless, Mosley and Zimpfer (1978) suggest that evidence of complex response may be damped in natural basins, compared with the experimental catchments where it has been observed, as well as being obscured by the effects of Quaternary climatic change. Absolute time scales in Figure 1.6(a) again vary according to the force:resistance ratios of fluvial systems, and although relatively rapid adjustments occur to abrupt environmental changes (which are often man-made) these may involve time lags, which can be measured by application of a rate law and calculation of a half-life (Graf, 1977) defined as the time required for half of the morphological adjustment to be made between two equilibrium states.

Measurement of the equilibrium state

To define the nature of equilibrium it is necessary to consider means of measuring its existence (Dury, 1966a). A number of approaches are discussed below.

(a) *Temporal variation of channel form*   Blench (1969, p. 2) defined 'regime' channels as those which 'adjust themselves to average breadths, depths, slopes and meander sizes that depend on (1) the sequence of water discharges imposed on them, (2) the sequence of sediment discharges acquired by them from catchment erosion, erosion of their own boundaries, or other sources, and (3) the liability of their cohesive banks to erosion or deposition'. While these factors are steady, constant average values of channel form variables occur which define a regime state analogous to climatic averages for a period of time. Thus, 'the relationship between process and form is stationary and the morphology of the system remains relatively constant over time' (Knox, 1975, p. 179). This constancy may be measured to define the steady state, although continuous data on stream morphology are rarely available over a long period. Ferguson (1977b) notes that discrete measurements of channel pattern from different map editions can give misleading evidence. Along an actively migrating reach, occasional bend cut-offs abruptly reduce the reach sinuosity, but amplitude growth on other bends then increases it again. A reach may have constant average sinuosity (Figure 1.7a), but a few random 'snapshots' on old maps may seem to imply a time trend. Identification of steady state is thus difficult in active streams, where equilibrium is dynamic in the two senses that stable average forms persist during migration and that stability is maintained by dynamic processes of sediment transport and deposition. The term 'steady state', however, is here preferred for a state of temporally constant average morphology.

(b) *Continuity of sediment transport*   Equilibrium of form is associated with equilibrium of process, implying temporal and spatial sediment

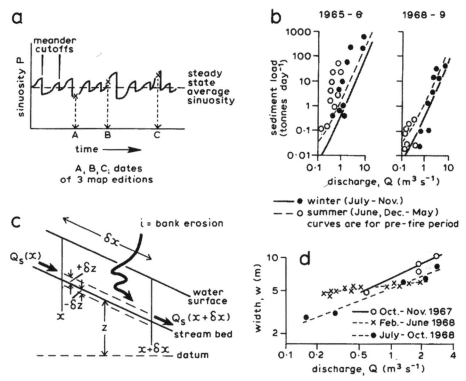

Figure 1.7 (a) Schematic illustration of temporal variation of sinuosity about steady stage average; measurements from maps suggest a time trend (after Ferguson, 1977b). A, B and C are the dates of three map editions. (b) Sediment-rating curves for an Australian basin, and sediment yield after a bushfire in 1965 (after Brown, 1972). (c) Definition diagram for continuity equation of sediment transport. (d) Temporal variation in width–discharge relationship at a single section (after Knighton, 1975a).

transport continuity. Brown (1972) analysed the effect of a bush fire on sediment yield from an Australian catchment monitored prior to this disturbance. Figure 1.7(b) shows the vastly increased erosion immediately after the fire, reflecting the bare surface and accelerated runoff, and the rapid return to the pre-existing equilibrium sediment rating after a few years of vegetation regrowth. Temporal stability of the process equilibrium is here more evident than morphological equilibrium, since gullies take more time to 'heal' than to erode. Mackin's (1948, p. 471) definition of a 'graded' stream stresses spatial continuity of transport, the graded stream being 'one in which, over a period of years, slope is delicately adjusted to provide, with available discharge and with prevailing channel characteristics, just the velocity required for the transportation of load supplied from the drainage basin.' This emphasizes average conditions over a period as in Blench's

(1969) definition of regime, but assumes channel gradient to be the crucial adjustable variable, with cross-section characteristics being predetermined rather than mutually adjustable with the long profile and plan form. Figure 1.7(c) illustrates the manifestation of this form of sediment transport continuity along an incremental reach δx in length and of elevation z at time t. Sediment inflow is $Q_s(x)$, and i is the input from bank erosion and slope runoff within the reach, per unit length; the output is $Q_s(x + \delta x)$. Over a period of time δt, total sediment gain within the reach is the difference between $Q_s(x)$ and $Q_s(x + \delta x)$ over that period, plus the total input from the banks. This would result in a change of bed elevation, or sediment storage, over the reach, and thus a change of channel form. This can be defined as $\delta_z \delta_x$ if the reach is treated two-dimensionally. Thus

$$[Q_s(x) - Q_s(x + \delta x)]\, \delta t + i\delta x\delta t = \delta z\, \delta x. \tag{1.5}$$

This is a two-dimensional sediment continuity equation. Dividing by δx and δt throughout and making them vanishingly small gives

$$-\frac{\mathrm{d}Q_s}{\mathrm{d}x} + i = \frac{\mathrm{d}z}{\mathrm{d}t}. \tag{1.6}$$

At equilibrium $\mathrm{d}z/\mathrm{d}t = 0$ (i.e. no change in storage and hence in bed elevation). Thus sediment transport varies along a reach by as much as is required to compensate for bank erosion. If this is also zero, the continuity equation demands a constant rate of transport along the reach, which can be empirically measured as a test of local reach equilibrium.

(c) *Efficiency of channel form*   A system in equilibrium might be expected to operate with maximum efficiency, but component relationships of complex systems may require conflicting characteristics. The most efficient cross-section for conveyance of flow, for given cross-section area and sediment roughness, is semi-circular since this has minimum hydraulic radius. However, unconsolidated sediments cannot sustain this shape, which is not optimal for bank materials. Natural cross-sections thus represent a compromise, so it is difficult to establish optimal forms. Given the variability of cross-section forms necessitated by different bank materials, all else being equal, several equilibrium slopes can maintain transport continuity – hence the use of the term 'quasi-equilibrium' to define the variable, shifting equilibrium of river channels (Langbein and Leopold, 1964). Kirkby (1977) argues that, in complex systems involving many processes, the maximum efficiency criterion only applies to a single limiting process, which for an alluvial river is the transportation of the load imposed on each reach with the available discharge. With load and discharge imposed, and hence sediment concentration (load per unit discharge) an independent variable, channel forms adjust to maintain this concentration on the minimum possible gradient, which implies minimum rate of energy loss at constant discharge (see Equation 1.3). Transport of sediment at a minimum rate of energy loss implies maximum efficiency, although a weakness of this model is the fact that work done in transportation only accounts for part of the total energy loss; as much as 96% may be dissipated in overcoming frictional resistance (Rubey, 1933b), and adjustments of form roughness may change rates of energy loss without altering slopes (Martinec, 1972).

(d) *Correlation of system properties* Strong intercorrelation of system variables (Figure 1.1) may be diagnostic of equilibrium, although relationships between channel form and process variables are often equally strong in aggrading and degrading streams. Knighton (1975a) measured stream width at sections on the River Bollin, Cheshire, at various discharges over 18 months, and showed that distinct width–discharge relationships occurred in the data for separate periods at a given section (Figure 1.7d). This may be interpreted in various ways. Firstly, it could reflect equilibrium in a multivariate system, with an unmeasured variable taking different values between each period and causing different width–discharge relations. Secondly, random bank collapse and small-scale temporary changes in channel form – scatter about a steady state – could be responsible. Thirdly, it could be a systematic adjustment in response to changing environmental conditions. This dynamic equilibrium interpretation is supported by Mosley's (1975a) evidence of widening of this river because of increased flood magnitudes caused by upstream land-use changes. Clearly these various interpretations of scatter in Figure 1.7(d) make it difficult to justify the definition of weak correlation as indicative of disequilibrium, the corollary of the initial premise.

Each criterion of steady state thus suffers ambiguity and needs to be qualified by the other criteria. Such practical problems in testing the existence of equilibrium explain the semantic difficulties (Dury, 1966a) which have characterized theoretical definitions since Davis (1902a, p. 86) applied the term 'grade' to a quantitatively undefined 'balance between erosion and deposition attained by mature rivers'. Steady state involves multivariate adjustment of three dimensions of channel form to allow transportation of sediment load from upstream with discharges imposed by the catchment, which implies ultimate control by environmental catchment properties. Discharge of water and sediment vary continually, which necessitates time-averaging of morphological and process variables. Perfect instantaneous 'balance' is inconceivable (Kesseli, 1941) but also irrelevant; steady state is identifiable over the 'modern' time scale (Schumm and Lichty, 1965). Furthermore, although 'grade' and 'slope' are not synonymous, post-Davisian concentration on long-profile adjustment (Kesseli, 1941, Mackin, 1948) fails to acknowledge the interrelationships of channel form (p. 20) recognized in Blench's (1969) definition of regime.

Indeterminate alluvial channel systems

Theoretical appraisal of basic system types should accompany the practical application of systems analysis to alluvial channels. 'Open' (non-isolated) systems are distinguished from 'closed' (isolated) systems by the passage of energy, or energy and matter, across their boundaries, and by the inter- action between the system and its environment. Chorley (1962) argued that long-term, Davisian models of river system development implied closed system thinking, since the end-product of a cycle of erosion, the peneplain, is compatible with the maximum entropy state achieved by thermodynamic decay in a closed system (Leopold and Langbein, 1962). After initial uplift, the hypothetical progressive denudation in the geomorphic cycle levels down potential energy differences in the landscape and reduces the capacity for work in the system. This interpretation has, however, been challenged (Chisholm, 1967, Conacher, 1969), partly because environmental systems at a sub-global scale are necessarily open (Dooge, 1968) and partly because quantitative evidence is lacking which demonstrates that the *relative* energy distribution is less uneven in low-relief than in high-relief landscapes.

However, alluvial rivers are clearly open systems over the intermediate 'modern' time scale (pp. 18–20) when continual energy inputs maintain the system in an equilibrium state and prevent decay to maximum entropy. Rivers result from the concentration of overland flow in the terrestrial phase of the hydrological cycle, and precipitation provides continual renewal of potential energy at higher altitudes, which is then dissipated as the water travels downslope, being converted to kinetic energy and work (in trans- porting sediment). At the catchment scale sediment inputs only occur as solutes in precipitation, which may represent a signficant element in geo- chemical budgets. At the river reach scale (Figure 1.7c), sediment input and output are both evident, while the disposition of sediment stored within the reach defines the channel morphology, and the imbalance of input and output results in changes of storage and channel form (erosion occurring if output exceeds input, deposition if input exceeds output). The major problems involved in treating the channel open system at this scale are to determine: (i) the catchment changes likely to disturb channel system equi- librium and cause imbalance in sediment transport rates from reach to reach; and (ii) how change in sediment storage within a reach leads to changes in channel form depending on the disposition of eroded or deposited sediment.

A general solution to these problems is inhibited by the multivariate nature of the catchment process controls and the channel geometry (pp. 9–17), as well as by the indeterminate behaviour of alluvial channels with mobile, erodible boundaries. Indeterminacy can be illustrated by the simple case of a straight, single-thread channel reach with uniform cross- section. The independent variables controlling the geometry of this reach are the discharge Q, the bed material size D_{50} (which influences the skin resistance), the slope s_b which is defined by the direction of the reach across the valley surface, and the input of sediment load at the upstream end (which in an equilibrium reach equals the output). With load and discharge imposed, the sediment concentration C (load per unit discharge) is fixed. The adjustable dependent variables in the reach are thus width w, depth $\bar{d}$ and mean flow velocity $\bar{v}$, and hydraulic roughness of the section f_f (the Darcy–Weisbach friction coefficient; see p. 64). These are the four 'degrees of freedom' of the river reach (Langbein, 1964a) and they can only be defined uniquely if there are four independent equations containing them, with the known values of the independent variables. It is only possible to define three equations: the continuity equation of flow

$$Q = w\bar{d}v; \tag{1.7}$$

the Darcy–Weisbach equation specifying a flow resistance law

$$f_f = \frac{8g\bar{d}s_b}{\bar{v}^2} \tag{1.8}$$

in which g is the gravitational acceleration (9.81 m s^{-2}); and an expression for sediment concentration. There are many transport equations, but Leopold *et al* (1964, p. 269) suggest one based on the Bagnold (1966) theory that transport rates reflect power expenditure per unit bed area, and the negative influence of bed material size and skin resistance. Thus

$$C \propto \frac{\bar{v}^{1/2} s_b^{3/2}}{f_f^2 \bar{d}^{1/6}}. \tag{1.9}$$

In applying these equations to a particular reach Q, g, s_b and C are given and w, $\bar{d}$, $\bar{v}$ and f_f are unknown. As there is one less equation than unknown variables, simultaneous solution is impossible, the system is indeterminate, and the reach geometry cannot be predicted. More generally, natural rivers are not straight and uniform, and Hey (1978) suggests that they have nine degrees of freedom, which allows adjustment to w, $\bar{d}$ (or wetted perimeter, p, and hydraulic radius, R), $\bar{v}$ and d_{max}. Cross-section shape is not specified by w and $\bar{d}$ alone; for example, rectangular and semi-circular channels could have the same widths and mean depths but different maximum depths. In addition to these cross-section variables, a sinuous channel may adjust sinuosity P, meander arc wavelength λ, and channel slope s_b. Slope here is not independent, since although valley slope represents a constraint, being the steepest possible channel slope, a sinuous channel has a lower gradient between two points because the sinuosity $P = s_v/s_b$, the ratio of valley to channel gradients. The final two adjustable variables are the wavelength l and amplitude h of sand bedforms, which reflect the rate of sediment transport and influence the 'form resistance' of the bed. Indeterminacy therefore reflects the lack of definitive relationships of (i) the friction factor to composite skin and form roughness effects (Maddock, 1970); (ii) the channel cross-section shape to bed and bank material properties; and (iii) the sinuosity and meander wavelength to stream power and sediment properties.

Figure 1.8 is a speculative representation of the alluvial channel system, with the indeterminate links between variables suggested by double arrows and signs. (The reader might care to evaluate this diagram critically after

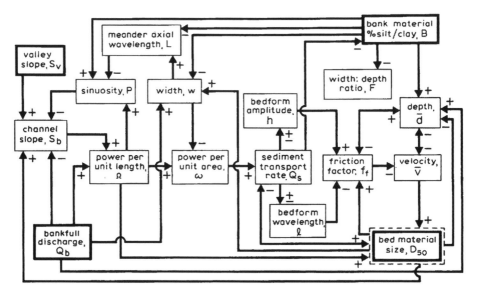

Figure 1.8 The alluvial channel system. Independent variables have heavy outlines; bed material size, though ultimately controlled by lithology, is semi-independent as it is affected by sediment transport. Direct relationships are shown by +, inverse by −, placed by arrow showing direction of influence. Some links are reversible; for example, as friction increases so does depth of flow, but increasing flow depth drowns skin resistance and reduces friction.

reading Chapters 4, 6, 7 and 8!) As well as inhibiting development of a general theory of equilibrium channel form and process, indeterminacy raises philosophical issues. Four sources of variance are inherent in river channels (Leopold and Langbein, 1963): random effects such as tree throws, measurement error, lack of physical relationships, and deviation of individual streams from the mean condition of steady state (Figure 1.7a). Thus, alluvial channel behaviour is stochastic (probabilistic), and although probable tendencies (average states) may be predicted, individual instantaneous conditions may not. Hey (1979) argues qualitatively that deterministic prediction is theoretically possible given the necessary additional, as yet undiscovered, physical laws. A pragmatic compromise (Watson,

1969) argues that deterministic modelling would be feasible even at the microscale, but that practical data-gathering limitations preclude this. The precise track of a gully is physically determined, but only predictable with detailed data on rainfall distribution, antecedent moisture, the distribution of soil and vegetation characteristics, and microlithological variation over a slope surface. Thus it may be necessary to utilize some broad assumptions about average river behaviour in order to limit the degrees of freedom.

One such assumption is provided by the 'minimum variance theory' (Langbein, 1964a), which argues that on average dependent variables adjust simultaneously to accommodate changes of stream power, with all variables changing equally within the limits imposed by the known physical relationships. For example, stream power changes downstream, and it may be represented by the discharge–slope product Qs_b (Equation 1.3). Thus

$$Qs_b = wd\bar{v}s_b \qquad (1.10)$$

demonstrates that four adjustable variables can accommodate downstream changes of power. The theory argues that this is achieved by mutual adjustment within constraints imposed by Equations (1.7)–(1.9), rather than by adjustment of any single dependent variable. Minimum total variance occurs when all variables adjust as equally as possible. A second set of basic assumptions relate to specific energy criteria. For example, Langbein and Leopold (1964) argue that equilibrium average channel forms result from a compromise between those adjustments of channel geometry needed to minimize total work in the system (and therefore equalize power expenditure per unit length) and those which minimize variation in power expenditure per unit bed area (see pp. 227–8). Alternatively, Yang (1971a) suggests that adjustments of dependent variables occur so that the imposed water discharge and sediment concentration can be transported with the minimum unit stream power (power expenditure per unit weight of water) under given geological and climatic constraints. The rationale for this hypothesis is that a given sediment experiences a threshold power expenditure above which it is subjected to entrainment and transport. Excessive rates of power expenditure locally therefore result in changes in channel form because of this transport, until power expenditure is reduced and the channel stabilized at a minimum power compatible with the threshold for the sediment, which is a geological constraint (Bull, 1979). For example, power expenditure per unit bed area ($\omega = \rho g Q s / w$) is high in a narrow channel, so the channel may be unstable and subject to widening until a smaller value of ω is reached and the channel stabilized. The negative feedback loop involving channel slope, stream power (Ω) and sinuosity in Figure 1.8 also illustrates this form of adjustment. However, minimum energy criteria have three important limitations: first, the minima are not always clearly defined, so a wide range of channel forms approximately satisfy a specific criterion; second, the actual minimum is an undefined function of geological constraints, so the method merely side-steps the indeterminacy issue; and third, minimization of an energy criterion is itself not the physical cause of the adjustments made. The processes of sediment transfer actually cause the adjustments, so that evident minimization may be interpreted as an effect of adjustment as much as a direct cause.

2

The drainage basin: environmental controls of the river channel

The drainage basin forms a fundamental spatial accounting unit in hydrological studies, since the catchment represents an open system (Dooge, 1968) within which moisture inputs, outputs and storages can be monitored to define the water balance equation:

$$\begin{array}{llll}
\text{Input} & = & \text{Output} & \pm & \text{Storage changes} \\
[\text{Rainfall}] & = & [\text{Streamflow} + & \pm & [\text{Groundwater and soil} \\
& & \text{Evapotranspiration}] & & \text{moisture storage changes}]
\end{array} \qquad (2.1)$$

(Penman, 1950). In fluvially eroded landscapes, weathering of bedrock and transfer of sediment both occur through the action of water moving through the terrestrial phases of the hydrological cycle, and as a result the drainage basin has also been considered a fundamental geomorphological unit (Chorley, 1969). The water and sediment yield of a catchment influence alluvial river morphology and process, and are focused on the stream by the catchment topography and the effect of gravity. However, continuity of sediment transfer within the drainage basin is rare, particularly where a floodplain exists and the river is a true alluvial channel, because limited direct contact occurs between the river and the slopes, and the products of slope erosion accumulate as colluvial deposits at the junction between floodplain and valley side. This lack of continuity is measured by the 'sediment delivery ratio' (Roehl, 1962), which is the percentage of total catchment erosion transported from the basin. In large catchments there are proportionally more locations for temporary sediment storage, so the delivery ratio decreases with catchment size (Figure 2.1a), but for a given

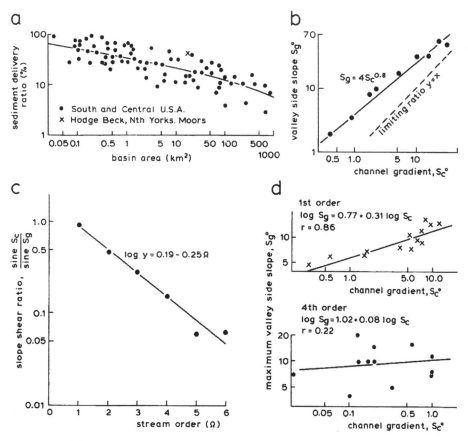

Figure 2.1 (a) The relationship between sediment delivery ratios and basin area (after Roehl, 1962). (b) The relationship between mean valley side slope and mean basal stream gradient (after Strahler, 1950). (c) The channel slope:valley side slope sine ratio as a function of stream order (after Carter and Chorley, 1961). (d) The channel slope–channel gradient relationships for streams of first and fourth order.

size it increases with overall catchment slope. In the Hodge Beck, North Yorkshire, 18,000 t yr^{-1} were eroded from unvegetated moorland in 1967, and 3500 t yr^{-1} of river bank erosion occurred (Imeson, 1974). Suspended sediment yield of 9100 t implied a delivery ratio of 42%, a slight underestimate since bedload yield was not measured.

Mutual adjustments of the morphological components of the drainage basin occur – in its slopes, river channels and drainage network structure – to maintain continuity of transfer of water and sediment, but these are less apparent where a floodplain exists. Strahler (1950) argued that high sediment yield from steep, valley side slopes (of angle S_g) demands a steep channel gradient for continuity of transport, resulting in a direct relationship between regional *average* slopes and gradients, that is

$$S_g = 4S_c^{0.8} \tag{2.2}$$

where channel gradient S_c is here measured in degrees (Figure 2.1b). This adjustment occurs because aggradation in the valley bottom steepens the channel gradient until the slope sediment yield can be transported. Carter and Chorley (1961) find no evidence of this relationship for *individual* streams and their adjacent slopes, and several factors account for this failure. First, streams can adjust their cross-sections as well as gradients to maximize sediment transport capacity, and secondly, the sediment yield from a slope depends on the transport process (creep or wash), so that for a given slope angle and area different channel gradients may be required for each slope process (Richards, 1977a). Thirdly, the ability of the stream to transport sediment depends on stream power, which is a function of discharge and gradient. Discharge increases downstream and the Strahler stream order (see pp. 32–3) is an imperfect surrogate index of discharge. The ratio of the sines of channel and valley side slopes decreases systematically with stream order (Figure 2.1c), reflecting a mutual adjustment in which large streams (sixth order) can transport sediment supplied from valley sides of given slope on a lower channel gradient than small streams (first order). Finally, the correlation between gradients may be strong for headwater streams of low Strahler order, but much weaker for higher-order streams in which floodplain development isolates the stream from the valley side slope (Figure 2.1d). Absence of basal erosion 'fossilizes' slope profiles, which retain features imposed by prior morphogenesis (e.g. periglacial erosion in mid-latitude areas), since present erosion by soil creep or wash is extremely slow; for instance, denudation rates of 0.8 mm and 0.03 mm per 1000 years have been attributed to these two slope processes following measurements in southern Scotland (Kirkby, 1967). Alluvial channel morphology adjusts to contemporary processes of water and sediment discharge, but present slope process rates reflect inherited profile gradients. In each case, adjustment of form and process is evident at the local scale (Slaymaker, 1972), but the directions of causal linkage between form and process differ.

Reversal of cause and effect at different scales thus complicates the relationship between a river and its catchment. For example, in the long term the spatial patterns of runoff and sediment production determine the evolution of drainage network structure. However, over the intermediate time scale of the river-channel steady state the catchment water and sediment delivery represent direct environmental controls of the river channel which are focused onto the river by the drainage network. The purpose of this chapter is therefore to concentrate on those aspects of the drainage basin which are of particular relevance to the alluvial channel equilibrium. The essential underlying control of the drainage network is considered in terms of both its structural arrangement, which controls exploitation of the spatially-distributed runoff and sediment sources, and its density, which influences the intensity of runoff and sediment yield to the channel. Secondly, drainage-basin processes which convert rainfall into runoff are examined. Slope hydrological processes, which vary regionally in relation to climate, relief, soils and vegetation, determine the form of the quickflow hydrograph which is then transformed as it is routed through the network. Thus downstream reaches are influenced by flood hydrographs which reflect the yield of storm runoff from the hillslopes modulated by network properties. Finally, the quantity and quality of river sediment load are discussed. These reflect the prevailing slope hydrological processes and the varying spatial distribution of sources for different sediment types. The longer-term history of sediment yield is, of course, an ultimate cause of the alluvial valley fill deposits across which the modern river flows, and in which its morphology has developed

Drainage network structure and density

Evolution and incision of a drainage network creates the essential framework which integrates the processes of water and sediment transfers through a fluvially eroded landscape. The network consists of a hierarchical system of 'links', classified by Strahler order (Strahler, 1952) or Shreve magnitude (Shreve, 1966, 1967), and connecting nodes which are 'sources', 'junctions', or the basin outlet. Exterior links (headwater streams) are initiated at sources and terminate at the first junction, and have both an order (Ω_s) and magnitude (n) of unity (Figure 2.2a). The magnitude of an interior link, which is a stream segment between two junctions, is simply the sum of the magnitudes of the tributaries at its upstream junction. It follows from this that a network of magnitude n consists of n exterior links, $n - 1$ interior links and $2n - 1$ links in total; all exterior links contribute in the definition of magnitude. A Strahler stream segment, however, may consist of several links. Strahler orders are defined as follows: if two Strahler streams of order Ω_s

Figure 2.2 (a) Definitions of network properties. (b) Topologically distinct channel networks of magnitude 9; illustrating the concept of 'generations' and varying structures (d is diameter, $\bar{p}_c$ is mean source height). (c) Mean source height ($\bar{p}_c$) increasing with magnitude along the mainstream; maximum, minimum and expected values, and a plot for the White Esk River, Scotland (after Jarvis and Werritty, 1975). (d) Cumulative catchment area downstream along the main channel of a Spanish ephemeral stream (after Butcher and Thornes, 1978). (e) Hydrograph form and the frequency distribution of distances to sources from the outlet of the basin concerned (after Rogers, 1972). (f) Network width as a function of generation number for networks in (b). (g) Relationship of paired laminae in ephemeral stream deposits to upstream tributary junctions (after Frostick and Reid, 1977).

join, they create a link of order $\Omega_s + 1$; if two streams of different order join, the resultant link has the order of the tributary of higher order. The relationship between numbers of links (y) and Strahler segments (x) in a network of order Ω_s is

$$y = 2x - (2^{\Omega_s} - 1) \tag{2.3}$$

(Coffman *et al*, 1972), which relates order and magnitude, since $y = 2n - 1$. Magnitude is a more consistent measure of basin size than order. Figure 2.2(b) shows three networks of similar size in terms of

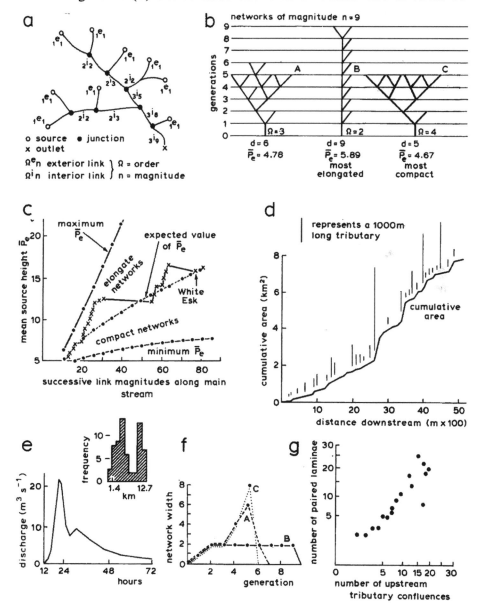

numbers of links, but whose varying structural organization results in different Strahler orders, because low-order 'adventitious' streams fail to change the higher order of the main streams they join.

All analyses of drainage network properties suffer from three related problems of network identification. First, networks are normally defined from secondary sources such as air photographs and topographic maps, preferably at scales of 1:24,000 (USA) and 1:25,000 (UK). Omission of headwater streams occurs with decreasing map scale, so network order is a function of map scale (Scheidegger, 1966). Furthermore, the degree of omission on maps of a given scale appears to increase with the ruggedness of the terrain (Eyles, 1966). Werritty (1972) suggests that only 1:2500 maps represent a satisfactory substitute for field mapping in relation to measurements of exterior link lengths. Second, the network is temporally dynamic, expanding and contracting during and after rainfall (Gregory and Walling, 1968). The static representations on various topographic maps may relate differently to the dynamic variations of network extent, which is particularly difficult to define when the successive integration and disintegration of discontinuous headwater channels occurs (Day, 1978). Longer-term temporal contraction of the network produces dry valleys, and it may be considered preferable to use contour crenulations to define a static valley network which measures total fluvial dissection, rather than the arbitrary blue line channel network which relates in an unknown way to variations of network extent (Gregory, 1966). However, this introduces the third problem, namely, that of operator variance (Chorley, 1958). The stream source is rarely apparent from the contours alone and it may be necessary to define arbitrarily the head of an exterior link using a valley floor slope criterion (Shreve, 1974).

Drainage network structure

Horton's (1945) 'laws' of drainage network composition relate numbers, lengths and catchment areas of stream segments systematically to stream order, but have little physical significance, being mathematical abstractions identifiable for any branching network subjected to the *ordering* scheme outlined above (Milton, 1966). Furthermore, the relevant structural indices such as the bifurcation ratio ($N_{\Omega_s}:N_{\Omega_s+1}$, the ratio of numbers of streams of ascending order), and equivalent length and area ratios, fail to correlate with hydrological and sedimentological phenomena (Smart, 1978). The alternative model assumes that network topology and geometry evolve at random in the absence of climatic, structural or topographic constraints. First, given *n* sources, there is a range of unique network arrangements (topologies) that can be created by joining the exterior links they initiate; these are Topologically Distinct Channel Networks (TDCNs; Shreve, 1966). The topology of the network in Figure 2.2(a) is reproduced as a simple connected graph in Figure 2.2(b) with two alternative TDCNs of the same magnitude. In the absence of constraints, each TDCN is equally probable, but some network 'types' (of a specific Strahler order) are more likely than others. The number of TDCNs of magnitude n is

$$N(n) = \frac{1}{2n-1} \left(\frac{(2n-1)!}{n!(n-1)!} \right) \qquad (2.4)$$

(Shreve, 1966), giving 1430 TDCNs of magnitude 9. Table 2.1 shows that there are fourteen ways of creating a fourth-order network from nine sources (by varying the position of the 'excess' tributary in Figure 2.2b), but the most probable type of network is of third order. The second element of the model is the assumption that interior and exterior links have lengths and

Table 2.1 The probabilities associated with various Strahler stream number sets generated from nine sources according to the random network topology hypothesis.

$\Omega_s =$	Stream number set				Number of TDCNs		Probability	
	1	2	3	4				
	9	4	2	1	14		0.0098	
	9	4	1		56		0.0392	
	9	3	1		560	1288	0.3916	0.9007
	9	2	1		672		0.4699	
	9	1			128		0.0895	
					1430		1.0000	

catchment areas which are random variables drawn from separate probability distributions (Smart and Werner, 1976). Thus, in the absence of physical constraints, network structures appear to develop by a random combination of links of randomly varying geometry, and there is no systematic variation of link length or area with position in a basin in which environmental conditions are uniform. The topological structure underlies the geometry of a network, and thus for a network of magnitude n, mean link length $\bar{l}$, and link catchment area $\bar{a} = k\bar{l}^2$ (where $k \approx 1.5$), the total basin area A is

$$A = \bar{a}(2n-1) = 1.5\bar{l}^2(2n-1). \qquad (2.5)$$

Figure 2.2(b) shows that Strahler order depends on network size and structure; ideally, structural indices should be scale-free and designed to provide information of greater hydrological significance. For example, Figure 2.2(b) illustrates the definition of successive 'generations' as the network branches headward; the network 'diameter' (d) is the maximum generation number, or the maximum link distance from source to outlet.

Mean source height $\bar{p}_e$ (mean exterior path length) is the mean number of links between the sources and the outlet. Both indices clearly relate to the travel time of floodwater through the network; they are scale-dependent, but for any network magnitude the maximum (most elongated), minimum (most compact) and mean expected values can be calculated (Werner and Smart, 1973). Figure 2.2(c) shows how the observed mean source height of networks which are tributary to the main-stream links of a single river can be compared with these theoretical values as a means of defining the downstream changes in network structure as tributary networks combine (Jarvis and Werritty, 1975).

The network structure influences hydrological and sediment transport processes, particularly in small drainage basins, and these in turn affect channel forms. Peak discharges imposed on river channels at various points within a network depend on the distribution of contributing area along the main stream. Thus, although there is an *average* relationship between catchment length (L) and area (A) which, over a wide range of catchment sizes, is

$$L = 4.63A^{0.47} \tag{2.6}$$

(Mueller, 1972), within a catchment the area–length relation is strongly stepped, with large increases in contributing area where major tributaries enter the main stream (Figure 2.2d). Rogers (1972) correlated the frequency distribution of metric distances from headwater-stream sources to the basin outlet with the hydrograph shape. It was found that unimodal distributions generate hydrographs with single peaks as long as the storm area exceeds the basin area, while a bimodal source–distance distribution produces a double-peaked hydrograph (Figure 2.2e). This correlation partly reflects the choice of class interval in the histogram, but a topological distribution can also be shown to control hydrograph form. 'Network width' is the number of links in successive generations from the outlet (Figure 2.2f). Kirkby (1976) shows that the input of a hillslope hydrograph to the network produces an output hydrograph whose shape mirrors the headward distribution of links: a long, narrow basin creates a flat hydrograph, while the time of rise of the hydrograph from a compact basin reflects the distance headward to the maximum number of links. Sedimentological influences are less apparent, but Frostick and Reid (1977) show that paired laminae of light and heavy grains in deposits of an arid stream in Kenya are generated by pulses of flow from tributaries, so that the number of laminae at a sampling site correlate with the number of tributaries joining the main channel upstream (Figure 2.2g). Douglas (1975) provides evidence of pulses of suspended sediment concentration reflecting tributary inputs. Leopold and Miller (1956) show that channel width is related to stream order, which is a scale effect, rather than a function of network structure. However, in Chapter 6 a width-link magnitude correlation is suggested which implies that channel width along the main stream increases particularly where large increases of magnitude occur at major tributary junctions.

Network density

Under constant climatic and hydrological conditions, the stream network density adjusts to an equilibrium conditioned by a balance between channel head stream erosion and depositional infill by slope processes (Calver, 1978). Woldenberg (1971) suggests that the equilibrium density reflects a compromise between the minimization of energy losses of flow over slopes and in channels. High drainage densities minimize energy loss during overland flow or throughflow on the resultant short slopes, while channel flow benefits from economies of scale which reduce frictional losses and favour a low density of large streams, which are best created by rotund basins which concentrate flow rapidly. Establishment of the equilibrium is shown by drainage densities on dated tills in Iowa (Ruhe, 1952), which imply rapid, initial network expansion followed by stabilization after about 20,000 years. This inference from space–time substitution assumes that the tills have similar runoff–production characteristics, and places considerable weight on the validity of point A in Figure 2.3(a). Surface runoff generation is the crucial control of drainage density D_d, which is equal to the total channel length divided by the basin area. High runoff rates cause headward extension and bifurcation during channel-cutting events which increase channel density; a small area is required to sustain a stream, and drainage density is high. The amount of rainfall therefore strongly influences drainage density, and Figure 2.3(b) shows the variation with mean annual rainfall based on the work of Abrahams (1972) and Gregory and Gardiner (1975). Fine-textured networks with densities of 15–150 km^{-1} occur in semi-arid environments where high-intensity, localized rainstorms combine with thin soils and vegetation cover to encourage high runoff rates. Schumm (1956) measured ultra-fine textures in badlands developed on land-fill at Perth Amboy and found that densities ranged from 310–820 km^{-1}. Humid temperate environments have coarse-textured networks ($D_d < 5$ km^{-1}), although Abrahams' (1972) curve suggests minimum densities in areas with a mean annual rainfall of about 450 mm. Alternative rainfall indices have been considered as controls of network density, including a measure of rainfall intensity (Chorley and Morgan, 1962) and a precipitation-effectiveness (P-E) index defined as

$$P\text{-}E = 115 \sum_{}^{12} [0.02P/(T+12.2)]^{1.11} \tag{2.7}$$

which adjusts monthly precipitation (P, mm) by a function of mean monthly temperature (T, °C) to account for evapotranspiration loss (Melton, 1957, 1958). The relation between drainage density and the P-E index (Figure

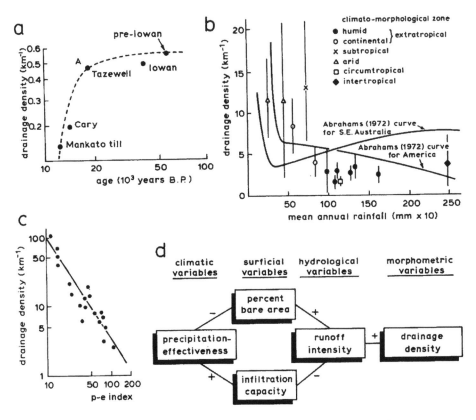

Figure 2.3 (a) Establishment of equilibrium drainage density inferred from measurements on tills of varying age (after Ruhe, 1952, and Leopold *et al.*, 1964). (b) Global variation of drainage density as a function of mean annual precipitation (after Gregory and Gardiner, 1975 and Abrahams, 1972). Each vertical line represents the range of values reported by a particular data source. (c) Drainage density as a function of precipitation-effectiveness (*P-E*) (after Melton, 1957). (d) Correlation structure relating climatic, surficial, hydrological and morphological variables. Based on Melton (1958).

2.3c) is a summary of a sequence of links between climatic, surficial, hydrological and morphological variables (Figure 2.3d). High effective precipitation results in a dense vegetation cover (low per cent bare area) and high infiltration capacity; this means low runoff intensity and, as a result, low drainage density. Lithology also influences drainage density, and for a given rainfall, runoff rates and densities are higher in areas with impermeable soils derived from clays and shales than where permeable soils are developed from sandstones. Although network density is controlled by runoff production in the long term, it in turn exerts some control over the form of the flood hydrograph in the short term, independently of the effect of network structure. Short slopes reduce the travel time of water moving to the river channels. and peak discharge (mean annual flood) per unit area increases approximately as the square of drainage density (Carlston, 1963, Howe *et al*, 1967). In addition, dense networks encourage rapid hydrograph rise, thereby reducing the time-to-peak.

Runoff processes and the flood hydrograph

Peak discharge events, through which the catchment hydrology influences channel form, result from a concentration of storm rainfall or snowmelt by catchment processes. Flood hydrographs (Figure 2.4a) consist of quickflow and delayed flow, and although much of the former component originates close to the basin outlet, both components are often genetically associated with specific flow processes. Rapid overland flow concentrates slope runoff into the quickflow peak, while delayed flow is sustained into dry periods by slow subsurface processes (throughflow and groundwater flow). Processes generating the quickflow peak vary, however, according to environmental conditions of soil, vegetation and topography (Figure 2.4b), as described below.

(a) *Infiltration–excess overland flow*   When storm rainfall intensity (i, in mm h^{-1}) exceeds soil infiltration capacity (f_p, mm h^{-1}), surface runoff occurs simultaneously over the catchment slopes (Horton, 1933, 1945). Initial infiltration rates (f_0) are high but decline to a limiting value (f_c) when the soil is saturated, after an initial time delay (t_0), while depression storage is filled and the soil is wetted, so that

$$f_p = f_c + (f_0 - f_c)\ \exp^{[-k(t - t_0)]} \tag{2.8}$$

(Johnson *et al*, 1980). The excess rainfall (Figure 2.4c) is assumed to become overland flow, which travels at velocities of up to 200–500 m h^{-1} in a thin film to the slope base. Within an hour on slopes of 200–300 m, therefore, a steady hillslope discharge per unit contour length (q_0) of

$$q_0 = (i - f_c)a \tag{2.9}$$

occurs (a is the area drained per unit contour length). This process, which feeds the stream discharge hydrograph rapidly, is thought to occur mainly in semi-arid environments (Kirkby and Chorley, 1967) or where natural vegetation has been cleared.

(b) *Partial area streamflow generation*   Thick, well-vegetated soils in humid temperate regions usually have high infiltration capacities: Weyman (1975) quotes values of 200 mm h^{-1} in the Mendip hills, England, where the maximum recorded rainfall intensity is 70 mm h^{-1}. The generation of infiltration-excess overland flow from an entire drainage basin is therefore improbable.

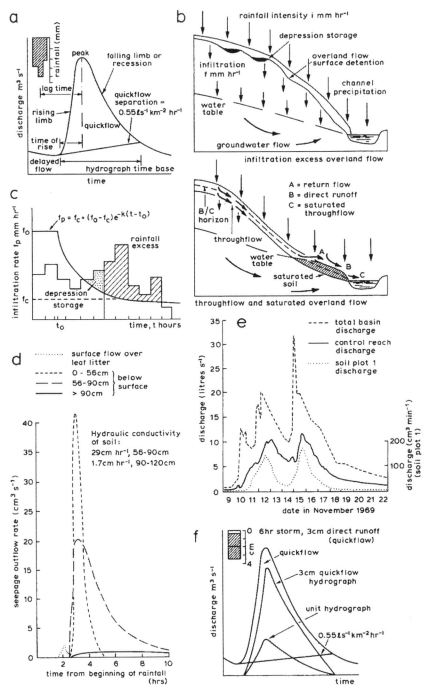

Figure 2.4 (a) Basic properties of a flood hydrograph. (b) Hydrological pathways over a slope: infiltration excess overland flow, throughflow and saturated overland flow. (c) Rainfall excess resulting from comparison of hourly rainfall intensity and infiltration rates as they vary through a storm. (d) Subsurface stormflow from an experimental laboratory slope (after Whipkey, 1965). (e) A flood event in the East Twin Brook, together with slope throughflow variation along a control reach and from a soil plot (after Weyman, 1970). (f) Derivation of a 6-hour unit hydrograph.

However, Betson (1964) suggested that surface runoff may be generated from a small proportion of the total topographic catchment (as little as 4%), particularly from areas with impermeable soils having thin organic A-horizons and from areas with high antecedent moisture content close to streams and in topographic hollows (Anderson and Burt, 1978). Rainfall therefore encounters a far more variable spatial distribution of infiltration capacity and antecedent moisture than the infiltration theory of runoff envisaged.

(c) *Saturation overland flow*   Surface runoff occurs at the slope foot where soils become saturated early during storms (Figure 2.4b). This runoff includes 'return flow' (upslope throughflow emerging at the surface) and direct runoff of precipitation onto the saturated soil (Dunne and Black, 1970a,b). As rainfall continues, the saturated area expands upslope, and thus the contributing area is dynamic or variable, as well as partial. This mechanism of supplying quickflow occurs particularly in humid areas with thin soils and in gentle concave footslopes and wide valley bottoms. It is a significant contributor to quickflow because soil saturation occurs at the slope foot (but may expand upslope), and the runoff created travels straight to the stream rather than infiltrating in more permeable soil downslope.

(d) *Subsurface stormflow*   Whipkey's (1965) experiments demonstrated the occurrence of lateral subsurface flow at a permeability discontinuity within the soil profile (Figure 2.4d), and in spite of the slow speed of throughflow (up to 400 mm h^{-1}) it has been considered a major contributor to peak flow generation in humid environments, especially on steep slopes where the soils are thick, permeable and have marked soil horizon development (Weyman, 1973). The resultant peak events are inevitably more delayed and attenuated than those generated by overland flow, unless non-capillary structures and macro-pores in the soil become integrated into a system of natural subsurface pipes (Jones, 1971) within which flow velocities of 200 m h^{-1} are possible. As the water table rises and the saturated area expands during rain, the discharge of saturated throughflow per unit length of slope foot (q_s) increases according to Darcy's law

$$q_s = -Kh\tan\theta, \qquad (2.10)$$

where K is the hydraulic conductivity, h the saturated zone thickness and tan θ the water-table gradient (negative in the direction of decreasing elevation). Figure 2.4(e) shows that increased saturated throughflow can respond on a similar time scale to that of the streamflow hydrograph (probably because of displacement of existing soil moisture; Hewlett and Hibbert, 1967), but it is quantitatively insufficient to account for the flood peak. Its main role is in increasing the potential area adjacent to the stream where saturated overland flow can occur, although it clearly maintains the prolonged flow recession during dry weather as the saturated wedge steadily contracts (Hewlett and Hibbert, 1963, Anderson and Burt, 1977).

Whichever processes generate the quickflow component of the hydrograph, it is often useful to separate quickflow and delayed flow to estimate the quickflow (flood) volume. Most separation methods are arbitrary, and a velocity-based method using a separation line inclined at $0.55 \, l \, s^{-1}$ per km^2 of catchment area per hour (Hewlett and Hibbert, 1967; Figure 2.4f) is often used in flood studies in small catchments. Process-based separation can be achieved using water chemistry measurements (Kunkle, 1965, Pinder and Jones, 1969). Total discharge at time t (Q_{Tt}) may be considered to be the sum of groundwater and surface runoff, which are each characterized by relatively constant, measurable, solute concentrations (C_G and C_S). Total solute load is the product of concentration and discharge, and is also the sum of groundwater and runoff loads. By measuring C_{Tt} and Q_{Tt} during a flood, separation into unknown groundwater (Q_{Gt}) and runoff (Q_{St}) component discharges is achieved by solving the simultaneous equations

$$\left.\begin{array}{c} C_G Q_{Gt} + C_S Q_{St} = C_{Tt} Q_{Tt} \\ Q_{Gt} + Q_{St} = Q_{Tt} \end{array}\right\} \qquad (2.11)$$

Hydrograph analysis falls into three phases: the explanation and prediction of (i) the quickflow volume resulting from particular rainfall input and infiltration and evapotranspiration losses; (ii) the time distribution of runoff in the quickflow hydrograph; and (iii) the progressive modification of hydrograph form as it travels downstream through the drainage network.

Storm rainfall–runoff relationships

After separation, total quickflow volume is expressed as an average depth after division by catchment area. Direct comparison with storm rainfall is then possible using a 'response factor' (Hewlett and Hibbert, 1967) such as

$$R_a = \frac{\text{Quickflow}}{\text{Rainfall}} \quad \text{or} \quad R_b = \frac{\text{Quickflow}}{\text{Rainfall} - \text{Evapotranspiration}} \qquad (2.12)$$

By employing annual totals to characterize the average catchment hydrological response, Walling (1971) quotes values of R_a of 0.06 to 0.23 for six Devon catchments, and Hewlett and Hibbert (1967) values of R_b of 0.07 to 0.33 for fifteen forested basins in the USA. High response factors reflect relatively impermeable conditions favouring quickflow generation, whereas low values of R_b arise if considerable infiltration 'losses' occur during storm rainfall (for any catchment, R_b exceeds R_a because it takes evaporation losses into account). The response in a single storm deviates from the catchment average according to antecedent moisture conditions prior to the storm; greater quickflow volumes are generated from 'wet' catchments in given rainstorms. Walling (1971) used step-wise multiple regression to

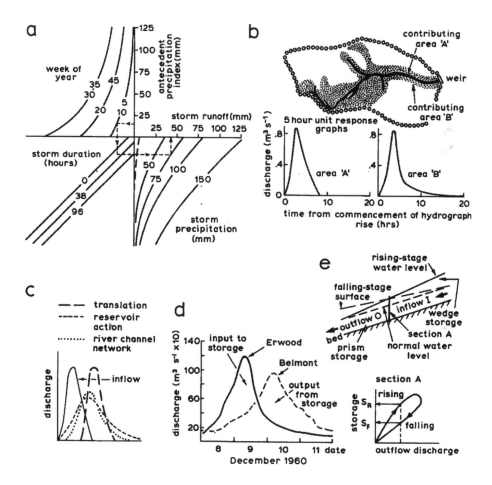

Figure 2.5 (a) A coaxial graphical correlation for predicting storm runoff for the Monocacy River, Maryland (after Kohler and Linsley, 1951). (b) Variation in the unit hydrograph according to antecedent catchment wetness and contributing area (after Walling, 1971). (c) Modification of hydrograph form during travel through a channel reach or reservoir. (d) Attenuation and delay of the hydrograph of an overbank flood passing between Erwood and Belmont on the River Wye, December 1960 (after Price, 1973). (e) Hysteresis in the storage–discharge relationship for a channel reach during a flood event.

identify 'soil moisture deficit' (S_{md}) as the main variable modulating storm rainfall–quickflow (R_s–Q_{fl}) relationships. An example is

$$Q_{fl} = 0.005 R_s^{2.67} S_{md}^{-0.52} \qquad (r^2 = 0.90) \qquad (2.13)$$

for one Devon catchment, which indicates greater quickflow with low soil moisture deficit. Graphical multiple (coaxial) correlation (Linsley *et al.* 1949) often employs an antecedent precipitation index (I_t) which defines a logarithmic recession during dry periods, starting with an index value of I_0 to give

$$I_t = I_0 K^t \qquad (2.14)$$

after t days; K is normally between 0.85 and 0.95. The index for a wet day ($t + 1$) is increased by the amount of rainfall, and renewed recession begins on day ($t + 2$). Figure 2.5(a) shows a coaxial diagram which defines quickflow as a function of antecedent wetness, calendar date (a surrogate for evapotranspiration) and storm duration and rainfall amount, which together define rainfall intensity.

The quickflow hydrograph

Whereas quickflow volume is determined by rainfall intensity and duration, and by losses to infiltration and evaporation, the hydrograph shape, defined by its peak, time base, lag time, time of rise, and recession (Figure 2.4a), reflects concentration of quickflow by catchment properties. Controls of the time distribution of quickflow are both transient and permanent (Rodda, 1969). The transient factors include storm characteristics of intensity, duration, area and migration; a short, intense storm creates a peaked hydrograph, which is accentuated by the storm moving downstream with the hydrograph. Permanent controls are catchment properties such as geology, soil type, and vegetation and land use (which are transient, but on a longer time scale than the storm characteristics). These determine the nature of quickflow generation. Catchment topography and morphology – the basin, network and channel properties – are also important. Numerous indices describe basin morphometry, but qualitative reviews (Gray, 1964) and quantitative investigation by principal components analysis (Wong, 1963, Gardiner, 1978) suggest that basin area, shape, relief (altitude and slope) and network density and distribution are major independent features. Area (or channel length) is a scaling factor, while compact shape, steep slopes and high drainage density or stream frequency encourage a peaked hydrograph. These factors therefore enter multiple regression models in which the peak discharge with a specific probability of occurrence is the dependent variable (pp. 134–5). These models summarize the influence of various independent variables on a peak discharge index, and permit prediction of that index in ungauged catchments.

The 'unit hydrograph' (Sherman, 1932) allows a standardized comparison of overall hydrograph shape between basins. It is defined as the hydrograph of 1 cm of quickflow from a storm of specified duration. Storms of, say, 6 h duration are assumed to create hydrographs of a similar time base within a given catchment. Quickflow hydrographs generated by several such storms are defined, their ordinates are reduced proportionally to create 'unit' hydrographs (Figure 2.4f), and the average of these is computed. A unit hydrograph is approximately triangular and can be defined by its peak discharge, time to peak (t_p) and time base (Reich, 1962). These parameters are dependent on catchment properties, so 'synthetic hydrographs' (Snyder, 1938) can be defined for ungauged catchments from appropriate empirical relationships such as

$$t_p = C_t(LL_c)^{0.3} \qquad (2.15)$$

which shows that lag time increases with catchment size (L and L_c are two basin length measurements) and decreases with basin slope, since the co-efficient C_t increases as slope decreases. Nash (1960) adopted a different approach, treating the unit hydrograph as the skewed probability distri-bution of travel times to the basin outlet by quickflow, characterized by mean travel time (m_1) and variance (m_2) and defined by

$$m_1 = 20.7A^{0.3}S_L^{-0.3} \qquad (2.16)$$

and

$$m_2 = 0.43L^{-0.1} \qquad (2.17)$$

(A is the basin area, S_L the overall catchment slope, and L the mainstream channel length). A 6-hour unit hydrograph can be used to define a hydro-graph of 5 cm of quickflow generated by a 12-hour storm simply by summing two unit hydrographs lagged by 6 hours (giving a 2 cm hydrograph) and by multiplying the ordinates by 2½. This method suffers severe limitations, however, in that it assumes that catchments effect a linear transformation of storm rainfall. Amorocho (1963), however, shows that successive equal bursts of rain produce floods whose peaks exceed those predicted by simple summation of a series of hydrographs equivalent to that generated by a single burst. Furthermore, the unit hydrograph is time variant. Figure 2.5(b) shows that the time base of a 5-hour storm hydrograph is longer when the catchment is wet because travel times from the more widely distributed contributing area are longer. Finally, 'parametric' models such as those in Equations (2.15) and (2.16) incorporate 'lumped' catchment averages of independent variables (e.g. catchment slope) which do not satisfactorily represent the spatially varied production of quickflow over the catchment surface. This remains a limitation of more advanced overall catchment models, such as the Stanford Watershed model, which transfer rainfall

inputs through a series of conceptual stores in order to simulate the hydrological response (Crawford and Linsley, 1966).

Modification of the hydrograph by the channel network

The spatial arrangement of contributing areas reflects the network
structure, and a successful model of catchment hydrological response needs
to accommodate a 'distributed' set of parameters defining variable infiltration and soil moisture characteristics in subcatchments (Beven and
Kirkby, 1979). The hillslope hydrograph enters the channel system, and the
catchment output hydrograph results from modifications which occur as it is
routed through the network; network-based models even account for the
interaction between network geometry and the storm area, speed and
direction of movement (Surkan, 1974). In small basins, downslope travel
time controls hydrograph form, but in large basins travel time through the
network dominates (Calver *et al*, 1972), particularly if catchment area
exceeds storm area and the hydrograph travels beyond its source. A hydrograph entering a dry reach with an impermeable bed may be translated
through the reach without substantial modification, while inflow (I) to a lake
or reservoir increases storage (S) which then 'forces' increased outflow (O).
Normally, the effect of channel reaches on hydrograph form is intermediate
between translation and reservoir effects (Figure 2.5c). The storage of
inflow within the reach (especially on the floodplain surface if overbank flow
occurs) causes attenuation and delay of the peak, as in the example for a
70 km reach of the River Wye shown in Figure 2.5(d).

   A simple continuity equation can be used to describe and predict this
modification by budgeting hydrological quantities at successive times,
$t, t+1 \ldots$,

$$\left(\frac{I_{t+1} + I_t}{2} \right) - \left(\frac{O_{t+1} + O_t}{2} \right) = \left(\frac{S_{t+1} - S_t}{2} \right). \qquad (2.18)$$

This equation can be solved if a relationship can be found between storage
and outflow, since both inflow and the initial outflow values are known. The
Muskingum method (McCarthy, 1938) uses the approximation

$$S = K[\epsilon I + (1 - \epsilon)O] \qquad (2.19)$$

in which K is the mean hydrograph travel time through the reach, which can
be calibrated using injected salt solution (Beven *et al*, 1979), and ϵ is a factor
weighting the relative influence of inflow and outflow on the storage. It is
theoretically shown to approximate a measure of wave diffusion and attenuation (Price, 1973) and is defined by examining the modification of observed
hydrographs: it is equal to 0.5 if the hydrograph is translated and to zero if
reservoir action occurs. Hydrograph modification involves conservation of

momentum and energy as well as mass, and frictional effects are accounted for by hydraulic routing methods based on the Saint-Venant gradually varied flow equations (Price, 1973, NERC, 1975). Mein *et al* (1974) used a simple storage relation

$$S = aO^n \qquad (2.20)$$

which subsumes frictional effects because the coefficient a includes channel width, slope, and the Manning coefficient. However, simple storage–outflow relations can only give approximate solutions to the continuity equation because storage is hysteretic (Figure 2.5e), being greater on the rising limb for in-channel flood flows, and greater on the recession limb especially when overflow occurs onto the floodplain (Hughes, 1980).

Sediment sources and sediment yield

The sediment and solute loads transported by rivers are significant in different contexts. Solute yield often represents the dominant mode of overall catchment denudation, not only in basins underlain by calcareous rocks. Erosion, transport and deposition of sediment, however, exert a major influence over channel morphology, the quantity, nature and balance of the types of sediment carried being important. Sediment load includes bedload, suspended load and wash load. The immersed weights of bedload particles are supported by the stream bed, and their motion involves frequent or continuous contact with the bed, the size and shape of particle determining whether movement is by saltation, rolling or sliding. Suspended sediment is carried within the body of flowing water, and turbulent fluctuations enable the fluid itself to support the immersed weight of the particles. Wash load consists of particles finer than those found in appreciable quantities in the mobile portion of the channel bed; it is derived from slope erosion rather than from the channel perimeter. Classification of the sediment load cannot be based solely on particle size, since a given particle may travel by rolling, saltation or in suspension at successively higher discharges through a channel reach.

The balance of the components of the total load depends on the distribution of appropriate source areas within a catchment and on the hydrological processes which transport sediments and solutes to the stream.

(a) *Solute sources* Solutes are derived from chemical weathering of bedrock and soil, whose significance as sources reflects the dominant hydrological pathways and the relative roles of groundwater flow and throughflow. Walling and Webb (1975) show that solute concentrations in the Exe basin, Devon, are highest on Permian marl and lowest on Devonian sandstone. The solute load, and its ionic composition, reflect the weathering susceptibility and mineralogy of bedrock. However, soils in regions beyond

Quaternary glacial limits reflect bedrock control, so geology appears to be the control although throughflow may supply most of the stream's solute load. Burt (1979) demonstrated that the storm hydrograph from a small catchment includes an instantaneous peak caused by channel precipitation and saturated overland flow, during which dilution reduces solute concentrations, and a delayed throughflow peak when solute concentrations increase as stored solutes are flushed from the soil profile, particularly from topographic hollows. Experimental evidence indicates that throughflow travel times are sufficient for chemical equilibration to occur with soil water (Smith and Dunne, 1977).

Non-denudational solute sources are often significant. Rainfall inputs, in particular, may account for 25–75% of total solute loads in Eastern Australia (Douglas, 1968a), so that failure to adjust for this source can result in overestimates of denudation rates by 1.4–2.4 times (Goudie, 1970, Janda, 1971). Ionic concentrations in bulk precipitation (rainfall and dry fallout) derive from distinctive sources; for example, in mid-Wales 50–60% of Mg^{2+}, Na^+, K^+ and Cl^- losses in runoff originate in precipitation from maritime sources (Cryer, 1976). These are non-denudational cyclic ions, whereas the Ca^{2+} precipitated from non-maritime air masses may originate from land sources, therefore representing denudation outside the catchment boundary. Geochemical cycling of solutes can result in an underestimation of denudation; Edwards (1973) suggests that 15% of the silica load of East Anglian rivers is seasonally assimilated by diatoms and fixed in an insoluble form. Finally, pollution loading is often significant, both from phosphate-rich point sources such as sewage works, and from the areally-distributed input of nitrogenous soil water derived from farmland subjected to fertilizer application. Ideally, therefore, complete geochemical budgets are necessary before chemical denudation rates can be established. A good illustration of this is the study of a small Maryland catchment by Cleaves *et al* (1970) where the total solute budget is

$$
\begin{array}{c}
\text{Rainfall input + Mineral weathering =} \\
\text{(soil and rock)} \\
5.95 \quad + \quad 9.75 \quad = \\
\\
\begin{array}{ccc}
\text{Solute load +} & \text{Solutes} & + \text{Solutes} \\
\text{of stream} & \text{stored in} & \text{stored in} \\
& \text{groundwater} & \text{plants} \\
8.90 \quad + & 2.37 \quad + & 4.43
\end{array}
\end{array} \qquad (2.21)
$$

(figures are in kg ha^{-1} yr^{-1}). The solute load in streamflow in this case underestimates the actual 'denudation' by mineral weathering.

(b) *Sediment sources* Sediment sources are generally more closely related to the river channel. Bedload is derived from the stream bed itself, and equilibrium is maintained by inputs of material of comparable calibre from bank erosion (especially of banks with coarse basal units of channel lag gravels; see pp. 163–5), or from bluffs of terrace gravels or solifluction deposits infilling the valley (Lewin *et al*, 1974). Gully erosion is important on steep upland valley sides. Gully erosion by runoff, mudflow and free-fall of debris generates 400 t ha^{-1} yr^{-1} of sediment which builds debris fans in Grains Gill, in the Howgill Fells in north-west England (Harvey, 1977). Erosional events which contribute to this fan building require a daily precipitation in excess of 13 mm, which occurs about thirty times per year. Even minor river erosion of debris fan bases demands more than 41 mm of daily precipitation, which occurs about once a year. Thus debris fans build up in the gullies until a flood occurs which is large enough to incorporate some or all of the coarse sediment supply into the channel sediments. Suspended sediment sources are more widely dispersed. Fine sediments trapped in pores between coarser bed materials are resuspended at high discharges, but bank collapse is a dominant source, as suggested by the concentration of source areas along migrating main streams and by comparable bank and suspended sediment properties (Carson *et al*, 1973). Wash load, however, enters the channel after catchment slope erosion and is independent of the hydraulics of streamflow except in that its continuing transport requires the appropriate conditions of turbulence. Wash load is recognized by its distinctive size properties; for instance, in a Pennsylvania catchment, median diameters are 9 mm for bed material, 0.4 mm for bank material and 0.005 mm for the wash-derived suspended load (Rendon-Herrero, 1974). In the context of denudation, organic and clastic (inorganic) suspended loads must be distinguished, particularly when a volume of erosion is computed from the sediment weight by applying an appropriate density. In the humid tropics, total organic loads of 50–120 kg ha^{-1} yr^{-1} occur (Brinson, 1976), and in sixteen catchments on the North Yorkshire Moors in England, suspended organic sediment of 5–11 kg ha^{-1} yr^{-1} represents up to 50% of total suspended load (Arnett, 1978). In the context of channel form, finely divided organic matter in suspension can encourage flocculation of clays whose incorporation in the channel sediments increases channel stability and influences channel shape. This is also true of the dissolved bivalent cations (Ca^{2+}, Mg^{2+}) which encourage cation bridging between clay particles.

Sediment yield

Sediment yield data are difficult to compare. Extreme annual surface-lowering rates of 1–2 cm have been measured on slope plots in semi-arid badlands, but delivery ratios decrease with basin size to produce much lower average areal sediment yields. Standard areal measuring units are therefore necessary. Temporal variations over storm-period, seasonal and annual scales occur in sediment concentration and load, and thus require that

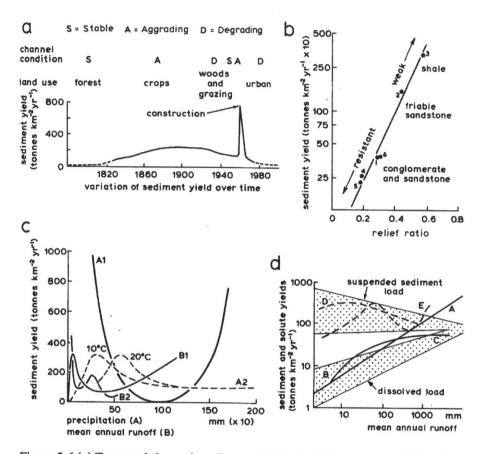

Figure 2.6 (a) Temporal change in sediment yield in the Piedmont region, USA (after Wolman, 1967). (b) The sediment yield–relative relief relationship (after Schumm, 1954). (c) Various relationships of sediment yield to mean annual precipitation (Fournier, 1960, curve A1), effective precipitation (Langbein and Schumm, 1958, curve A2) and runoff (Douglas, 1967, curve B1, and Judson and Ritter, 1964, curve B2). Note that runoff values represent precipitation less evapotranspiration and infiltration. (d) The balance between sediment and solute yields. Curve A, Livingstone (1963); curve B, Van Denburgh and Feth (1965); curve C, Langbein and Dawdy (1964); curve D, Langbein and Schumm (1958); curve E, Douglas (1967).

representative sampling periods are chosen. The sampling frequency also influences the estimates of sediment yield. A varying and usually unknown percentage of sediment yield is transported as bedload, but most estimated yields are based solely on measured suspended loads. Natural erosion rates are accelerated by man (Douglas, 1967, Meade, 1969), and Figure 2.6(a) illustrates the effects of deforestation and urban construction in the Eastern USA on erosion and stream channel behaviour. Furthermore, sediment yields are controlled by interacting relief, land-use, soil resistance and climatic factors whose individual effects are obscured by covariation. The relationship between erosion rates and relative relief in Figure 2.6(b) (Schumm, 1954) occurs partly because the steep relief occurs in weaker rocks.

Langbein and Schumm (1958) evaluated climatic controls using data from suspended sediment gauging stations in the USA (average basin area, 4000 km^2) and reservoir sedimentation studies (average basin area, 80 km^2), and defined sediment yield as a function of effective precipitation (adjusted for evapotranspiration loss). Maximum erosion rates occur at an annual effective precipitation of 300 mm for a mean annual temperature of 10°C (Figure 2.6c). Sediment concentrations in drier climates may be very high, but total runoff is too small for sediment yield to be significant. In wetter climates, the protective vegetation cover reduces rainsplash and runoff and inhibits erosion. Fleming (1969a) also showed that, although sediment yield always increases with stream discharge, it is highest at a given discharge in scrub desert areas and lowest in mixed broadleaf–coniferous woodlands. Sediment yields may increase again with runoff or precipitation in very wet climates not represented in the American data, particularly the seasonal tropics (Douglas, 1967, Fournier, 1960). The general trends in Figure 2.6(c) may be compared with that of drainage density in Figure 2.3(b).

Climate cannot be adequately represented by annual quantities; seasonality of rainfall is an important control of erosion (Wilson, 1973). Fournier's (1954) index p^2/P accounts for both total annual precipitation (P, mm) and seasonality, where p is the average precipitation of the wettest month. Regional sediment yields are distinguished by this parameter, which is highest in semi-arid and seasonal tropical environments, but a relief factor is necessary to rationalize the scatter in a multivariate relationship (Fournier, 1960)

$$S = 0.03(p^2/P)^{2.65}(R^2/A)^{0.46} \qquad (2.22)$$

where S is suspended sediment yield ($t\,km^{-2}yr^{-1}$), R is basin relief (m) and A basin area (km^2). More locally, the factors responsible for wash load production (S_{WL}) are summarized by the Universal Soil Loss Equation (Smith and Wischmeier, 1957)

$$S_{WL} = F_R F_K F_L F_S F_C F_P \qquad (2.23)$$

where the controlling factors measure the effects of rainfall intensity, soil erodibility, slope length, slope angle, cropping, and management practices and conservation measures (F_R to F_P respectively).

Sediment calibre, as well as quantity, is an important control of alluvial channel morphology. Chemical weathering in warm, wet climates favours production of silts and clays, while cold, dry climates generate coarse sediments. Steep slopes and mountainous relief generally produce coarser sediments. Sediments on the inner continental shelves reflect climatic controls of terrestrial sediment yield (Hayes, 1967). Gravels are most common (>15%) where median annual temperatures are cold (<0°C) and precipitation is relatively low (<1000 mm), while muds exceed 40% in warm (>25°C), wet (>1500 mm) climates. Sands are ubiquitous, but are most important in hot, dry areas and areas of moderate rainfall and temperature.

Solute concentrations are highest in dry climates where they are increased by evaporation (Durum *et al*, 1960). but total solute yield is also dependent on runoff quantities, which increase rapidly with rainfall to more than compensate for lower concentrations. Thus, solute yields are highest in wetter climates (Langbein and Dawdy, 1964, Van Denburgh and Feth, 1965). Figure 2.6(d) shows this general trend of increasing solute yield with runoff, based largely on data which do not compensate for cyclic atmospheric salts. Total and specific ion concentrations reflect rock type as well as climate, especially in small catchments. For example, in the southern Pennines, England, $CaCO_3$ concentrations increase from 140–220 ppm as the proportion of catchment area underlain by limestone increases from 30–100% (Pitty, 1968). However, large catchments usually contain a representative sample of rock types and therefore have a similar range of components in their solute loads. The importance of relief is illustrated by sediment and solute loads in the Amazon basin subcatchments (Gibbs, 1967). Solute concentrations are actually higher in areas of steep relief, perhaps because the soils have greater transmissibility. However sediment concentrations are much higher, especially in the wet season, and therefore dominate the total load. It is evident from Figure 2.6(d) that, notwithstanding variations between studies, sediments generally comprise about 90% of total load (erosion) in dry environments, while solutes are increasingly important (>50% of total load) in wetter climates. The balance between sediments and solutes reflects the dominant, operative hydrological process: subsurface flow processes encourage chemical weathering and dominant solute yield, while surface runoff is associated with sediment yield. Solute concentrations are diluted during peak flows (pp. 90–6) and although solute load (concentration times discharge) usually still increases with discharge (Douglas, 1964), lower, more frequent discharges are relatively more important than extreme floods as transporters of solutes. Thus, solutes tend to be more important relative to sediments in uniform, non-flashy hydrological regimes, and the sediment–solute balance is a reflection of the magnitude–frequency properties of the flows (Chapter 5).

The history of sediment yield and valley fill evolution

Alluvial channels form in unconsolidated sediments which represent the accumulated deposition of sediment yield. The products of catchment erosion are not transported immediately from the basin, but are temporarily stored in a floodplain which extends downstream from the point of initiation of valley alluviation. In semi-arid river valleys in Spain, alluviation begins at the point of maximum network width where converging tributaries deliver sediment to the main valley (Butcher and Thornes, 1978). However, many valley fill deposits result from net aggradation after successive phases of scour and fill during the Quaternary. Aggradation occurs because the load:discharge ratio is increased and excess sediment load is deposited. This arises if transport capacity is reduced by a decrease in gradient, spreading of the flow over a wide area, or loss of streamflow by evaporation and infiltration, or if excessive sediment supply results from the introduction of glacial outwash, or accelerated slope and tributary valley erosion because of climatic change or de-vegetation by man. An aggrading state is difficult to identify because sediment movement is discontinuous. Chains in the stream bed may be buried by accumulating sediment (Miller and Leopold, 1963), but this may be temporary fill which is scoured in the next major flood. In some small Welsh catchments, the sum of slope and river bank erosion exceeds basin sediment yield (Slaymaker, 1972), but if sediment delivery ratios are low this may not imply aggradation; the erosion, storage and release of sediment from a drainage basin is a complex and discontinuous process, in which soil and sediment are progressively weathered during lengthy storage periods on slopes or in valley fill (Dietrich and Dunne, 1978).

The depth of accumulated valley fill is measured by auger or by seismic exploration, which demonstrate that bed-rock surfaces may form symmetrical cross-sections in straight valley reaches but asymmetrical sections at bends (Figure 2.7a), suggesting that the valleys were occupied by larger rivers prior to valley filling (Dury, 1962, 1964b). Palmquist (1975), however, argues that normal scour depth is approximately twice the bank height, so the upper surface of valley fill or floodplain is reworked to this depth by migrating meanders which create characteristic sedimentary structures (pp. 206–11). Unless the fill is extremely thick, asymmetric bedrock profiles at valley bends may simply reflect frequent scour by a channel in its preferred position on the outside of the valley bend.

Depositional structures reflect the nature of the aggradation control (Mackin, 1948, Smith, 1973). Downstream control results from a rise in base level (Figure 2.7b) and is associated with backfilling as a wedge of deposition extends headwards. Deposition is mainly by overbank flows, and therefore tends to be of fine sediment, and the progressive *decrease* of gradient caused by backfilling results in the general fining of sediment up through the deposits. Fisk (1939) interpreted terraces in the lower Mississippi valley as follows: high interglacial sea levels resulted in aggradation, but relative sea levels in successive interglacials were slightly lower, because of isostatic uplift inland from Baton Rouge to compensate for deltaic accumulation. The result was a flight of terraces.

Upstream control occurs when sediment enters the head of the system, particularly in the form of glacial outwash, to cause downfilling as a wedge of

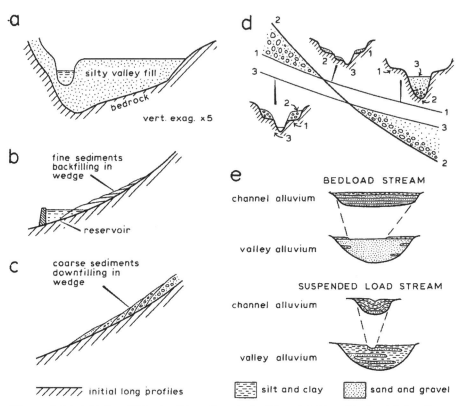

Figure 2.7 (a) Valley fill in the River Evenlode, Warwickshire, England (after Dury, 1964b). (b) Downstream control of aggradation; base level rise because of reservoir impoundment. (c) Upstream control of aggradation; input of glacial outwash (after Smith, 1973). (d) Complex long-profile and cross-section development during a glacial cycle (after Baulig, 1935). (e) Valley fill by bedload and suspended load streams (after Schumm, 1968b).

sediments extends downstream (Figure 2.7c). Sedimentation of coarse sand and gravels in channel bars gradually *steepens* the slope and a coarsening-upwards sequence occurs at a particular location. The outwash plain of the Bossons glacier aggraded rapidly during glacial advance from 1968 to 1974 (Maizels, 1979), the intensified braiding and steepening gradient being a hydraulic adjustment to the heavy sediment load. There is no general rule, however, that aggradation occurs during short-term glacial advance and degradation during retreat, since the load:discharge ratio is critical. In the Columbia icefield outwash, buried volcanic ash which has been dated indicates that aggradation has been the rule for 2450 years (Smith, 1973). The combination of upstream (glacial) and downstream (sea level) controls in large rivers results in a complex pattern of deposits and terraces, as summarized in Figure 2.7(d) for a *single* glacial cycle (Baulig, 1935). A preglacial long profile is steepened by outwash aggradation upstream, and degradation to low sea level downstream, during the glacial period, after which the reverse pattern occurs when a high postglacial sea level drowns the lower valley and the glacial sediment source retreats fron the valley head. Such patterns are evident in the Ohio–Mississippi, Rhone–Durance and many other river systems. Of course, the succession of glacial epochs during the Quaternary has created an even more complex pattern of aggradation and degradation in these river valleys than this single cycle example illustrates. Finally, basin control and vertical filling result from accelerated slope erosion. This may be a natural consequence of steepening slope angles during valley incision (Tinkler, 1971), or of the coupling of tributary and main valleys which result in 'complex responses' (Schumm, 1973) to initial external stimuli such as rejuvenation (pp. 18–20). Alternatively, slope erosion may increase because of accelerated runoff caused by overgrazing or by some other form of man-induced vegetation removal (Strahler, 1956). Subtle climatic changes may be the cause of complex cut and fill cycles in semi-arid valleys (Cooke, 1974). Changes in the balance of runoff and sediment yield from hillslopes may influence the occurrence of aggradation and degradation in the valley bottoms. In addition, there may be variations in the erosional capacity of flow over the valley floor, and in the resistance to erosion of valley floor sediments especially as a result of changes in the protective effect of the vegetation cover.

Whether upstream, downstream or basin controls influence aggradation, the detailed sedimentary structures of the accumulated fill reflect the type of sediment being deposited and the rivers transporting it. Coarse sand and gravel carried by braided streams forms a fill dominated by lenticular channel fill deposits, while finer sediments in meandering streams are characterized by fining-upwards sequences (Allen, 1965, Schumm, 1968b; Figure 2.7e). These depositional structures are considered in detail in relation to channel patterns in Chapter 7.

3

The mechanics of flow and the initiation of sediment transport

Quickflow is generated by processes ranging from infiltration excess overland flow to subsurface stormflow and is concentrated into storm hydrograph form by the basin topography and network geometry. A succession of such hydrographs arriving over a period in the main valley bottom carves in the alluvium a channel whose form reflects the interaction between the local sedimentology and the physical mechanisms of channel flow. Particularly important physical processes are those determining the distribution and magnitude of bed shear stress exerted by the flow, the momentum transfer from rapid to slow zones of flow, the velocity distribution with depth and across the channel, and the rate of transformation of potential energy to kinetic energy and work. At local scales, that is from the few centimetres closest to the stream bed to the single short channel reach, it is possible to apply to these processes theoretical models derived from fluid mechanics (the science of the motion of fluids, including air and water) and hydraulics (the science of the conveyance of liquids in conduits, including pipes, canals and natural channels). Such 'fundamentalist' models (Allen, 1977), which are based on physical principles and supported by laboratory scale-model investigations in flumes, also provide an insight into the initial interactions of flow and sediment which form the prerequisites for sediment transport (Chapter 4). However, the restrictive assumptions of many theoretical models limit their application to a narrow range of conditions. For example, much of the theory of hydraulics derives from studies of flow through pipes with rigid boundary roughness. The velocity (mean and distribution) of fully developed turbulent flow under such conditions is essentially controlled by the fixed roughness height. However, the loose boundary of a sandy alluvial channel is moulded into bedforms such as dunes during rapid sediment transport, and these augment the grain roughness with form roughness. In essence, the additional form roughness which develops under high-energy flow conditions with rapid sediment transport provides a negative feedback which helps to dissipate the excess energy by introducing an additional source of flow resistance. Flow velocity is controlled by the changing roughness of the mobile boundary, but is often indeterminate in loose boundary channels because the process of bedform development remains poorly understood.

The hydraulics and mechanics of open channel flow

The various theoretical models of open channel flow involve simplifying assumptions concerning the spatial and temporal variations of flow properties. The important spatial classification into uniform, gradually varied and rapidly varied flow is defined by the pressure distribution, which in turn reflects the streamline pattern. (Streamlines are imaginary lines to which the local velocity vector is tangent.) In uniform flow the depth is constant from section to section, and the bed, water surface, energy grade line and all streamlines are parallel (Figure 3.1a). The pressure distribution is everywhere hydrostatic, and at any point the pressure is dependent on the depth beneath the free surface. In Figure 3.1(a), the head h is equal to $y\cos^2\theta$, but for small ($<10°$) bed slopes, $\cos\theta$ may be taken as unity with little error, and the head equals the depth y. In gradually varied flow the streamlines are not parallel because the channel cross-section, and hence the depth and

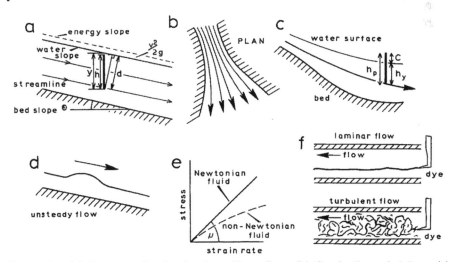

Figure 3.1 (a) Pressure distribution in uniform flow. (b) Gradually varied flow. (c) Pressure distribution in curvilinear flow; h_y is hydrostatic head, h_p total piezometric head, and c the correction due to streamline curvature. (d) Unsteady flow: passage of a flood wave along a channel. (e) Stress–strain rate relation in a Newtonian fluid. (f) Laminar and turbulent flow as revealed in Reynold's (1883) experiment.

velocity, change along the reach (Figure 3.1b). Bed, surface and energy gradients are not necessarily equal. Divergence or convergence of the streamlines results in a deceleration or acceleration of flow velocity, and this curvilinear flow produces centrifugal forces so that the pressure distribution is not hydrostatic. Figure 3.1(c) shows that 'concave flow' results in a downward-acting centrifugal force which augments the gravitational effect to give pressures in excess of hydrostatic. The converse is true of 'convex flow', and similar effects occur because of curvature in plan. However, streamline curvature is sufficiently slow with distance that, for short channel reaches, the flow may be assumed uniform. This distinguishes gradually varied from rapidly varied flow, where streamline curvature and non-parallelism are extreme. Rapidly varied flow involves phenomena such as hydraulic jumps and drops, including the sudden increase in depth downstream from a throated flume (Figure 5.2h) or the free overfall of a weir. These sudden changes in water level usually reflect contractions or expansions of cross-section geometry, and are exploited in control sections for stream gauging (Chow, 1959). The associated rapid energy dissipation inhibits the continued existence of significant examples of rapidly varied flow in natural channels with erodible boundaries.

Natural rivers usually experience gradually varied flow, since the bed slope and cross-section vary even between tributary junctions. However, the total energy loss over a reach experiencing gradually varied flow is usually approximately the same as that over a reach experiencing uniform flow with the same (average) velocity and depth, so that uniform flow formulae (see below) can be used locally.

Temporally, flow conditions are defined as 'steady' or 'unsteady'. Steady flow involves constant discharges, and therefore constant depth and velocity, through time at each section. Stream discharge (Q) is the product of cross-section area (A) and mean velocity ($\bar{v}$), and under continuous steady flow at successive sections $1, 2, \ldots, n$,

$$Q = A_1\bar{v}_1 = A_2\bar{v}_2 = \ldots = A_n\bar{v}_n. \tag{3.1}$$

This is the continuity equation of flow. Unsteady flow is associated with temporally changing discharge, depth and velocity, as when a flood wave passes along the channel reach (Figure 3.1d). Since the water surface slope changes as the wave passes a section, uniform unsteady flow is impossible. Unsteady gradually varied flow is the rule in natural channels, although rapid unsteady flow variation occurs if sluice operation causes surges. Nevertheless, in large rivers flood passage is often sufficiently slow that steady conditions can be assumed for short periods.

At a more detailed level, the flow may be classified according to its energy, momentum, inertia and viscosity. Chapter 1 emphasized the fundamental importance of continual potential energy renewal by the hydrological cycle, and the conversion of potential energy to kinetic energy and work as a volume of water travels downstream. The stream power measures the rate of work done by the flowing water in overcoming bed and internal flow resistance and transporting sediment (units of measurement are outlined in the Appendix). The momentum of flowing water is the product of its mass and velocity, and it defines the difficulty of stopping a body of water already in motion. Stream discharge is volume per unit time, and so the momentum per unit time of water passing a section is density $\times$ discharge (=mass per unit time) $\times$ velocity. For a small unit volume of water within the body of flow, momentum is simply density (ρ_w) $\times$ velocity (= unit mass $\times$ velocity). The mass of a body of water is a measure of its inertia and is therefore a measure of the difficulty of initiating motion, or of changing the motion of a steadily moving body. Again, mass per unit time is density $\times$ discharge (Q), and dividing by the width of flow we have

$$\rho_w Q/w = \rho_w w dv/w = \rho_w dv \qquad (3.2)$$

or the inertia of flow per unit width per unit time. Finally, water is a typical Newtonian viscous fluid; when subjected to a stress, the rate of strain or

Table 3.1 Density and viscosity of water at various temperatures.

Temperature (°C)	Fluid density, ρ_w (g cm^{-3})	Dynamic viscosity, μ (g cm^{-1} s^{-1})	Kinematic viscosity ν (cm^2 s^{-1})
0	1.000	0.01800	0.0180
10	1.000	0.01310	0.0131
20	0.998	0.00998	0.0100

deformation of the fluid increases linearly with the applied stress (Figure 3.1e). The 'dynamic viscosity' (μ) is the gradient of the stress–strain rate curve and it measures the force per unit area (stress) required to maintain a unit difference of velocity (strain rate) between two parallel layers separated by a unit distance. Viscosity values for water, which vary with temperature, are given in Table 3.1. The 'kinematic viscosity' (ν) is also listed. This is a measure of the interference between adjacent layers of fluid, and is simply μ/ρ_w; it is measured in units of area/time (see Appendix). When water contains a high concentration of suspended sediment (>5% by weight), it ceases to behave as a Newtonian fluid (Simons et al, 1963), and the coefficient relating stress and strain rate is an 'apparent viscosity' which varies as a function of the rate of shear.

Using these criteria, the state of flow may be defined by two dimensionless numbers, namely the Reynolds and Froude numbers.

(a) *The Reynolds number (Re)* This distinguishes 'laminar' and 'turbulent' flow on the basis of the ratio between inertial and viscous forces. It is defined as

$$\text{(i) } Re = \frac{\rho_w v L}{\mu}, \quad \text{or} \quad \text{(ii) } Re = \frac{\rho_w v R}{\mu} = \frac{v R}{\nu} \qquad (3.3)$$

In (i) L is a 'characteristic length' taken as the hydraulic radius (R) in (ii) to give the Reynolds number of the total flow. In wide, shallow channels the hydraulic radius approximates to the mean depth and is referred to as the 'hydraulic mean depth'. A 'particle Reynolds number' is obtained by replacing L by the grain diameter and using the viscosity and density of the fluid through which the particle is moving. The numerator in Equation (3.3) (ii) is the inertia per unit width per unit time. Laminar flow experiences smooth, linear streamlines with water apparently 'sliding' in layers, and dye

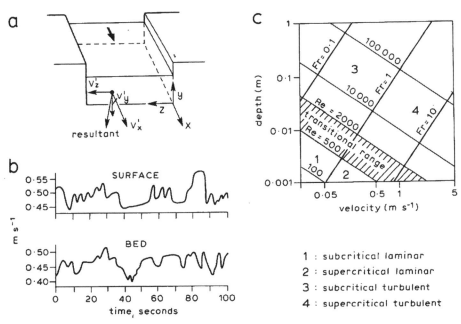

Figure 3.2 (a) Definition of orthogonal coordinates for measurement of turbulent velocity fluctuations. (b) Velocity fluctuations (v'_x) in turbulent flow (after Embleton and Thornes, 1979). (c) Classification of flow types by Reynolds and Froude numbers.

injected at a point forms a straight coherent thread (Reynolds, 1883). Viscous forces are significant, and here Re is normally less than 500. In turbulent flow, inertial forces predominate and Re exceeds 2000. The flow involves random secondary motions (Figure 3.1f) which are superimposed on the main downstream movement and which can be identified from the distortion of injected dye. The instantaneous velocity vector at a point varies as random eddies of fluid pass. Velocity components may be measured parallel to orthogonal x, y and z co-ordinates (Figure 3.2a), and at any instant and point location, these consist of a time-average mean velocity (marked with a bar in Equation 3.4) and turbulent fluctuating velocity components (denoted by primes):

$$v_x = \bar{v}_x + v'_x$$
$$v_y = \bar{v}_y + v'_y \quad \quad \quad \text{(3.4)}$$
$$v_z = \bar{v}_z + v'_z$$

(Simons and Şentürk, 1977, Chapter 3). The mean values of v'_x, v'_y and v'_z are zero, but as their standard deviations are non-zero the intensity of turbulence is therefore defined by

$$\{\sqrt{[(v'^2_x + v'^2_y + v'^2_z)/3]}\}/\bar{v}_x. \quad \quad \text{(3.5)}$$

Those instantaneous components of velocity acting parallel to the y axis are particularly important in providing the means of sustaining the suspension of sediment, which cannot occur in true laminar flow. Kalinske (1943) showed that the relative intensity of turbulence is greater over rough beds and at high Reynolds numbers, and measurements by a hot-wire anemometer (Blinco and Simons, 1974) demonstrate that turbulent fluctuations of velocity are greater close to the bed, relative to the lower mean velocity, than near the surface (Figure 3.2b).

(b) *The Froude number (Fr)* This dimensionless number relates the inertia of a unit mass of streamflow to the celerity of a shallow gravity wave:

$$Fr = \frac{v}{\sqrt{(gd)}} \quad \quad \text{(3.6)}$$

where d is the depth of flow. If $Fr < 1$, the flow is subcritical, or tranquil, and the wave celerity ($\sqrt{gd}$) exceeds the flow velocity so that ripples on the water surface are able to travel upstream. When $Fr > 1$, the flow is supercritical, or rapid, and gravity waves cannot migrate upstream. Furthermore, surface waves are unstable and may break, which results in a considerable energy loss. Free surface instability, such as the development of roll waves, generally tends to increase resistance to flow.

Figure 3.2(c) summarizes the variation of the states of flow, defined by these two criteria, for different depths of flow and velocities. Laminar flow requires depth and velocity values too small for most channel flows, but would be common in sheet wash if it were not for the disturbing effect of raindrops (Emmett, 1970). At a temperature of 10°C the Reynolds number of a 50 cm deep channel flow at 25 cm s^{-1} is $Re = 95,420$; clearly, turbulent flow is virtually inevitable in open channel flow. Mean Froude numbers rarely exceed $Fr \approx 0.5$ in natural, mobile-bed channels (Leopold *et al*, 1960), for the reasons discussed below. It is worth noting that these two dimensionless numbers are independent of the scale of river to which they are applied, and that flow conditions in laboratory flume studies should therefore aim to recreate the turbulence intensity and the Froude number of natural streams.

Uniform flow formulae

The mechanics of uniform flow were first treated mathematically by Chezy (1775), who utilized two assumptions. The first is that the force driving the flow – the downslope component of the weight of water – is exactly balanced by the total force of bed resistance. This follows from the definition of uniform flow, in which no acceleration of flow occurs, local gain of momentum being balanced by loss, and local energy loss due to boundary resistance balancing renewal of kinetic energy. The second assumption (evaluated below) is that the force resisting the flow per unit bed area, τ_0, varies as the square of velocity, according to

$$\tau_0 = kv^2 \tag{3.7}$$

where k is a roughness coefficient. The quantity τ_0 is also the mean bed shear stress, which is the drag exerted by the flow on the bed; this is equal and opposite to the resistance offered by the bed to the flow. The relevant forces are (Figure 3.3a):

$$\text{Resisting force} = \text{Total bed area} \times \text{bed shear stress}$$

$$= pL\tau_0 = pLkv^2 \tag{3.8}$$

$$\text{Driving force} \quad = \text{Total water weight} \times \text{sine of the bed slope}$$

$$= W \sin \theta = \rho_w gAL \sin \theta. \tag{3.9}$$

These forces are balanced if no acceleration occurs, and so these expressions may be equated to give:

$$pLkv^2 = \rho_w gAL \sin \theta. \tag{3.10}$$

For small angles, $\sin \theta$ equals the slope s; also the ratio A/p is the hydraulic mean depth R. Thus Equation (3.10) becomes

$$v^2 = (\rho_w g/k)Rs \tag{3.11}$$

or

$$v = C\sqrt{(Rs)} \tag{3.12}$$

where

$$C = \sqrt{(\rho_w g/k)} \tag{3.13}$$

and C is the 'Chezy coefficient'. Equation (3.12) is the Chezy velocity formula. Since $\tau_0 = kv^2$, the balance equation (3.10) may be written as:

$$pL\tau_0 = \rho_w gAL \sin \theta \tag{3.14}$$

and therefore

$$\tau_0 = \rho_w gRs = \gamma_w Rs \tag{3.15}$$

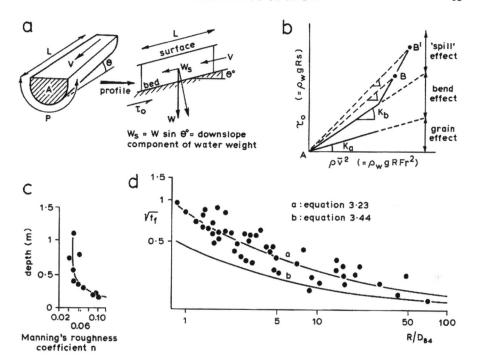

Figure 3.3 (a) Definition diagram for derivation of the Chezy equation. (b) Relationship of bed shear stress (τ_0) to the velocity ($\bar{v}$) squared, illustrating elemental friction coefficients (after Leopold *et al*, 1960). (c) Changing roughness at Seven Springs, Brandywine Creek, as depth changes (after Wolman, 1955). (d) The friction factor-relative roughness relationship; empirical (after Limerinos, 1969) curve a; and theoretical, curve b.

which defines the mean bed shear stress as the product of the unit weight of water γ_w, the hydraulic mean depth (often simply approximated by the average water depth) and the slope.

Alternative velocity formulae exist, of which that due to Manning (1891) is well known. This formula is

$$v = k_1 R^{2/3} s^{1/2} n^{-1} \qquad (3.16)$$

and is an empirical formula in which the exponent of R is an average of the range of exponents (0.65 to 0.84) obtained for various channel shapes and roughnesses. The constant, k_1, is taken to be 1 in SI units, 1.49 in fps units and 4.64 in cgs units, absorbing the units of Manning's roughness coefficient n so that it is possible to use a constant 'dimensionless' value of n for a given channel, regardless of the units used to measure R. The Darcy–Weisbach roughness formula for frictional head loss (h_f) in circular pipes is

$$h_f = f_f \frac{L}{d_0} \frac{v^2}{2g} \qquad (3.17)$$

where f_f is a friction factor, L the length of pipe, and d_0 its diameter (which equals $4R$). This may be rearranged, noting that $d_0 = 4R$ and $h_f/L = s$, to give the Darcy–Weisbach friction factor

$$f_f = \frac{8gRs}{v^2} \qquad (3.18)$$

which is also applicable to flow in open channels. There are thus three commonly used velocity formulae (Equations 3.12, 3.16 and 3.18) with separate friction coefficients. These coefficients are interrelated, however, in the following way, in SI units with k_1 from Equation (3.16) therefore equal to 1:

$$C = R^{1/6}/n; \quad f_f = 8gn^2/R^{1/3} \quad \text{and} \quad \sqrt{f_f} = [\sqrt{(8g)}]/C \qquad (3.19)$$

Note that n and f_f are directly related to bed roughness, while C is more correctly an index of flow conductance than resistance. The American Society of Civil Engineers Task Force (1963) recommends use of Equation (3.18), as it has a sound theoretical basis and the friction factor is dimensionless.

Langbein (1966) justified the square-law resistance–velocity assumption (Equation 3.7) by using a random walk model for particles of water and by showing that the rate of absorption of random walks at the channel boundary (analogous to frictional head loss) is proportional to the square of the velocity. However, laboratory studies (Leopold et al, 1960) provide more detailed insight into the range of its validity. Four sources of flow resistance are considered to exist: skin (grain) resistance, form resistance of bedforms, internal distortion resistance caused by channel bends, and spill resistance. Spill resistance occurs where local flow acceleration or deceleration is rapid. It may be associated with 'tumbling flow' (Peterson and Mohanty, 1960), where local hydraulic jumps occur in the lee of coarse particles, or with acceleration of flow between large-scale roughness elements such as boulders which project above the surface (Herbich and Shulits, 1964). Experimental data from straight flume channels show a linear relationship between τ_0 and $\rho_w v^2$, the gradient k_a being a grain roughness coefficient. In sinuous channels with the same flow depth and bed material, a new roughness coefficient k_b measures the sum of grain and internal distortion resistances (Figure 3.3b). The additional resistance due to internal distortion is greater for more sharply curved bends. However, runs at high velocities (and hence at high Froude numbers) give data which lie on a steeper line, reflecting an increased energy loss due to the onset of 'spill resistance'. This indicates a breakdown in the square law. At low Froude numbers, resistance components are additive with elemental coefficients identifiable for each source. At high Froude numbers, spill resistance occurs and the roughness coefficient, defined by the gradient of a line from the origin (AB) in Figure 3.3(b), is a function of the Froude number. This departure from the square law occurs at Froude numbers of 0.45–0.61, and significantly data from natural streams suggest a well defined limit on the bankfull Froude number at $Fr \approx 0.5$. The excessive rate of energy dissipation associated with spill resistance caused by obstructions such as collapsed bank blocks leads to the removal of the obstruction and the reversion to stable flow conditions.

An important requirement in uniform flow analysis is the definition of the roughness coefficient for a channel section or reach. This may be achieved by inverting the relevant velocity formula and estimating the coefficient from the measured velocity, depth and slope (Wolman, 1955), though an objective estimation from independent controls is preferable. In practice, the numerous effects on roughness mitigate against this, and tabulated

Table 3.2 Average resistance coefficients for straight channels in various conditions.

Type of channel and description	n	f_f[†]	C[†]
Artificial channel, shuttered concrete	0.014	0.016	71
Excavated channel, earth	0.022	0.039	45
Excavated channel, gravel	0.025	0.049	40
Natural channel <30 m wide, clean, regular	0.030	0.072	33
Natural channel <30 m wide, some weeds, stones	0.035	0.093	29
Natural channel <30 m wide, sluggish weedy pools	0.070	0.400	14
Mountain streams, cobbles and boulders	0.050	0.196	20
Major streams >30 m, clean, regular	0.025	0.049	40

[†]for hydraulic radius, R, of 1 m.

values (Table 3.2) or photographs of representative reaches of known roughness (Chow, 1959, Barnes, 1967) are used. Cowan (1956) defined elemental coefficients for the effects of bed material calibre, bed configuration, cross-section variability, obstructions such as trees and boulders, and channel vegetation growth, added these and then applied a multiplier m_5 which varied with channel sinuosity ($m_5 = 1.3$ if sinuosity exceeds 1.5). Thus,

$$n = (n_0 + n_1 + n_2 + n_3 + n_4)m_5. \qquad (3.20)$$

Scobey (1939) suggested an extra, additive, coefficient for the effect of sinuosity, increasing Manning's coefficient n by 0.001 for each 20° of channel curvature per 30 m of channel. Suspended sediment load tends to dampen turbulence (Vanoni and Nomicos, 1960), and Buckley (1923) produced a version of the Manning equation in which the resistance coefficient decreased with increasing suspended sediment concentration. At a given section, the roughness coefficient generally decreases with increasing water depth (Figure 3.3c), at least until overbank flow encounters the high resistance of floodplain vegetation. This decrease obscures the effects of two counteracting influences, however, namely the decrease in relative grain resistance as deeper water 'drowns' the grain roughness, which is the dominant effect in Figure 3.3(c), for a gravel-bed stream, and the change in bedforms that may occur in sand-bed streams (pp. 84–9). The most useful 'objective' methods of estimating a roughness coefficient are based on measurements of bed material size. Strickler's (1923) data on gravel-bed streams in Switzerland may be summarized as

$$n = 0.0151D_{50}^{1/6} \tag{3.21}$$

and Scott and Culbertson (1971), defining a dimensionless Chezy coefficient as $C_* = C/\sqrt{g}$, showed that plane sand beds experiencing sediment transport yield

$$C_* = 10.2D_{50}^{-1/2} \tag{3.22}$$

(in Equations (3.21) and (3.22), D_{50} is measured in millimetres). However, the empirical relationship developed by Limerinos (1969) is the most useful because it employs a relative 'smoothness' measure, and is close to the theoretical relationship expected from consideration of the two-dimensional vertical velocity profile (Boyer, 1954; pp. 68–71). This is (Figure 3.3d)

$$1/\sqrt{f_f} = 1.16 + 2 \log (R/D_{84}) \tag{3.23}$$

where D_{84} is the 84th percentile of pebble intermediate-axis measurements, in the same units as R, the hydraulic radius. A similar relationship was defined by Leopold and Wolman (1957), and approximated by a power function over a narrow range of relative roughness. The spacing and arrangement of larger roughness elements are also important on a hetero-geneous gravel bed, and Kolosseus and Davidian (1961) have studied the roughness effects of cubic elements and introduced a measure of their projected area exposed to the flow, as a proportion of total bed area, to a roughness equation similar to (3.23). Mirajgaoker and Charlu (1963) also show, in experiments with natural cobbles, that cobble density influences their frictional effect.

Boundary layer theory and the velocity distribution

Water flowing in river channels is retarded by the resistance of the bed and the banks. A very thin layer of water adjacent to the solid boundary is slowed to a stop, but shear resistance between adjacent fluid layers is less effective and the retardation diminishes with distance from the boundary. The influence of boundary drag is restricted to the 'boundary layer', within which a velocity gradient occurs (Figure 3.4a). In practice the boundary layer extends to the surface in rivers. Fluid remote from the boundary has a greater momentum per unit volume ($\rho_w v$) than fluid close to the boundary, but momentum transfer occurs from 'layers' of high momentum to those of low momentum. The amount of retardation of a faster layer reflects the shear resistance between layers, which in turn depends on the degree of interference between layers and the contrast in momentum between them (Carson, 1971, Chapter 2).

The nature of the velocity profile, which is initially considered in two dimensions for an infinitely wide flow with no side-wall effects, depends on the character of the flow. If the flow is laminar, layers of fluid appear to glide

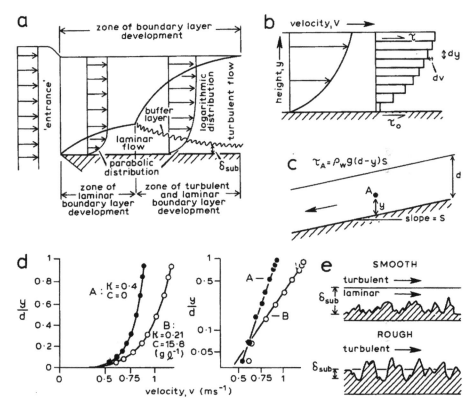

Figure 3.4 (a) Boundary layer development in a free stream passing over a solid surface (δ_{sub} is the laminar sublayer). (b) Velocity profile in laminar flow. (c) Definition diagram for shear stress variation with depth. (d) Velocity profiles in turbulent flow for different suspended sediment concentrations (after Vanoni and Nomicos, 1960). (e) Hydrodynamically smooth and rough boundaries.

over one another (Figure 3.4b) and interference between layers occurs at a molecular scale. Molecules move between layers, adding to the momentum of the slower layer and decreasing that of the faster layer, at a rate measured by the kinematic viscosity ν. As shown in Figure 3.4(c), the shear stress at any point is determined by the overlying water depth, according to

$$\tau = \rho_w g(d-y)s \tag{3.24}$$

where y measures height above the bed. It is also defined, as suggested above, by the momentum gradient and the degree of molecular exchange. Thus

$$\tau = \nu \, \frac{d(\rho_w \nu)}{dy} = \rho_w \nu \, \frac{dv}{dy} . \qquad (3.25)$$

If these equations are combined and rearranged, the rate of change of velocity with distance above the bed is

$$\frac{dv}{dy} = (gs\nu^{-1}) \, (d-y) \qquad (3.26)$$

which, on integration, gives the velocity profile, or the variation of velocity with height above the bed. This is

$$v = (gs/2\nu)(2yd - y^2) + C_1 \qquad (3.27)$$

where C_1 is a constant of integration. This shows that the velocity profile in laminar flow is an approximate parabolic function of distance from the solid boundary.

In turbulent flow, momentum exchange between 'layers' is achieved by eddies which act over distances far beyond the molecular scale. Thus, the high momentum of the upper flow can be transferred close to the bed, resulting in a rapid increase of velocity immediately above the bed (a steep velocity gradient). The opposite effect – water of low momentum being carried upwards – results in slower increases of velocity near the surface than occur in laminar flow (Figure 3.4d). The steep gradient near the bed is an important cause of suspension of sediment and of hydrodynamic lift forces entraining bed sediment in turbulent flow. The shear stress at any point in a turbulent profile is

$$\tau = (\nu + \epsilon) \, \frac{d(\rho_w \nu)}{dy} \approx \epsilon \rho_w \, \frac{dv}{dy}. \qquad (3.28)$$

The molecular viscosity may be neglected, being insignificant relative to the 'eddy viscosity' ϵ. An analysis of the velocity profile of turbulent flow is complicated, partly because the eddy viscosity is not a constant but varies with distance from the boundary, and partly because the complete profile is composed of three elements, the laminar sublayer, the buffer zone, and the main turbulent profile. In a thin layer close to the bed, viscous forces are significant and laminar flow occurs. The thickness of the laminar sublayer, δ_{sub}, is

$$\delta_{sub} = 11.6\nu/\sqrt{(\tau_0/\rho_w)} = 11.6\nu/v* \qquad (3.29)$$

(Keulegan, 1938). Here, v_* is an important quantity which is known as the 'shear velocity' and which converts the mean bed shear stress to a quantity with the dimensions of a velocity. Equation (3.29) indicates that the laminar sublayer is thinner under conditions of high bed shear stress, when turbulence penetrates closer to the bed. This element in the velocity profile is important if the bed sediment is fine silt and clay, which may be 'protected' from the effects of turbulence if its linear dimension is less than δ_{sub}. Above a transitional buffer layer, fully turbulent flow occurs in which velocity is often observed to increase as the logarithm of height (Figure 3.4d), according to the relationship (Carson, 1971)

$$v = b(\ln y - \ln y_0) = b \ln \left(\frac{y}{y_0} \right) \tag{3.30}$$

which involves two parameters, b the gradient and y_0 the (imaginary) height where velocity is zero (the intercept). The boundary roughness projection height (k) determines y_0. The boundary is 'hydrodynamically smooth' if $k < \delta_{sub}$ (Figure 3.4e), when the roughness is contained within the laminar sublayer. Under these conditions the intercept is independent of the roughness height, but decreases as shear stress increases, according to

$$y_0 = \frac{\nu}{9 v_*} \tag{3.31}$$

For the 'hydrodynamically rough' boundaries most common in rivers, strictly defined when $k > 5\delta_{sub}$,

$$y_0 = \frac{1}{30} k = \frac{1}{30} D_{65} \tag{3.32}$$

where D_{65} is referred to as the 'Nikuradse sand roughness'. Equation (3.32) is derived from experimental studies of flow in pipes with sand grains stuck to the wall to form an artificial roughness and a rigid boundary.

The gradient, b, can be defined using the Prandtl–von Karman theory (Prandtl, 1952). This theory considers the eddy viscosity ϵ to be

$$\epsilon = l^2 \frac{dv}{dy} \tag{3.33}$$

where l measures the depth of eddy penetration (the 'mixing length') and the velocity gradient controls the frequency of penetration. The mixing length is dependent on proximity to the boundary and is experimentally defined as

$$l = \kappa y \tag{3.34}$$

where κ is von Karman's universal constant, $\kappa = 0.4$. Substituting (3.33) and (3.34) in (3.28), and assuming $\tau = \tau_0$ close to the boundary, gives

$$\tau_0 = \kappa^2 y^2 \rho_w \left(\frac{dv}{dy} \right)^2. \tag{3.35}$$

Given the definition of the shear velocity ($v_* = \sqrt{(\tau_0/\rho_w)}$) this equation can be rearranged to yield

$$\frac{dv}{dy} = \frac{v_*}{\kappa y} \tag{3.36}$$

which provides the velocity variation with height after integration:

$$v = \frac{v_*}{\kappa} \ln y + C_2. \tag{3.37}$$

The constant of integration, C_2, can be defined simply because $v = 0$ when $y = y_0$, and therefore

$$0 = \frac{v_*}{\kappa} \ln y_0 + C_2. \tag{3.38}$$

Defining C_2 from (3.38) and substituting in (3.37), we obtain

$$v = \frac{v_*}{\kappa} \ln y - \frac{v_*}{\kappa} \ln y_0 = \frac{v_*}{\kappa} (\ln y - \ln y_0) = \frac{v_*}{\kappa} \ln \frac{y}{y_0}. \tag{3.39}$$

Thus the gradient b in (3.30) is v_*/κ and a measure of velocity shear (steepness of the velocity gradient) close to the boundary. Equation (3.39) can be simplified by inserting $\kappa = 0.4$, converting to common logarithms using $\ln x = 2.303 \log x$, and adding the relevant expression for y_0. For hydrodynamically smooth boundaries, y_0 is defined by (3.31), giving

$$v = 5.75v_* \log (v_* y/v) + 5.5v_* \tag{3.40}$$

and for rough boundaries, using (3.32), the velocity profile is

$$v = 5.75v_* \log (y/D_{65}) + 8.5v_* \tag{3.41}$$

These are the universal velocity profiles which define the logarithmic 'law of the wall' for rough and smooth turbulent flows.

The velocity profile equations incorporate a measure of bed shear stress τ_0 in the shear velocity term v_*. Thus, the fitting of the logarithmic profile to current meter measurements in a single vertical is an alternative means of estimating local bed shear, often preferable to using Equation (3.15) which demands accurate measurement of local slope. However, problems exist in applying a two-dimensional rigid boundary theory to flow in a three-dimensional mobile-bed channel. Strictly, the 'law of the wall' only applies close to the wall where the assumption that shear in the fluid equals shear at the bed ($\tau = \tau_0$), used to form equation (3.35), is valid; Figure 3.4(d), however, suggests that it does fit throughout the flow depth. This is not necessarily surprising, since most of the turbulent energy is generated close to the bed in a layer where shear stress varies by less than 10%. Nevertheless, amendments to the mixing length theory have been made: the relationship $l = \kappa y$ generally applies only to fully developed turbulence in an 'inner region' in the bottom 20% of the flow, and the mixing length becomes constant in an outer region where an empirical 'velocity defect law' is often applied (Simons and Şentürk, 1977, Chapter 3). Over a deformable boundary two limitations are apparent. First, grain roughness height is irrelevant in estimating the imaginary height of zero velocity when bedforms occur (see pp. 88–9), and second, the von Karman constant becomes a variable when sediment occurs in suspension. The energy required to maintain suspension is derived from the turbulent eddies, which dampens the turbulence and reduces the mixing length (Yalin and Finlayson, 1972). For high concentrations of suspended sediment (10–15 g l^{-1}) von Karman's κ (Equation 3.34) reduces to 0.2 (Vanoni and Nomicos, 1960), thereby steepening the velocity gradient and increasing mean velocity (Figure 3.4d), although this may be particularly apparent close to the bed where concentrations are highest and result in a departure from the logarithmic velocity curve (Yalin and Finlayson, 1972).

Finally, channel flow is three-dimensional, and the effect of confinement in a channel can be shown as follows. Integrating the rough boundary velocity profile equation (3.41) and dividing by the flow depth gives mean velocity as a function of 'relative smoothness':

$$\bar{v}/v_* = 5.75 \log (d/D_{65}) + 6.00. \tag{3.42}$$

It may be noted that this equation implies that the mean velocity in the vertical profile should be found at 0.63 of the depth measured from the surface (Boyer, 1954); current meter measurements are normally made at $0.6d$ to obtain the mean (p. 127). Inserting $v_* = \sqrt{(\tau_0/\rho_w)}$ and $\tau_0 = \rho_w gds$ (for infinite width $d = R$) into Equation (3.42), we obtain

$$\bar{v}/\sqrt{(gds)} = 5.75 \log (d/D_{65}) + 6.00 \tag{3.43}$$

and on substituting $f_f = 8gds/v^2$, this yields

$$1/\sqrt{f_f} = 2 \log (d/D_{65}) + 2.12. \tag{3.44}$$

This theoretical relationship of friction factor to relative roughness is compared with Limerinos' (1969) empirical curve in Figure 3.3(d), where it is an envelope curve for infinitely wide channels. For a given relative roughness, the friction factor is higher for three-dimensional channel flow, where side-wall friction is significant (Graf, 1966). Tracy and Lester (1961) show that velocity profiles in a smooth rectangular channel are unaffected by wall friction in the central section (Fig. 3.5a). This constitutes about half the total width in channels whose width:depth ratio is about 20, but is less extensive in narrower and deeper sections. Outside this region in which the flow is essentially two-dimensional, the mean velocity and the velocity gradient are both reduced (Figure 3.5a), and flow retardation close to the channel side wall is associated with depression of the velocity maximum beneath the

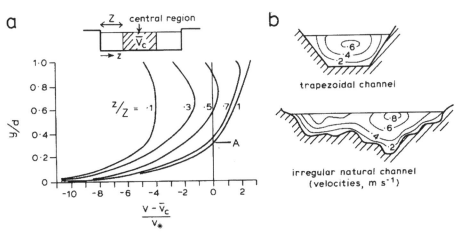

Figure 3.5 (a) Dimensionless vertical velocity profiles in the region of a smooth rectangular channel affected by side-wall friction (after Tracey and Lester, 1961). A is the mean velocity in the vertical in the central region. (b) Typical three-dimensional velocity distributions in channel sections.

surface because of secondary currents. The lower velocity gradient reduces bed shear stress near the wall (Cruff, 1965). *Rough* walls are even more effective and generate vortices which induce strong transverse secondary currents (Einstein and Shen, 1964). Furthermore, at the bank foot where curvature of the boundary is sharp a generalized law of the wall for three-dimensional boundaries predicts rotation of the velocity vector (Van Den Burg, 1975), and flow separation (see below) also occurs. These phenomena also help to encourage the transverse currents normal in a three-dimensional channelized flow, which are important precursors of meander patterns (pp. 189–91). In a channel, the velocity distribution involves slower moving water at the sides as well as at the bed (Figure 3.5b), so that *lateral* transfer of momentum also occurs. This is significant in the context of bank erosion (pp. 163–7).

Gradually varied flow

In gradually varied flow, streamlines are approximately parallel and the pressure distribution is hydrostatic. The total energy of a unit volume on a streamline is the sum of its potential energy of location (mgh) and its kinetic energy of motion ($mv^2/2$), where m is mass, h height and v velocity. At point A in Figure 3.6(a), the energy of a unit mass, E_m, is

$$E_m = \rho_w g h_1 + \rho_w g d \cos \theta + \rho_w v^2/2 \qquad (3.45)$$

where ρ_w, the density, is the mass of a unit volume, and the three terms on the right-hand side are the potential energy per unit volume, the pressure

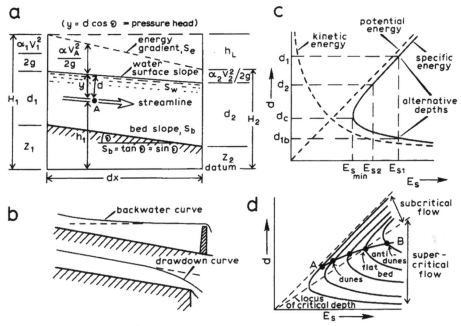

Figure 3.6 (a) Definition diagram for gradually varied flow. (b) Backwater and drawndown curves. (c) Specific energy curve for a given unit discharge. (d) Specific energy curves for a series of discharges.

energy, and the kinetic energy per unit volume respectively. For small slopes (θ), $y \approx d$, and the total energy of a unit volume at the bed in a single section (e.g. the upstream section in Figure 3.6a) is

$$E_m = \rho_w g \left(z_1 + d_1 + \frac{v_1^2}{2g} \right) \qquad (3.46)$$

The total energy per second for the discharge (Q) through this section is

$$E = Q\rho_w g \left(z_1 + d_1 + \alpha_1 \frac{\bar{v}_1^2}{2g} \right) \qquad (3.47)$$

where the coefficient α accounts for the non-uniform velocity distribution in the section. This is necessary because an average velocity head is required for the section, and the square of the mean velocity is less than the mean of the squares of the individual point velocities. Thus, α ranges from 1.03 for wide, deep sections to 1.36 in small sections where velocity is relatively more variable. The term in parentheses in (3.47) is the total head H, and

$$H_1 = z_1 + d_1 + \alpha_1 \frac{\bar{v}_1^2}{2g} \qquad (3.48)$$

is the Bernoulli equation, which defines the conservation of energy from section to section. In Figure 3.6(a), the energy, water surface and bed slopes are not parallel, the energy gradient being above the water surface (hydraulic gradient) by an amount equal to the velocity head. The fall of the energy gradient, which always slopes downstream, reflects energy losses caused by skin, form, internal distortion and spill resistances and sediment transport. Its relation to the hydraulic gradient reflects energy loss and the conversion between potential and kinetic energy implicit in the Bernoulli equation. If the hydraulic gradient is steeper than the energy gradient, accelerated flow occurs and potential energy is converted to kinetic energy, whereas in decelerating flow (Figure 3.6a) the hydraulic gradient is less steep and kinetic energy is converted to potential energy.

Between two sections separated by an incremental reach of constant discharge, the total energy (head) upstream equals that downstream plus the intervening head loss, according to the principle of conservation of energy. This results in an energy balance equation

$$z_1 + d_1 + \frac{\alpha_1 \bar{v}_1^2}{2g} = z_2 + d_2 + \alpha_2 \frac{\bar{v}_2^2}{2g} + h_L \qquad (3.49)$$

(Figure 3.6a), which applies to both uniform and gradually varied flows. Using the convention that decreasing elevation downstream gives a negative change of height, the head loss over the reach, $dH/dx = -\sin s_e = -s_e$ for small slopes. Similarly, $dz/dx = -\sin s_b = -s_b$. Thus, the energy balance equation may be expressed in differential form as

$$\frac{dH}{dx} = \frac{dz}{dx} + \frac{dd}{dx} + \frac{d(\alpha\bar{v}^2/2g)}{dx} \qquad (3.50)$$

and, using the bed and energy slopes, and noting that

$$\frac{d(\alpha\bar{v}^2/2g)}{dx} = \frac{d(\alpha\bar{v}^2/2g)}{dd} \frac{dd}{dx} \qquad (3.51)$$

this may be expressed as

$$\frac{dd}{dx} = \frac{s_b - s_e}{1 + d(\alpha\bar{v}^2/2g)/dd}. \qquad (3.52)$$

Since the energy slope, s_e, may be locally defined in gradually varied flow by a uniform flow equation, such as the Chezy equation ($s_e = v^2/C^2 R$), this can be inserted in (3.52), which becomes a non-linear differential equation solvable by a range of graphical and numerical integration techniques (Chow, 1959), to predict the unknown upstream depth if bedslope, width, roughness and downstream depth are known. Applications of this method are discussed in Chapter 6. Note that if dd/dx is positive, the surface diverges from the bed in a backwater curve, whereas when dd/dx is negative a drawdown curve is defined (Figure 3.6b).

The energy at the stream bed in a section is the specific energy, E_s:

$$E_s = d + \frac{\alpha \bar{v}^2}{2g} \tag{3.53}$$

Since $\bar{v} = Q/A$, and unit discharge q is Q/w (A is cross-section area, w is width), for a given discharge through a given section, this may be expressed as

$$E_s = d + \frac{\alpha Q^2}{2gA^2} = d + \frac{\alpha q^2}{2gd^2} \tag{3.54}$$

A specific energy curve (Figure 3.6c) sums the potential and kinetic energies and is a nonlinear function of depth for a given unit discharge, defining two 'alternative depths' for a given specific energy except where a single 'critical depth' occurs at the minimum specific energy. At this minimum, $dE_s/dd = 0$, and so the condition of critical depth is defined by differentiating (3.54) (with $\alpha = 1$) and setting the derivative equal to zero. Thus,

$$\frac{dE_s}{dd} = 1 - \frac{q^2}{gd^3} = 0 \tag{3.55}$$

and therefore

$$1 = \frac{q^2}{gd^3} = \frac{Q^2}{gA^2d} = \frac{v^2}{gd} = \frac{v}{\sqrt{gd}} = Fr. \tag{3.56}$$

Thus, the critical depth is that where the Froude number, Fr, is unity. Those depths on the upper limb of the specific energy curve are for subcritical flow ($Fr < 1$), while those on the lower limb occur in the alternative state of shallow depth, rapid velocity and supercritical flow ($Fr > 1$). A sudden change between alternative depths from the supercritical to subcritical states is a 'hydraulic jump'. If it is assumed that energy loss is negligible in a short reach, the total energy $H = E_s + z$ is constant and a specific energy curve defines the changes in water level that occur if bed elevation changes locally. If bed elevation z increases, specific energy E_s must decrease by a similar amount for H to remain constant (e.g. E_{s1} to E_{s2} in Figure 3.6c). In subcritical flow the water depth decreases (to d_2), whereas in supercritical flow it increases. This explains why water surface depressions occur over large dune bedforms in subcritical flow, but bedforms in supercritical flow are in phase with the fluctuations of the water surface (pp. 84–7). Dune bedforms tend to wash out into a flat bed if the Froude number is near unity, and Figure 3.6(d) suggests an interpretation of this behaviour. Separate specific energy curves are required for each unit discharge, and depth, specific energy and sediment transport rate all increase with discharge. The curve of increasing depth begins in subcritical flow at low discharges, but converges to the locus of the critical depth points; eventually, depth is critical at a certain discharge. This means that the sediment transport must be maintained by the minimum specific energy at that discharge, which is best achieved in the absence of bedforms which 'absorb' energy because of their form resistance. Thus, energy is not 'wasted' on scour, bedform troughs are filled, and the bed becomes plane.

The mechanics of flow–sediment interactions

The theoretical fluid mechanics and hydraulics models discussed above generally assume a rigid channel boundary. This is reasonable for gravel-bed streams when the finer sandy sediment size fraction is being transported over a static gravel surface at low discharges, but less so at the higher discharges responsible for gross channel morphology. In sand-bed streams, the mechanics of flow normally involve two-phase problems, with an interaction between the mobile channel perimeter sediments and the flow. Scour and fill during peak discharge events are pronounced, and at lower discharges sufficient sediment transport occurs to deform the boundary into ripples and dunes which strongly affect bed roughness, energy loss and velocity gradients. Initial mechanical interactions of the solid and fluid phases of the sediment transport problem include the determination of fall and traction velocities of individual sediment particles, and when sediment transport becomes significant, feedback to the flow properties occurs through the development of form roughness.

The fall velocity of particles in still water

In still water, a vertically falling particle briefly accelerates from rest until a constant terminal fall velocity is achieved. If this velocity exceeds that of the instantaneous vertical velocity components in turbulent flow, the particle will be deposited. The magnitude of the fall velocity reflects a balance between the downward-acting force due to the submerged particle weight and opposing forces due to viscous fluid resistance and inertia effects. A particle Reynolds number may be defined as

$$Re_p = \omega_0 D / \nu \qquad (3.57)$$

where D is the diameter and ω_0 the fall velocity. When Re_p is less than 0.1, for small particles in the silt–clay range (< 0.0625 mm diameter), viscous resistance dominates and inertia is negligible. The grains fall sufficiently slowly that the surrounding boundary layer is laminar and streamlines display 'fore-and-aft' symmetry with no separation from the particle and no wake development (Figure 3.7a). The fall velocity varies as the square of the particle diameter. For gravels (>2 mm), the particle Reynolds number exceeds 1000, and the high fall velocity induces a turbulent boundary layer, flow separation and a turbulent wake (Figure 3.7a). The fall velocity, which

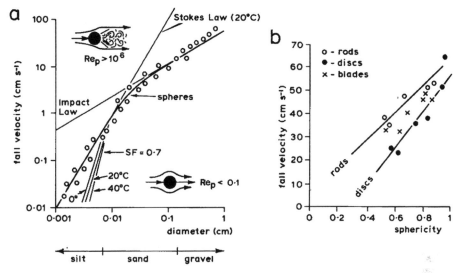

Figure 3.7 (a) Fall velocities for quartz spheres at 20° C (after Rubey, 1933a). Curves are shown for particles of shape factor 0.7 at various temperatures, and the insets show streamlines round particles at low and high fall velocity Reynolds numbers. (b) Fall velocities for rods, discs and blades compared to that for a sphere of equal size (after Krumbein, 1942).

varies as the square root of the diameter for these sizes, is controlled by inertia rather than by the negligible viscous resistance. Sand grains, however, in the intermediate 0.063–2 mm size range, are influenced by a combination of viscous and inertial forces.

The fall velocity of silts and clays is defined by Stokes' Law (Stokes, 1851), which balances a downward force due to the submerged particle weight (W) and a viscous resistance force (V). The submerged weight is the particle volume multiplied by the gravitational constant and submerged density. Assuming spherical particles

$$W = \frac{1}{6} \pi D^3 (\rho_s - \rho_w)g. \qquad (3.58)$$

The viscous resistance depends on the surface area, the dynamic viscosity, the relative velocity of water passing the particle (the fall velocity), and a drag coefficient which varies with the Reynolds number. These combine to give

$$V = 3\pi D \mu \omega_0. \qquad (3.59)$$

These forces are balanced when there is no acceleration and $W = V$. Equating the forces gives

$$\frac{1}{6} \pi D^3 (\rho_s - \rho_w)g = 3\pi D \mu \omega_0 \qquad (3.60)$$

from which the fall velocity can be defined as

$$\omega_0 = \frac{1}{18} D^2(\rho_s - \rho_w)g/\mu. \qquad (3.61)$$

In clear water at 20°C, by using the data from Table 3.1 (in cgs units) and assuming a sediment density (ρ_s) of 2.65 g cm^{-3}, this equation simplifies to

$$\omega_0 \approx 9000 D^2. \qquad (3.62)$$

Since the fall velocity of small particles is dependent on fluid viscosity, it varies with temperature, and at 0°C is approximately one-half that at 20°C. Furthermore, the presence of fine material in suspension increases the apparent viscosity of the water–sediment mixture (Simons et al, 1963), and 10% by weight of dispersed suspended clay reduces the fall velocity of individual sand grains by 30% to 80%, the latter for smaller grains whose 'true' fall velocity is 2.5 cm s^{-1}. An increase in viscosity, whether the result of lower temperatures or suspended sediment, has the effect of reducing the 'apparent' or 'effective' grain size, so that a sand grain of median diameter 0.28 mm appears to be of median diameter 0.11 mm according to its fall velocity in a 10% bentonite suspension (Simons et al, 1963). In streams, suspended sediment concentrations are highest near the bed, and thus a falling grain slows as it experiences increased apparent viscosity. However, silt and clay in suspension in natural stream water tend to flocculate, and the floccules settle at the velocities associated with large individual particles. The viscosity effect is, of course, the major reason for careful temperature control when Stokes' Law is used in sedimentation methods of particle size analysis.

   Particles coarser than 2 mm encounter resistance from the inertia of the water as they fall, and viscosity is unimportant. Rubey (1933a) equated the case of a grain falling at a constant velocity with that of an identical grain held stationary by the impact of a rising column of water. This 'impact force' (I), which balances the submerged weight force, is the momentum per unit time of the cylindrical column of water whose cross-section area equals the projected area of the spherical grain. The momentum is mass (volume × density) muliplied by velocity, and the volume per unit time is the cross-section area multiplied by velocity. Thus

$$I = \frac{1}{4} \pi D^2 \rho_w \omega_0^2 \qquad (3.63)$$

where the water velocity is equivalent to the particle fall velocity in still water. By equating I and W, the 'impact law' for fall velocities is

$$\omega_0 = \sqrt{\left[\frac{2}{3} Dg(\rho_s - \rho_w)/\rho_w \right]} \qquad (3.64)$$

which is the square root law for large grains. A composite law for sand grains (0.063–2 mm) affected by both viscous and inertial resistance is defined by equating W with the sum of I and V and solving for ω_0 (Rubey, 1933a) (Figure 3.7a). Particle shape is the predominant cause of departures from the fall velocities predicted for spheres in the larger size categories, although it is also significant for smaller grains. Using the shape factor $c/\sqrt{ab}$, based on axis measurements (Figure 1.5b), fall velocity curves for different shapes can be drawn, and since natural sediment particles have an average shape factor of about 0.7, the relevant curve is shown in Figure 3.7(a) for comparison with the theoretical curve for spheres. Krumbein (1942) classed test particles of constant weight as blades, discs and rods (Figure 1.5c), and Figure 3.7(b) shows that discs have the lowest and rods the highest fall velocity for a given sphericity. Heavier discs fall erratically: at low Reynolds numbers, discs fall steady and flat, but as the Reynolds number increases this action changes progressively from a regular lateral oscillation about a vertical axis to a gliding sideways-angled fall, and a tumbling motion. All these actions have been revealed by movie photography (Stringham *et al*, 1969). In the latter cases, the fall velocity over the vertical extent of fall is significantly less than over the longer path actually followed.

The initiation of particle movement

In natural streams the initial bedload transport process – rolling or sliding, depending on grain shape – requires the exceedance of a threshold flow intensity, since grains must be lifted from bed roughness 'hollows'. The flow intensity controlling initial particle movement and eventual sediment transport rate is measured by shear stress, velocity or stream power, and the critical flow has the minimum intensity capable of initiating motion of a particular sediment. Flow *competence* is the maximum particle size transportable, so at the initiation of bed material movement the flow is critical for the sediment and the sediment size is the competence of the flow. However, competence to *maintain* particle motion is usually greater than competence to *initiate* motion (Nevin, 1946), and at any time the apparent competence (maximum particle size in the prevailing load) may be less than the true competence, in the absence of larger sediment sizes.

The critical stage of initiation of transport is difficult to measure. Flume experiments suggest that increasing flow intensity over a smoothed bed eventually causes particles to vibrate with a frequency just less than the dominant frequency of turbulent oscillations (Lyles and Woodruff, 1972), then entrains firstly a few small grains, and finally a general movement of even the larger grains. In turbulent flow, fluctuations of velocity can cause spatially and temporally random movements at any flow intensity, and initial particle motion is a statistical phenomenon whose probability never becomes zero until still water occurs (Paintal, 1971). Visual identification of the threshold is therefore highly subjective. An alternative method, especially in rivers, involves relating bedload transport rate to shear stress or velocity and projecting the curve to zero transport; if the relationship is a power function with no zero, however, this fails. Curvature often occurs close to the threshold because of the statistical aspect of particle motions and the transport through the measuring section of imported particles already in motion (Figure 3.8a). Tagged particles, either painted (Keller, 1970) or wrapped in aluminium wire and traced by metal detector (Butler, 1977), may be used in gravel-bed streams. Distances travelled by particles of each given size are related to the peak shear stresses in a series of floods. This relationship is then used to identify the threshold stress for each particle size, this being the shear stress at which movement just occurs. These threshold stresses may then be shown to increase with particle size (Wilcock, 1971). However, the tagged particles travel through different reaches in which bed shear stresses and velocities vary, and it is difficult to correlate distance travelled to conditions in the initial section.

It is convenient to relate incipient motion to easily measured time-averaged properties of the whole flow, such as mean bed shear stress ($\tau_0 = \rho_w gds$) or velocity ($\bar{v}$). Hjulström's (1935) empirical curve (Figure 3.8b) predicts that the lowest threshold mean velocity occurs for well sorted 0.2–0.5 mm sands, while higher critical velocities are needed to entrain larger, heavier gravel and pebbles and the smaller, cohesive clays which are protected by submergence within the laminar sublayer. However, clays are normally eroded as floccules rather than as individual particles, and it is primarily the cohesion which discourages detachment. The critical mean velocity for a given sediment size varies with flow depth and slope (i.e. shear stress) and with sediment sorting and consolidation (Sundborg, 1956, 1967), and for gravels critical conditions exist for sliding and overturning (Novak, 1973). The velocity at which deposition occurs is about two-thirds of the threshold velocity for gravels, because of particle inertia. Generally it approximates to the fall velocity, but Francis (1973) found theoretically that the flow velocity at which bedload transport ceases is dependent on the shear velocity v_* ($= \sqrt{(\tau_0/\rho_w)}$) and on the fall velocity. Transport of suspended silt and clay is maintained over a wide range of flow velocities between the threshold and fall velocities. Figure 3.8(b) thus defines approximate conditions for the initiation of transport, deposition, and transport as bedload or suspended load, as well as the bedform regimes occurring in sands at progressively higher transport rates (Menard, 1950). The mean bed shear stress criterion of Shields (1936) recognizes that critical bed shear (τ_{0c}) increases with particle size, but also depends on the bed roughness condition. Thus, a dimensionless critical shear stress

$$\theta_c = \tau_{0c}/(\rho_s - \rho_w)gD \qquad (3.65)$$

is employed, relating the critical mean bed shear stress to the weight per unit

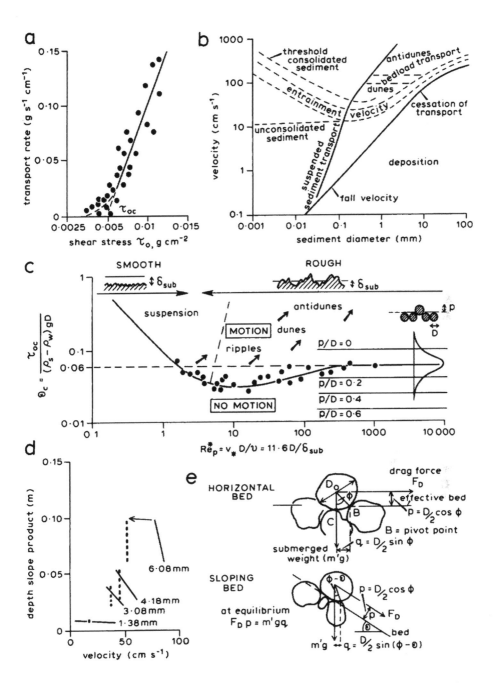

Figure 3.8 (a) Definition of critical shear stress using a sediment transport relation (after Johnson, 1943). (b) Threshold velocities (after Hjulström, 1935), and transport and bedform regimes. (c) The Shields entrainment function, showing the effects of relative particle exposure and the concept of probabilistic entrainment (after Shields, 1936, and Fenton and Abbott, 1977). (d) Depth–slope product and mean velocity at incipient motion for different sediments. Vertical dotted lines indicate constant bed velocities required for entrainment (after Rubey, 1938). (e) Forces acting on grains on flat and sloping stream beds.

area of a single layer of submerged grains. This is a function of bed roughness, defined by a particle shear velocity Reynolds number,

$$Re_p^* = v_* D/\nu = 11.6 D/\delta_{sub} \qquad (3.66)$$

(using Equation 3.29). On a smooth surface, when Re_p^* is less than 3.5, grains are well submerged in the laminar sublayer ($D/\delta_{sub} < 0.3$), and for a given grain size a larger shear stress is needed to initiate movement on progressively smoother surface ($Re_p^* \ll 3.5$; Figure 3.8c). On hydrodynamically rough beds, θ_c is independent of Re_p^* and becomes constant at $\theta_c = 0.06$. Thus,

$$\tau_{0c} = 0.06(\rho_s - \rho_w)gD \qquad (3.67)$$

on a rough surface. As bed shear stress decreases over a rough bed, the maximum sediment size movable decreases, but the laminar sublayer thickens until eventually $D = \delta_{sub}$. This occurs between 0.18 mm (Inman, 1949) and 0.7 mm grain sizes (Carson, 1971, p. 56). Finer particles than these are associated with smooth boundaries, which need higher critical shear stresses. Thus, the Hjulström and Shields curves are basically similar.

Experimental data reveal considerable scatter about the Shields curve (White, 1970), possibly reflecting differential grain exposures. Although the curve is based on evidence of entrainment on flat beds, this initial condition is difficult to achieve in fine sands, and Fenton and Abbott (1977) define a family of rough-bed critical states depending on relative grain projection (Figure 3.8c). They recommend using $\theta_c = 0.01$ for rough natural stream beds on which easily moved perched grains are inevitable. This mobile, unconsolidated type of stream bed is referred to as 'over-loose', while an imbricated, well-packed gravelly sediment would present an 'under-loose' bed with θ_c approaching 0.3, from which detachment is inhibited (Carling, 1983). Yang (1973) has simplified the practical application of this approach by introducing the parameter $\theta_c' = \bar{v}_c/\omega_0$ as an alternative criterion of sediment particle stability. Over rough beds, θ_c' is a constant implying $\bar{v}_c = 2.05\,w_0$ (i.e. critical velocity is approximately twice the sediment fall velocity).

In palaeohydraulic studies it is desirable to 'reconstruct' the flow intensity responsible for a particular deposit, and Baker and Ritter (1975) suggest a purely empirical relationship

$$D_{max} = 65\tau_0^{0.54} \qquad (3.68)$$

derived from data on maximum sizes moved in large floods (units are mm for D and kg m^{-2} for τ_0). This relationship predicts larger sizes to be mobile at lower shear stresses than the Shields curve, and smaller maximum sizes to be mobile at higher stresses. This is probably because bed velocity, shear and turbulence are greater in shallow streams and less pronounced in deep streams. Thus, local conditions at the bed control entrainment, and a general criterion based on mean flow properties, can only be an approximation. Data compiled by Rubey (1938) clarify this. Fine sediment entrainment seems to depend on a critical shear stress, over a range of mean velocities, while coarse sediment moves at a constant mean velocity over a range of shear stresses (Figure 3.8d). However, bed velocity is the crucial control, and this varies with shear velocity and bed roughness for a given mean velocity. A constant, critical bed velocity occurs for each sediment size. This measure of competence can be applied by fitting the logarithmic velocity profile to current meter data, and projecting it to the bed level. This method is of considerable utility in sediment transport studies as a means of obtaining bed velocity and *local* rather than mean bed shear stress (Bridge and Jarvis, 1977).

The stability of an average non-cohesive grain on a stream bed may be analysed using the turning moments acting about the downstream fulcrum (White, 1940). Roughly spherical, equally sized particles with an angle of internal friction $\phi°$ are arranged on a horizontal bed such that angle BOC *averages* $\phi°$ (Figure 3.8e). At limiting equilibrium moments due to fluid force and submerged particle weight just balance. If the number of grains per unit bed area is $n = \eta/D^2$, where η is a packing coefficient, the exposed area per grain is D^2/η and the drag force acting on each grain is the shear stress multiplied by this area, or $F_D = \tau_0 D^2/\eta$, assuming that the entire fluid drag is transmitted to the grains alone (i.e. no form drag due to bedforms). At limiting equilibrium,

$$F_D p = m'gq \tag{3.69}$$

(Figure 3.8e), and in terms of shear stress, with $\tau_0 = \tau_{0c}$, this is

$$\frac{\tau_{0c}D^2}{\eta}\frac{D}{2}\cos\phi = \pi\frac{D^3}{6}(\rho_s - \rho_w)g\frac{D}{2}\sin\phi \tag{3.70}$$

which reduces to

$$\tau_{0c} = \eta\frac{\pi}{6}(\rho_s - \rho_w)gD\tan\phi. \tag{3.71}$$

This may be modified for steep bed slopes (Figure 3.8e) and also requires adjustment for the additional downslope component of the gravity force for non-cohesive particles on the channel banks (see p. 283). Equation (3.71) provides the theoretical justification for the Shields entrainment function, θ_c. Furthermore, since shear stress is proportional to the square of velocity, it implies that critical velocity varies as the square root of the diameter. Thus the impact law is equivalent to Equation (3.71) and may be applied equally to entrainment of bed material and derivation of fall velocity (Rubey, 1938). Particle weight varies as the cube of the diameter, and by equating the concepts of competence and critical bed velocity, and noting that $v_c \propto \sqrt{D}$, it follows that the maximum particle weight transported by a current varies as the sixth power of velocity ($D^3 \propto v^6$); this is the so-called 'sixth power law'. However, in White's experiments, mean shear stresses at incipient motion calculated from $\tau_{0c} = \rho_w gds$ were about one-half those expected, for two reasons.

First, White (1940) neglected lift forces. Because of the steep velocity gradient at the bed, the velocity above a particle is markedly greater than beneath, and by the Bernoulli equation (Equation 3.45) this results in a pressure difference which causes a lift force (Einstein and El-Samni, 1949), which may be of the same order as the drag force (Chepil, 1958). Helley (1969) successfully predicted incipient motion of ellipsoidal pebbles using the principle of moments (3.71) to define drag force, but including an expression for the lift-force effect. On a smooth boundary, within the laminar sublayer, lift forces are negative and help to resist erosion. The second phenomenon ignored in White's experiments is the instantaneous velocity variation which occurs in turbulent flow. Kalinske (1947) estimated that normally distributed instantaneous velocities may be 2–3 times the mean bed velocity for about 5% of the time, and instantaneous drag and lift forces are therefore 4–9 times the mean. Cheetham (1979) estimated apparent bed shear stress in braided streams by lowering to the bed a plate mounted with fixed hemispheres on which loose spheres of varying density were placed. The 'competence' was defined by the heaviest sphere moved, and the apparent shear stresses implied were up to ten times those defined by $\tau_0 = \rho_w gds$, because of the combined effects of peak lift and drag forces induced by instantaneous turbulent velocity fluctuations.

The Hjulström or Shields curves are thus difficult to apply to natural rivers, where grain size heterogeneity, variable grain exposure and instantaneous velocity variations render incipient motion a probabilistic phenomenon (Yalin, 1977; Figure 3.8c). Keller's (1970) experiments on pebble movement indicate that travel distance varies with pebble size and shape and bed velocity, with pebble parameters being dominant controls of distance moved by pebbles starting from the deeper water of pools, and bed velocity dominating in the shallower, more turbulent flow of riffles. Butler (1977) also emphasizes initial location as a control of pebble movement in natural streams with non-uniform velocity distributions across the section. Bed relief, microtopography, sediment fabric and packing, and shielding effects are all sedimentological properties that render classic concepts of competence and critical motion inapplicable to heterogeneous natural bed sediments (Laronne and Carson, 1976). In addition, the erosion of cohesive bank material reflects aggregate and bulk mechanical properties rather than grain size and shape, and these are considered in Chapter 6 (pp. 163–9).

Bedform development

The friction factor is controlled by relative roughness in gravel-bed streams (Figures 3.3c,d), but mobile sand beds experiencing increasing flow intensity and sediment transport are moulded into a sequence of bedforms which introduce form resistance. Figure 3.9(a) summarizes the succession of forms identified in flume experiments (Simons *et al*, 1961, Simons and

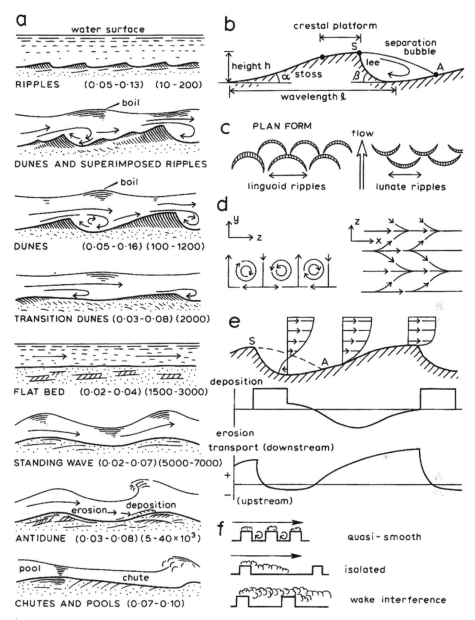

a
water surface

RIPPLES (0·05-0·13) (10 - 200)

boil

DUNES AND SUPERIMPOSED RIPPLES

boil

DUNES (0·05-0·16) (100 -1200)

TRANSITION DUNES (0·03-0·08)(2000)

FLAT BED (0·02-0·04) (1500-3000)

STANDING WAVE (0·02-0·07)(5000-7000)

erosion→ deposition
ANTIDUNE (0·03-0·08) (5-40×10³)

pool chute
CHUTES AND POOLS (0·07-0·10)

b
crestal platform
separation bubble
S
height h
α stoss
lee
β
A
wavelength ℓ

c PLAN FORM
flow
linguoid ripples   lunate ripples

d
y
z
z
x

e
S
A
deposition
erosion
transport (downstream)
+
−
(upstream)

f
quasi - smooth
isolated
wake interference

Figure 3.9 (a) The sequence of bedforms caused by increasing flow intensity, to observed in flume studies with sands of D_{50} ranging from 0.19 to 0.93 mm (after Simons and Richardson, 1966). Figures in parentheses are: first, the Darcy–Weisbach friction factor; second, bedload transport (ppm). (b) Form elements of ripples and dunes. (c) Linguoid and lunate three-dimensional ripples. (d) Flow patterns in Taylor–Görtler vortices, in section and plan (after Allen, 1969). (e) Velocity profiles over a dune, and associated patterns of sediment transport, erosion and deposition (after Bridge and Jarvis, 1977, and Allen, 1969). (f) Roughness due to cubic blocks on a plane bed.

Richardson, 1966, Guy *et al*, 1966) with associated friction factors and transport rates, the former being composite grain and form resistance measures whose lower values generally occur in the finer sands tested (Richardson *et al*, 1962).

Sediment transport occurring at shear stresses or bed velocities just above the threshold for motion deforms the bed into ripples, immediately in coarse sands but after a further increase in flow over fine sand. Ripples are from 2–100 grain diameters high, of triangular cross-section, and with height $h < 4$ cm and length $l < 60$ cm. Small or absent crestal platforms result in average steepness (h/l) values of 0.1, with maxima of 0.2. Larger ripples are steeper, since

$$h = 0.074l^{1.19} \quad (h, l \text{ in cm}) \tag{3.72}$$

(Allen, 1968). At higher shear stresses a transitional phase occurs of ripples superimposed on dunes, suggesting bimodality of bedform sizes, with two distinctive features. Dunes are larger, longer and flatter, with an average steepness of 0.06 and maxima of 0.1 (Yalin, 1977). Large dunes are less steep, since

$$h = 0.074l^{0.77} \quad (h, l \text{ in m}), \tag{3.73}$$

and may be up to 10 m high and 250 m long in major rivers such as the Mississippi. Their scale is such that their presence affects the water surface, which is depressed over the dune summit (see p. 75). Ripples and dunes have gentle stoss (upstream) slopes ($\alpha = 1 - 8°$, Figure 3.9b) and steeper lee (downstream) slopes which are approximately at the angle of repose. The lee face is formed by the avalanching of material transported up the stoss side to the crest, and because of this process the entire bedform migrates downstream. Individual ripples and dunes constitute an assemblage of randomly varying forms, and the variation of bed elevation along a rippled bed may be treated as a stochastic phenomenon (Nordin and Algert, 1966, Nordin, 1971). Ripple and dune bedforms may be essentially two-dimensional features, with straight crest lines transverse to the flow direction. However, a variety of three-dimensional forms also occur. These may have long sinuous crests, or take short-crested lunate or linguoid forms in which the crest lines have peak heights separated by troughs. Such bedforms thus have a *transverse* wavelength (Figure 3.9c) which is normally larger in flows with a high Reynolds number (Allen, 1969).

Ripples and dunes are *lower-regime* bedforms. At Froude numbers of 0.3–0.8 dunes elongate and flatten, washing out to a transitional flat-bed regime before *upper-regime* bedforms develop at high discharges and shear stresses. These are average transitional Froude numbers. In natural rivers particularly, upper-regime forms occur when the mean Froude number is significantly below 1, and ripples coexist in marginal, slower-flowing water, while antidunes occur at the thalweg where the local Froude number exceeds 1. During the flat-bed phase, roughness is reduced markedly, being controlled by grain resistance alone (Scott and Culbertson, 1971). A further increase in flow intensity leads to standing wave and antidune development and renewed form roughness. These roughly sinusoidal waves are in phase with water surface waves 1.5–2 times their amplitude, as expected in super-critical flow. Erosion on the downstream face and deposition on the adjacent upstream face results in upstream migration (hence, *anti*dunes) in some cases. Rapid energy dissipation occurs when the waves break, which destroys the bedform temporarily and gives way to a pattern of chutes and pools.

Bedform regimes may be defined empirically in relation to stream velocity (Figure 3.8b), shear stress (Figure 3.8c) or stream power (Figure 1.3a), and to sediment mobility, defined by sediment size or fall velocity. Criteria based on mean flow properties experience the limitations noted in an analysis of incipient motion, since bed conditions determine the changes between bedform types. Fluid viscosity controls effective sediment size via the fall velocity, and so variations in temperature and suspended sediment concentration are important and result in complex adjustments of form roughness. If the temperature drops when the bed is rippled, increased sediment mobility results in larger bedforms offering more flow resistance, but over a dune bed the result may be that the bedforms wash out and flow resistance decreases. This example illustrates the complex mechanics of bedform processes, which preclude deterministic understanding. However, it is evident that the transitional flat-bed is stable, becoming plane by washing out any temporary discontinuity, whereas ripples, dunes and standing waves/antidunes result from the instability of plane beds at certain flow stages when discontinuities are amplified. Ripples develop from disturb-ances in the laminar sublayer under conditions of relatively low bed shear stress. Chance discontinuities in the sublayer, with a height (h') of a few grain diameters, cause flow separation eddies of up to $100\ h'$ in length. Sedimentation in these eddies results in incipient ripple formation (Liu, 1957, Williams and Kemp, 1971). Dunes involve more funda-mental disturbance of the velocity gradient, with large-scale 'roller eddies' producing alternate flow acceleration and deceleration, which causes bed scour and accretion, with a wavelength between successive scour points of $2\pi d$, which is the average dune wavelength (Yalin, 1971, 1977). Once initiated, dunes disturb the flow to cause flow separation. Streamlines at the bed are skin-friction lines, along which the bed shear acts and the bed sediment travels. The skin-friction line departs from the bed at the 'separation point' (S, Figure 3.9b) and re-attaches at the 'attachment point' (A), enclosing a separation bubble where reverse eddies occur. These are 'rollers', with axes transverse to the flow. Eddies with axes parallel to the flow are Taylor–Görtler vortices (Allen, 1968), which cause the longi-tudinal flow disturbances responsible for three-dimensional bedforms

(Figure 3.9d). Separation roller eddies may provide a strong enough reverse flow in the lee of large dunes to form counter-current ripples facing upstream, which are buried by the advancing dune. The final instability process, causing standing waves, is the result of free surface instability in supercritical flow. Water surface waves above a flat bed cause an acceleration of flow in the wave troughs and a deceleration beneath the wave itself; the resulting erosion and accretion generates a bed wave in phase with the surface wave. Kennedy (1963) used hydrodynamic stability analysis to show that standing waves require average Froude numbers of 0.84 and develop wavelengths of

$$l = 2\pi v^2/g. \tag{3.74}$$

Since $Fr = v/\sqrt{(gd)} \approx 1$, this is approximately $2\pi d$.

Bedform development has several important effects. First, bedforms locally distort the vertical velocity profile (Bridge and Jarvis, 1977), with reverse flow in the separation bubble and increasing velocity shear towards the crest up the stoss slope (Raudkivi, 1963) as flow acceleration occurs (Figure 3.9e). When the logarithmic velocity profile is fitted to measurements over a dune bed, the imaginary height at which velocity is zero is found to be much greater than one-thirtieth of the grain size (Equation 3.32), because the bed roughness is controlled by dune size rather than grain dimensions. Secondly, the bed shear stress generated by the flow is balanced in uniform flow by the sum of grain and form resistances to flow. As a result of flow separation the form drag is caused by pressure differences between the dune stoss and lee. Only that part of total bed shear associated with grain resistance is responsible for bedload transport (Einstein, 1950), and as a result the threshold shear stress for incipient movement is higher on an initially rippled bed. Thirdly, form roughness affects energy loss and, as a result, flow depth and velocity. Ripples have an effect on energy loss in the River Jizera in Czechoslovakia equivalent to a doubling of channel length by meandering (Martinec, 1972), and dune-bed friction factors may be six times the flat-bed grain roughness values (Raudkivi, 1967). Thus, depth and velocity depend on the bedform regime. Discontinuities occur in depth–discharge relations when roughness decreases sharply as dunes wash out, causing velocity to increase sharply and depth to become almost constant as discharge increases (Dawdy, 1961, Beckman and Furness, 1962). However, hysteresis occurs in bedform development, which lags behind the discharge changes (Simons and Richardson, 1962). Bedforms in rivers with varying discharge are often out of equilibrium with prevailing flow characteristics, and are therefore rather different from the stabilized equilibrium forms developed in flume experiments (Allen, 1976a). For example, during the rising limb of a flood peak, small partially developed dunes lag behind the increasing discharge, and large dunes inherited from the peak discharges occur on the falling limb.

Bedform resistance effects are suggested by experiments using rigid triangular or cubic roughness elements. Closely spaced cubes behave like an elevated smooth bed, and widely spaced cubes also present little flow resistance. However, at intermediate cube height:spacing ratios of about 0.1, the separation wake of each cube interferes with that of the next downstream (Figure 3.9f) and roughness is maximized (Johnson, 1944). At this spacing, roughness increases with increasing cube height:flow depth ratio, or relative form roughness (Robinson and Albertson, 1952). Thus, the height:length ratios of ripples and dunes appear appropriate to maximize their resistance to flow (Davies and Sutherland, 1980, Yalin, 1977), which suggests that their development represents a negative feedback which permits efficient energy dissipation. By adding pegs to cubic roughness elements Einstein and Banks (1950) simulated grain and form roughness, and found their effects to be additive. This leads to the concept of separate friction factors

$$f_f = f'_f + f''_f \tag{3.75}$$

where $f_f = 8gds/\bar{v}^2$ and is the total flow resistance, f'_f is the grain roughness and f''_f is the form roughness. Grain roughness can be estimated using a relative roughness relation such as Equation (3.23) (Chang, 1970), and form roughness is a function of dune height, projected area and flow depth. For example, Vanoni and Hwang (1967) developed the relationship

$$1/\sqrt{f''_f} = 3.5 \log (R_b/\epsilon h) - 2.3 \tag{3.76}$$

where R_b is the hydraulic radius of the bed, ϵ is the horizontally projected area of lee faces as a proportion of total bed area, and h is the dune height. This represents a 'relative form roughness' relation similar to the 'relative grain roughness' relation of (3.23) (Figure 3.3d). Since an increase of roughness results in greater depth and steeper energy gradient, it is possible to partition the hydraulic radius (Einstein and Barbarossa, 1952) or energy slope (Engelund, 1966) into components resulting from grain and form roughness, so that, for example,

$$\tau_0 = \tau'_0 + \tau''_0 = \rho_w g R's + \rho_w g R''s. \tag{3.77}$$

This divides bed shear stress into those components 'carried' by grain and form resistance, of which only the former provides the drag force on the grains which causes sediment transport.

4

Sediment transport processes

The dynamics of sediment transport complicate attempts to quantify both the influence of transport mode and intensity on channel form and the total denudational yield of solutes and sediments from a catchment. Solute, suspended sediment and bedload transport are successively more discontinuous, responding variously to changes of stream discharge and sediment supply during storm events. Transport of solids is controlled not by stream discharge but by associated flow characteristics such as bed velocity, shear stress or stream power, which vary along a reach of constant discharge to cause spatial variations in competence and discontinuous bedload movement. However, empirical 'rating curves', usually simple power functions, are commonly used to define the general variation with *discharge* of the concentration (c) and load (G) in each transport mode. Concentration measurements in $mg\,l^{-1}$ (weight per volume) or ppm (weight per weight) are interchangeable up to concentrations of about $7000\,mg\,l^{-1}$, but concentrations in $mg\,l^{-1}$ are strictly dependent on temperature and on excessive solute concentrations both of which affect fluid density. The load is the product of concentration and discharge ($G = cQ$) and is measured in units of $kg\,s^{-1}$ if concentrations in $mg\,l^{-1}$ are multiplied by discharges in $m^3\,s^{-1}$ and divided by 1000. Rating curves for solute transport (subscript d) are

$$c_d = a_1 Q^{-b} \quad \text{and} \quad G_d = a_1 Q^{1-b} \tag{4.1}$$

where a_1 and b are empirical constants. Solute concentrations normally decrease with increasing discharge, but if $|b| < 1$ the load still increases. Suspended sediment ratings (subscript s) for a section or reach are

$$c_s = a_2 Q^b \quad \text{and} \quad G_s = a_2 Q^{b+1} = a_2 Q^j. \tag{4.2}$$

The exponent j normally lies in the range 2–3 (Leopold and Maddock, 1953a), so concentrations commonly increase with discharge at a section.

A bedload rating would normally be

$$G_b = a_3(Q - Q_c)^b \tag{4.3}$$

because a threshold flow intensity is required to initiate bed material transport. Concentrations are less meaningful and difficult to measure because transport is confined to a restricted bed layer.

These empirical models have commonly been based on infrequent, random water and sediment samples. Walling (1975), however, has emphasized that frequent systematic sampling is an essential basis for an understanding of sediment and solute dynamics. Automated sampling techniques, direct monitoring and continuous recording provide a data base which reveals the many limitations of both simple empirical rating curves and complex, but often unrealistic, theoretical models. Theoretical approaches to sediment transport invariably assume full capacity loads controlled by flow hydraulics. Continuous data reveal that catchment hydrological controls of sediment supply are often equally important influences of the load carried at a particular moment.

Solute transport dynamics

Solute transport has little effect on alluvial channel form, which reflects the disposition of solid sediment. However, interaction continually occurs between solutes and suspended sediments as dissolved ions are adsorbed onto the surfaces of fine suspended particles, and partial dissolution of mineral particles provides a source of solutes (Lemmens and Roger, 1978). Solutes may dominate the denudational process in humid climates, but catchment losses of specific ions only represent products of geochemical weathering after adjustment is made for rainfall and pollutant inputs and biological uptake in non-equilibrium vegetation systems (Likens *et al*, 1967). Inverse solute concentration–stream discharge relationships reflect the dilution of baseflow (groundwater and interflow) by stormflow of low ionic concentration. However, the solute concentration of a given discharge varies seasonally partly because the quickflow:delayed flow proportions differ and partly because of seasonal biological influences, such as the secretion of silica by diatoms in summer. Solute dynamics thus reflect nutrient cycling processes, with separate nutrients behaving uniquely both seasonally and in storm periods, so that the total solute concentration is a complex summation of these separate ion variations.

Sampling and measurement of solutes

Lateral dispersion of solutes from a mid-channel point source approximates to a turbulent diffusion process (Sayre and Chang, 1968) in which a symmetrical distribution about the channel centre line spreads and flattens with distance from the source. Cross-channel solute distributions become uniform over 'mixing lengths' of 25–250 m in narrow, fast-flowing Alpine streams (Day, 1975). Thus, in general, natural solutes from diffuse sources may be sampled with confidence by single withdrawals from mid-channel, since in turbulent streams cross-channel variation in concentration is normally less than 5% (Johnson, 1971). Even under non-steady flood conditions in the River Tyne, transverse bank-to-centre concentration gradients are less than 0.05 ppm m^{-1} (Glover and Johnson, 1974). Cross-channel sample integration is more appropriate in deep, sluggish reaches, or in wide, shallow streams where secondary current cells inhibit complete lateral mixing. For example, the Susquehanna River at Harrisburg, Pennsylvania, has an asymmetric solute distribution, with high east bank sulphate concentrations reflecting coalmine drainage and high west bank bicarbonate levels derived from tributaries draining limestone outcrops (Anderson, 1963). If the lateral concentration gradient is negligible, automatic pump sampling from a single point in the section is acceptable. However, the samples are exposed to atmospheric and container contamination; normally, samples should be filtered to prevent any chemical interaction with suspended solids, and stored for no more than 72 hours before laboratory analysis (Johnson, 1971).

Total solute concentrations may be measured by weighing the residues after evaporating known, filtered sample volumes over a steam bath. However, the systematic increase in electrical conductivity of water with increasing ionic concentration enables measurements of specific conductance (micromhos per centimetre of water, μS cm^{-1}) to be calibrated against known concentrations, and conductivity meter output currents may be continuously recorded in field installations. Calibration is slightly ion-dependent (Figure 4.1a) and strongly temperature-dependent; conductance increases by about 2% per degree Centigrade and is usually adjusted for comparative purposes to a standard temperature of 25°C. Specific cation concentrations (K^+, Na^+, Ca^{2+}, Mg^{2+}, Si^{4+}, and various metals) are determined by atomic absorption spectrophotometry (Fishman and Downs, 1966), although specific ion electrodes may be more accurate at low concentrations in glacial meltwaters containing rock flour particles (<1 μm) not removed by normal filtration (Reynolds, 1971). A range of laboratory techniques including colorimetry, spectrophotometry and titration (Rainwater and Thatcher, 1960) are used for anion determination (SO_4^{2-}, Cl^-, NO_3^-, HCO_3^-). Water hardness due to calcium and magnesium salts in solution is measured by titration with EDTA (ethylenediamine tetra-acetic acid) in the laboratory, but simple field determinations using hardness indicator tablets yield results comparable with laboratory measurements (Douglas, 1968b). These techniques are particularly useful in the analysis of river solute loads in limestone areas.

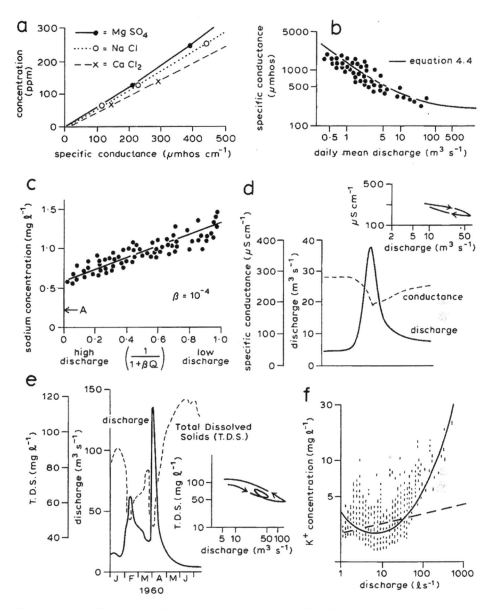

Figure 4.1 (a) Specific conductance–concentration calibration curves for three salts. (b) Solute rating, based on conductance measurements, for the San Francisco River, Arizona; rating curve derived using a three-component mixing model (Equation 4.4) (after Hem, 1970). (c) Variation of sodium concentrations in stream water as a function of discharge in a forested catchment. A is mean concentration in net precipitation (after Johnson *et al*, 1969). (d) Hydrograph, chemograph and clockwise looped rating for the River Creedy at Cowley (after Walling and Foster, 1975). (e) Hydrograph, chemograph and anticlockwise looped rating for Spring Creek, Georgia (after Toler, 1965). (f) Quadratic potassium rating curve for a small catchment in Devon; maximum baseflow discharge is $8–10 \, l \, s^{-1}$ (after Foster, 1980).

Mixing models: theoretical approaches to solute dynamics

Mixing models are simple mass balance equations for mixtures of component solutions of varying volumes. They assume ideal (complete) mixing, no chemical reaction on mixing, no evaporative concentration of solutes, and no biological uptake (Hem, 1970). Assuming distinct, constant solute concentrations for delayed flow and quickflow, a two-component mixing model describes the dilution effect of the latter and is used in hydrograph separation (Equation 2.11). For example, in glacial meltwater streams snow and glacier melt of low solute concentration ($0.1–10$ mg l^{-1}) mixes with the more concentrated solute load in baseflow, and streamflow concentrations reflect the relative volume and solute load of these two sources (Rainwater and Guy, 1961). A mass balance equation for multiple runoff sources is also possible. Hem (1970) identified three sources with distinctive discharge and conductivity properties contributing to the San Francisco River, Arizona: a saline spring of 0.06 m^3 s^{-1} constant flow and $16,000$ μS cm^{-1} conductance; a baseflow of conductance 500 μS cm^{-1} and discharge up to 2.83 m^3 s^{-1}; and storm runoff characterized by a specific conductance of 200 μS cm^{-1} and flows in excess of 2.89 m^3 s^{-1}. The concentration (represented by conductivity c_d') of the combined flow at any time is satisfactorily predicted by

$$c_d' = \frac{c_1' Q_1 + c_2' Q_2 + c_3' Q_3}{Q_1 + Q_2 + Q_3} \tag{4.4}$$

in which the sum of the three component loads is divided by the total discharge (Figure 4.1b). In the absence of data on solute concentrations of component flows, a mixing model may be derived by expressing the mass balance in terms of quickflow (subscript R) and delayed flow (D) volumes and assuming a storage volume–outflow relationship for discharge of the form $V_R = \gamma Q$, where γ is a measure of the residence time of storm runoff moving through the catchment. The mass balance is therefore

$$c_D V_D + c_R \gamma Q = c_d (V_D + \gamma Q). \tag{4.5}$$

If Δc is the difference between the two constant concentrations ($\Delta c = c_D - c_R$) and $\beta = \gamma / V_D$, this becomes

$$c_d = \left(\frac{1}{1 + \beta Q} \right) \Delta c + c_R. \tag{4.6}$$

Linear regression of c_d on the 'inverse discharge' term in parentheses yields Δc as the gradient and c_R as the intercept, with β being adjusted to maximize linear correlation. In a forested catchment, Johnson et al (1969) found this model appropriate for discharge-related variations of individual ion concentrations, with $\beta = 10^{-4}$ or 10^{-5} (Figure 4.1c). Mixing models can be further generalized by incorporating variable solute concentrations in the component flows (Nakamura, 1971, Walling, 1974). For example, empirically observed quickflow and delayed flow dilution processes for conductance (c_d') may be combined to give a complex mixing model (with discharge here in $1 s^{-1}$)

$$c_d' = [(204Q_R^{-0.04}Q_R) + (251Q_D^{-0.03}Q_D)]/(Q_R + Q_D) \qquad (4.7)$$

(Walling, 1974), but even this only explains 53% of the variation of conductivity, because it fails to account for flushing and hysteresis effects in storm runoff events.

Empirical evidence of solute dynamics

Mixing models describe a generally non-linear dilution process in which a near constant baseflow concentration exists at low flows and streamflow solute concentrations decline when quickflow begins. Ultimately a lower, steady concentration comparable to that in rainfall occurs at high discharges (Hem, 1970). The exponents in simple concentration–discharge relationships (Equation 4.1) thus reflect the discharge range sampled (Foster, 1978a). The dilution process results in a 'chemograph' trough (Figure 4.1d), but this is rarely a mirror image of the hydrograph peak, so hysteresis concentration loops occur which the mixing models cannot accommodate.

If the chemograph trough lags behind the hydrograph peak, clockwise hysteresis occurs, with higher concentrations on the rising limb (Figure 4.1d). This is particularly evident when a 'flushing' event occurs at the beginning of a storm, as chemically enriched soil moisture is displaced into the stream. Under these conditions solute concentrations may even increase with discharge on the rising limb (Oxley, 1974). However, dilution may occur if an initial hydrograph peak occurs as a result of direct channel precipitation; streamflow solute concentrations then increase during a subsequent throughflow discharge peak, when accumulated weathering products are flushed from the soil as the soil moisture deficit is replenished and the water table rises (Burt, 1979). Data on storm-period solute response at Cowley on the River Creedy, Devon, indicate that chemograph lag times are smaller in major floods and longer when dry antecedent conditions permit salt accumulation in the soil which intensifies the initial flushing effect (Walling and Foster, 1975). The chemograph lag is, however, a spatial as well as a temporal variable, and because a flood wave generated by a local storm travels downstream from its source faster than the river water itself, as a kinematic wave, the chemograph lag increases progressively downstream (Glover and Johnson, 1974). Changes in the storage-outflow discharge relation also induce hysteresis by changing the proportions of quickflow and delayed flow at different stages of a storm hydrograph; drainage of river water into bank storage at peak flow prevents baseflow discharge, for example, and causes a decrease in solute concentrations in streamflow (Hendrickson and Krieger, 1964). The time delay before solute-rich subglacial meltwater reaches a glacier snout may cause similar hysteresis in diurnal melt-controlled streamflow variations in glacier streams, again reflecting a changing balance between two components of flow (Collins, 1979). Anticlockwise hysteresis, with higher concentrations on the falling limb, is sometimes evident in limestone areas where good hydraulic connections between groundwater storage and the river allow increased groundwater discharge coincident with the flood peak, but Figure 4.1(e) indicates that this operates over relatively longer time scales (Toler, 1965).

Other controls of solute concentrations are numerous. Seasonal effects interact with geology and land use as expansion of the saturated runoff-producing area incorporates varying lithological and surface conditions (Walling, 1971). Accumulated salts are flushed in autumn, but supply is exhausted in spring. Some ions (K^+, Ca^{2+}, Mg^{2+} and $N-NO_3^-$) experience a seasonal change in their relation to discharge because they accumulate during the summer, while the Na^+ and Cl^- ions show no seasonal variation especially in maritime catchments where they originate from atmospheric sources (Foster, 1978b). Individual components of total solute load thus behave differently. In limestone areas, rating curves for Ca^{2+} and Mg^{2+} parallel that of total solute concentration, since these cations predominate. Dissolved organic concentrations tend to be independent of discharge, because of buffering in the soil and the cancelling effects of out-of-phase carbon production cycles (Arnett, 1978). Some ions even require polynomial ratings: for example, the potassium curve in Figure 4.1(f) demonstrates roughly constant concentration during baseflow and a sharp increase as quickflow detaches mobile potassium from the soil–clay colloids with which it is associated (Foster, 1980).

Suspended sediment transport

Suspended sediment, supported by the vertical velocity component of turbulent fluid eddies, includes wash load and suspended bed material load. Wash load originates in slope runoff, and its source area reflects the process of runoff generation. Bank erosion also contributes to the wash load, with loose weathered sediment removed by rising flood stages and bank collapse occurring on falling stages (p. 164). Wash load is generally fine sediment readily transported even at low discharges and at rates controlled by the supply of sediment to the stream. Suspended bed material load occurs at high 'transport stages', measured for example by the ratio v_*/v_{*c} (shear velocity/the threshold shear velocity defined by the Shields entrainment function of Figure 3.8c). Homogeneous bed material experiences changing transport modes (Figure 4.2a) at higher transport stages, with saltation being interrupted by grain collision or partial suspension at stages above 2,

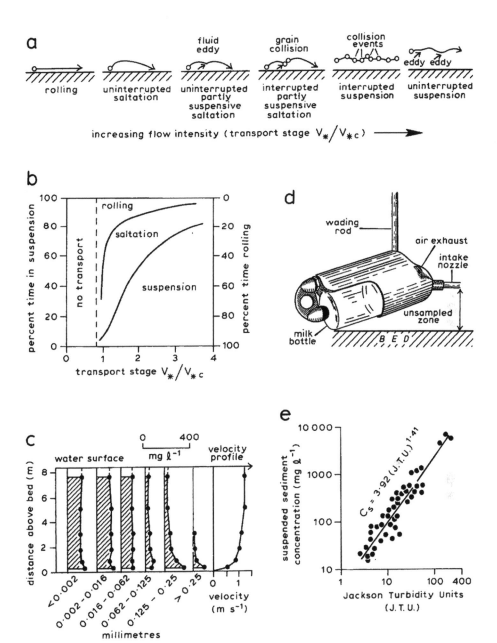

Figure 4.2 (a) Transport modes experienced under conditions of increasing flow intensity (after Leeder, 1979). (b) Percentage of time single particles experience rolling, saltation and suspension during transport at different transport stages (after Abbott and Francis, 1977). (c) Vertical profiles of suspended sediment concentrations of different size fractions, Mississippi River at St Louis (after Colby, 1963). (d) The US DH-48 suspended sediment sampler. (e) Calibration of turbidity units against suspended sediment concentrations, Sleepers River watershed, Vermont (after Kunkle and Comer, 1971).

when bed shear stress is four times the threshold shear stress (since $\nu_* = \sqrt{(\tau_0/\rho_w)}$). Fully suspensive transport occurs at even higher transport stages (Leeder, 1979), although experiments indicate that at $\nu_*/\nu_{*c} \approx 2$ individual grains introduced to flow over a static rough bed are suspended for 50% of their time in transport (Abbott and Francis, 1977; Figure 4.2b). The finest sediment in a heterogeneous bed may be suspended at relatively low shear stresses unless protected by coarse surface 'armour'.

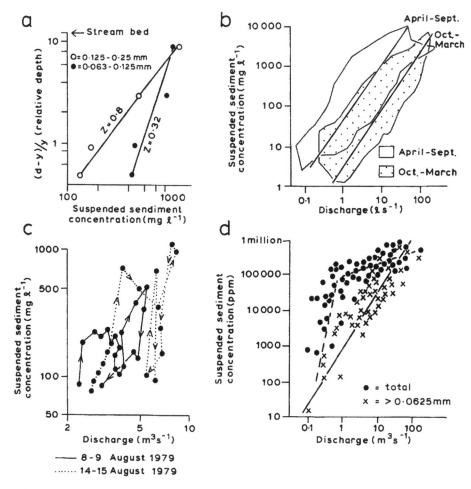

Figure 4.3 (a) Vertical concentration gradients of suspended sediment in the Rio Grande near Socorro (after Nordin and Dempster, 1963). (b) Seasonal variation in suspended sediment concentration in the Rosebarn catchment, Devon (after Walling, 1974). (c) Storm-period suspended load variation in Storbregrova, the meltstream draining the Storbreen glacier catchment, in the Jotunheimen, Norway. Two storm-period hysteresis loops are plotted. (d) Suspended sediment concentrations in the Paria River, Arizona (after Gregory and Walling, 1973). Coarser suspended bed material is closely related to discharge, but wash load exhaustion occurs at high discharges.

These two components are simultaneously sampled and cannot be separated; indeed, fine bed material may originally have been wash load in a prior flood. However, an arbitrary distinction based on sediment size may be justified by two criteria. First, the two components behave differently during floods. Bedload and suspended bed material load vary in phase with flood discharge (Figure 4.3d), whereas wash load transport peaks early in a flood while rain and surface runoff are still occurring (Einstein *et al*, 1940). In the Enoree River, transport of sediment coarser than 0.35 mm is closely related to discharge, while finer sediment – the wash load – is not. Modal bed material diameter in the Rio Grande is 0.5–1 mm, but suspended sediment is trimodal, and two of the modes are the clay (<0.002 mm) and silt (0.008–0.016 mm) components which represent wash load sediments unrepresented in the bed. The third sandy mode matches the dominant bed material size (Nordin and Beverage, 1965). The second general criterion distinguishing these two components of the suspended load is the vertical concentration gradient, which is commonly uniform for wash load but steeply decreasing with height above the bed (Figure 4.2c) for suspended bed material load (Colby, 1963). Conventionally, both of these criteria tend to associate wash load with the silt and clay (<0.0625 mm) sizes, but obviously in a cobble-bed stream the wash load may be sandy.

Measurement of suspended sediment concentration

Suspended sediment is non-uniformly distributed with depth and across the channel, especially where secondary currents are well developed, and vertical and transverse integration is therefore necessary during sampling. Sampling devices require streamlining to minimize turbulence, fins to permit correct orientation in the flow, and nozzle entrance velocities equal to stream velocity. Depth-integrating samplers, such as the US DH-48 sampler used in wadeable rivers (Figure 4.2d) or the US DH-59 cableway sampler, have continuous intakes and are employed in the Equal Transit Rate method, being lowered to the bed in several verticals at a constant velocity equal to about 40% of the maximum flow velocity in a midstream vertical, to give a sample weighted according to the discharge variation across the section. To define the concentration gradient in one vertical, point-integrating samplers (e.g. the US P-61) are used, and these open at the required depth for a finite time. These samplers all fail to sample close to the bed (Figure 4.2d) where suspended bed material concentrations are highest. The theoretical suspended sediment distribution of Equation (4.10) indicates that for fine sediment or turbulent conditions (low z exponent), the unsampled proportion is small because the distribution is uniform (Chien, 1952). Large errors arise for coarse sediment and weak turbulence (high z exponent) because the concentration gradient is steep and the load close to the bed is a large percentage of the total. Errors are also significant in shallow streams (<1.5 m deep) where the unsampled zone is a larger proportion of the depth.

Automatic pump samplers (Walling and Teed, 1971) can be programmed to sample at specified time intervals or at rates proportional to stream discharge (Fredriksen, 1969). With single intakes, vertical and lateral integration is impossible and calibration against hand samples is necessary. Pump pressures are constant and intake pipe flow velocities are fixed across the discharge range. Thus, coarse suspended sediment of high fall velocity may be systematically unsampled, an error not readily corrected by simple calibration. Indirect automated monitoring by turbidity meters, which measure the attenuation of an artificial light source by suspended sediment, is particularly useful where brief 'pulses' of high sediment concentration occur (Fleming, 1969b; Truhlar, 1978). Standard 'turbidity units' are calibrated against measured concentrations (Figure 4.2e), but this calibration varies with the optical properties of sediments, which are influenced by particle size, shape and mineralogy. Calibration can therefore only be considered valid for data from a single stream. Radioisotope methods based on attenuation of an x-ray beam are more expensive, but more reliable, since calibration appears to be independent of size and type of sediment and the beam orientation across the flow (McHenry et al, 1970). These varied automatic sampling and monitoring techniques have been instrumental in providing flood-period data on suspended concentrations (see pp. 103–6).

Laboratory analysis of samples requires vacuum filtration through pre-weighed and oven-dried filters of appropriate pore size, and reweighing the filter plus retained sediment after drying at 80°C; the suspended sediment concentration is the dry sediment weight divided by the filtrate volume ($mg\,l^{-1}$). Fine sediments passing through the filter pores must be treated as part of the solute load.

Other properties analysed in the laboratory include sediment size, mineralogy and organic fraction. The Coulter Counter (Fleming, 1967, Walker et al, 1974) provides a useful approach to size analysis of the small quantities of fine sediment usually sampled from suspensions, and X-ray diffraction (Kennedy, 1965, Wood, 1978) may be used to define mineralogy. A comparison of suspended sediment clay mineralogy with that of catchment soils and lithologies permits an assessment of source areas to be made. The organic fraction, which must be identified in any assessment of catchment erosion, is measured by wet oxidation and titration of excess potassium dichromate against ferrous sulphate (Finlayson, 1975).

The vertical gradient of suspended sediment concentration

Suspension occurs because of turbulence, and the concentration at any point in a cross-section therefore continually varies. However, in steady uniform flow the time-averaged vertical sediment distribution is constant, with equilibrium between rising and falling sediment at any level. This may be modelled as a diffusion process (O'Brien, 1933). The upward diffusion of particles through a horizontal plane of unit area is proportional to the concentration gradient at that height above the bed (dc/dy) and is $-\epsilon_s(dc/dy)$ by analogy with the momentum transfer (Equation 3.28). The sediment diffusion coefficient, ϵ_s, is analogous but not necessarily equal to the momentum transfer coefficient ϵ; it is negative to define diffusion towards lower concentrations at a higher level. Downward movement through the same unit area is $\omega_0 c$, where ω_0 is the settling velocity in clear, still water (as a first approximation). In an equilibrium concentration profile these balance, so that

$$\omega_0 c = -\epsilon_s \frac{dc}{dy}. \tag{4.8}$$

The true downward particle velocity, v_s, is less than ω_0 in turbid water, so that in high sediment concentrations the effect of neighbouring particles may be acknowledged by multiplying the left-hand term by $(1-c_v)$, where c_v is the volumetric concentration of the grains (Hunt, 1954). This differential equation is usually solved by assuming $\epsilon_s = \epsilon$ in order to define the variation with depth of the sediment transfer coefficient. From Equation (3.28) $\epsilon_s = \epsilon$ is defined as a function of shear stress and velocity gradient, and from Equation (3.24) (Figure 3.4c) a linear decrease in the shear stress with height above the stream bed (y) is obtained after dividing by $\tau_0 = \rho_w gds$ to give $\tau/\tau_0 = (d-y)d$. Finally, the velocity distribution of Equation (3.36) is assumed, giving

$$\omega_0 c = -\left(\frac{\tau}{\rho_w(dv/dy)}\right)\frac{dc}{dy} = -\left(\frac{\tau_0(d-y)}{\rho_w d(dv/dy)}\right)\frac{dc}{dy}$$

$$= -\left(\frac{\tau_0(d-y)\kappa y}{\rho_w dv_*}\right)\frac{dc}{dy} = -\left(\frac{v_*\kappa y(d-y)}{d}\right)\frac{dc}{dy}. \tag{4.9}$$

Integrating with respect to y yields an equation of the general form

$$c_y = \text{constant}\left(\frac{d-y}{y}\right)^{\omega_0/\kappa v_*} = \text{constant}\left(\frac{d-y}{y}\right)^z \tag{4.10}$$

(Vanoni, 1946, Carson, 1971, pp. 61–3), in which the concentration gradient is defined by the exponent z for an individual size fraction of fall velocity ω_0. If z is small, for fine particles (small ω_0) or rapid velocity shear and turbulence (high v_*), the distribution is uniform. Strong vertical concentration gradients (large z) occur for larger particles and weak turbulence (Figures 4.3a and 4.2c).

   Exponents (z_1) obtained by fitting power functions to measured concentration data are normally less than the values of z calculated from the sediment fall velocity and flow shear velocity, implying a more uniform distribution than predicted by Equation (4.10). This is especially apparent for coarser suspended sediment such as sand (Anderson, 1942, Nordin and Dempster, 1963). This departure arises because v_s, the actual settling velocity, is significantly less than ω_0 in turbulent and turbid water (Brush *et al*, 1962). Furthermore, von Karman's κ varies with suspended sediment concentration (Einstein and Abdel-Aal, 1972) as the suspension damps the mixing length of eddies, and it is also affected by secondary flow induced by bedforms. Generally the inertia of sediment grains causes the sediment diffusion coefficient, ϵ_s, to be less than the eddy diffusion coefficient, ϵ, especially for larger, heavier grains (Brush *et al*, 1962). However, conflicting results emerge in studies of the sediment diffusion coefficient because κ, v_s and ϵ_s all vary together and are interdependent. A satisfactory theory of the suspended sediment profile is inhibited by the dependence of these controls on the concentration itself: for example, v_s and κ both vary with height above the bed in response to changes of concentration.

Equation (4.10) may be rewritten in a form which incorporates a reference concentration, c_a, at a height a above the bed:

$$\frac{c_y}{c_a} = \left(\frac{a(d-y)}{y(d-a)} \right)^z = H^z. \tag{4.11}$$

The concentration profile of a narrow particle size range may be fitted to a single observed concentration value by plotting this value at its known relative height H and drawing the curve defined by the exponent z through it. Alternatively, Equation (4.11) may be used in a purely theoretical approach which defines the concentration profile of suspended bed material by relating the reference concentration at a height close to the bed to bed material or bed load properties. The wash load component in each size class may then be estimated by subtraction from the total measured suspended load. For example, Lane and Kalinske (1939) show that c_a/c_b, the ratio of the reference concentration immediately above the bed to the concentration of comparable particle sizes in the bed, decreases as relative turbulence intensity (ω_0/v_*) increases. Fine sediment of low fall velocity in a turbulent flow (low ω_0/v_*) is virtually all ejected from the bed into suspension, and so its concentration at height a is of the order of 100 times the concentration measured in the bed material. Alternatively, Einstein (1950) assumed the reference concentration, at a height a equal to two grain diameters above the static bed, was equivalent to the concentration of similar grains in the moving-bed layer. Thus, the bed material transported in suspension may be estimated from the bedload transport rate of similar particle sizes.

The suspended load transport rate at height y is the product of concentration and velocity at that height. Hence, the suspended load per unit width is obtained by multiplying the velocity and concentration profiles and integrating the product over the flow depth. This approach yields the suspended bed material load in steady, uniform flow, since the theoretical velocity and concentration profiles are derived on the assumption of this condition. In the unsteady, gradually varied flow of natural rivers, temporal and spatial variations of concentration occur throughout the flow depth and result in entrainment and deposition. The coarser suspended sediment is deposited when the shear velocity (v_*) decreases after the passage of a flood wave, or when slower zones are encountered in pools, backwater areas, or lakes. The probability of deposition increases as the inverse relative turbulence intensity index, ω_0/v_*, increases, and clays and silts carried into backwaters during overbank flow are deposited as mud drapes where turbulence is virtually absent in stagnant water. The Hjulström curve (Figure 3.8b) shows that transport of fine sediment is maintained over a wide range of flow velocities since the discrepancy between erosion and fall velocities is large. However, clay minerals may flocculate, particularly if solute concentrations increase downstream so that bivalent cations in solution can encourage cation bridging. The aggregates then behave as larger particles of higher fall velocity and may be deposited in more turbulent locations, such as at the channel sides. The type of clay mineral is significant in this context; kaolinite flocculates more readily than illite and montmorillonite (Edzwald and O'Melia, 1975).

Extreme suspended sediment concentrations approaching 1 kg 1^{-1} have been reported, for example from the semi-arid south-west USA (Beverage and Culbertson, 1964) and the loess areas of China (Stoddart, 1978); heavier concentrations are characteristic of mudflows. Generally, however, suspended sediment transport consumes little energy and is limited only by sediment supply, being invariably less than the transporting capacity of the flow. Satisfactory prediction of total suspended load from mechanical properties of the channel flow (bed shear stress, or shear velocity) is therefore impossible, and is a problem involving the consideration of catchment hydrology rather than channel hydraulics. Estimation of wash load generation must thus employ extensions of catchment hydrological models which predict hillslope rainsplash, runoff intensity and soil erosion (Negev, 1967, Meyer, 1971).

Empirical evidence of the hydrological control of suspended load

Coarser suspended bed material load responds directly to discharge changes (Figure 4.3d), but empirical data reveal that considerable variations in, and departures from, simple bivariate suspended sediment rating curves (Equation 4.2) arise because of catchment-controlled variation in wash load supply. Exponents in bivariate concentration–discharge relationships for Hungarian rivers range from 0.48 to 2.57, and have been shown by multiple regression and graphical coaxial correlation methods to increase with higher mean discharge and runoff per unit area, but decrease with increased catchment width and hydrological 'flashiness' (Bogardi, 1974, p. 495). Wide catchments present more opportunities for wash load deposition before runoff reaches the channel, and in a flashy regime relatively low discharges include high proportions of runoff. However, frequent systematic sampling or continuous recording generally demonstrate the inadequacy of bivariate relationships, notwithstanding the logic of these correlations.

Analysis of the mean concentration, total load, or discharge-weighted concentration of suspended sediment transported by quickflow in single storm events (Guy, 1964, Walling, 1974) indicates dependence on variables other than mean or peak discharge. Storm properties such as duration and rainfall intensity, and seasonal effects including antecedent discharge levels and water temperature, all influence storm-period sediment load, so that similar floods at different times are associated with different sediment yields. Furthermore, the distribution of sediment transport through a single storm is variable, and a wide range occurs between concentrations measured at similar discharges on rising and falling stages. Thus, multivariate rating curves are essential.

Seasonal variation is illustrated in Figure 4.3(b). Summer concentrations in a small Devon catchment are consistently four times the winter concentrations (Walling, 1974), reflecting the supply of surface dust which accumulates between storms in summer and the greater quickflow proportion in any storm discharge. The seasonal effect is often represented by an inverse correlation between concentration and the delayed flow discharge level prior to a storm (Q_{pr}), which tends to be low in summer. Storm characteristics are also partly seasonal and sediment concentrations are usually maximized in flashy events, which have a rapid rate of discharge increase on the rising limb (Q_{inc}). Suspension of sandy bed material is also strongly influenced by water temperature, since the higher viscosity at low temperature reduces the fall velocity and hence the 'effective' sediment size. The classic example of this effect is the three-fold increase of sediment concentration and load observed in the Colorado River at roughly constant discharge, and caused by a seasonal water temperature decrease from 30 to 15°C (Lane et al, 1949).

Hysteresis in the concentration variation through a single storm reflects two main influences. Firstly, sediment supply changes between the rising and falling flood stages, and in small catchments in particular wash load inputs continue as rain and runoff occur early in the flood but cease when hillslope runoff ends after the flood peak. Exhaustion of supply amplifies this, so that if suspended sediment is predominantly derived from bank erosion (Carson et al, 1973) or from the immediate channel margins (Bogen, 1980), the rising water level introduces new sources of sediment which are exhausted by the time the flood recedes. In glacier meltwater streams, whose sediment load is often of practical significance in the context of power generation and possible turbine damage, diurnal and storm-period discharge changes are both associated with hysteresis in concentration variations, although the complex pattern of sediment storage and sub- and en-glacial drainage may produce involuted concentration–discharge relationships (Collins, 1979). Clockwise hysteresis, with high sediment concentrations on the rising limb (Figure 4.3c), may be modelled in multivariate rating curves by using the rate of change of discharge per hour (ΔQ) or the time in minutes of sampling relative to peak discharge (T), which is negative on the rising stage (Richards, 1984). For the Storbreen meltwater stream in the Jotunheimen, Norway, ratings were

$$\left.\begin{array}{ll} \text{Diurnal:} & \log G_s = 1.98 + 0.96 \log Q + 1.23\Delta Q \\ \text{Storm period:} & \log G_s = 1.51 + 2.20 \log Q + 0.29\Delta Q \end{array}\right\} \quad (4.12)$$

in the 1979 summer season. In the urbanizing catchment in Devon monitored by Walling (1974), a multivariate rating of the form

$$\log c_s = 1.69 + 0.48 \log Q - 0.002T - 0.027 \log Q_{pr} + 0.17 \log Q_{inc} \quad (4.13)$$

was derived. In Equation (4.12), the load G_s is in units of g s^{-1} and the discharge is m s^{-1}, while in Equation (4.13) concentration c_s is in mg l^{-1} and the discharge is in l s^{-1}. These empirical models fail to account for exhaustion effects between successive closely spaced storms, when a given discharge in second and third events may be associated with lower concentrations because of insufficient time for renewal of sediment storage. Sediment yield from random catastrophic events, such as landslides, is also not predicted. Furthermore, these analyses of total suspended sediment data fail to reveal selective exhaustion of specific sediment sizes, such as the observed 'limiting' concentration of silt and clay wash load carried by the heavily laden Paria River in Arizona (Figure 4.3d).

The second major influence on storm-period variation in sediment concentration is the spatial effect of storm size relative to catchment area, and the associated lag effect of the downstream travel of sediment and water 'waves'. Heidel (1956) noted that flood waves on the Bighorn River travel downstream at speeds 25–67% faster than the stream water and suspended sediment load, so that the suspended sediment pulse is delayed by up to 36 hours after the flood peak to cause three- to four-fold variations of sediment concentration at near-constant discharges after the floodwater recession.

Suspended sediment yield over a period may be calculated by applying rating curve estimates of concentration to the observed discharge record. The seasonal, storm-period and lag effects all complicate the relationship between instantaneous sediment concentration and discharge, and considerable errors may arise if simple bivariate ratings are used rather than multivariate ratings accounting for seasonality and hysteresis (Walling, 1977, Loughran, 1976). Also, in stable geomorphological environments the catchment sediment yield may be dominated by the bulk input of dust in rainfall and as dry atmospheric fallout, so even *solid* denudation estimates must discount this sediment source (Finlayson, 1978).

Bedload transport

Whereas there is no practical limit to suspended sediment concentrations, a maximum bedload transport rate may be defined for a given flow and sediment, and bedload transport often occurs at this full capacity. Several reasons for this can be identified. Firstly, the source of bedload is confined to the channel bed and its immediate environment, so transport is directly controlled by conditions within the channel. Secondly, bed particle movement is brief and discontinuous, partly because a transport threshold exists and partly because particle velocities are only 2–15% of flow velocity, which inevitably means that the flood wave which initiates motion soon outstrips the particle, which is redeposited. Thirdly, bedload is normally less than 10% of the total solids transport, although in (non-alluvial) mountain streams it may reach 70%. It is generally more significant where the threshold discharge is a relatively frequent event. Exhaustion of supply is less likely to occur in a minor component of solids load. Finally, bedload transport utilizes much of the available stream energy, and exhaustion of supply is unlikely if the 'available power:threshold power' ratio is small.

Hysteresis in flood-period bedload transport may occur if a coarse surface armour is rendered mobile at peak flow to expose underlying sand which intensifies transport on the falling limb. Evidence is generally lacking, however, because measurement techniques tend to integrate transport over a period rather than monitor continuous variation. However, seasonal variation is apparent, and Nanson (1974), for example, identified a 95% reduction in bedload transport in a mountain stream after the peak snowmelt flood, as well as a doubling of the threshold discharge. These both reflect the removal of fines and loose, mobile bed material.

The existence of transport at full capacity should simplify the theoretical prediction of bedload transport. However, although bedload transport may relate closely to discharge variations, the precise threshold of motion and transport rate are usually defined by assuming uniform, spherical, unshielded grains, whereas natural stream bed material consists of heterogeneous, non-spherical, interlocked and 'hidden' particles. These sedimentological factors, which are difficult to quantify, inhibit the successful application of theoretical models. Furthermore, most bedload transport equations include empirical constants and demand testing in natural streams. The necessary measurement of bedload transport rates represents a major obstacle.

Measurement of bedload transport rates

The total solids load may be measured at contracted, highly turbulent sections where even the bedload becomes suspended. The difference between this total and the measured suspended load at a nearby normal section represents bedload plus the unmeasured suspended sediment near the bed. Attempts to measure the moving bedload at normal sections have often been of questionable value, because the sampling device disturbs and alters the process it is intended to measure.

Direct monitoring by permanent sediment traps – pits or slots in the bed – is successful if the slot width exceeds the saltation path lengths (~100–200 grain diameters). Simple pits merely fill with bedload during flood periods, and only indicate the total transport during any flood event. They may lose

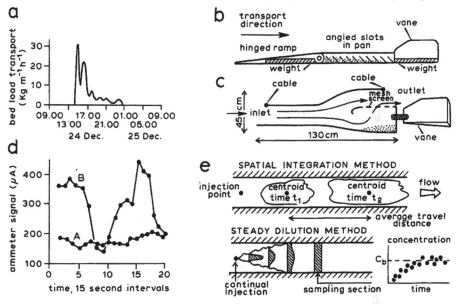

Figure 4.4 (a) Continuous record of flood-period bedload transport, including initiation, fluctuation and cessation, provided by a pressure pillow slot trap (after Reid *et al*, 1980). (b) Pan sampler of the Polyakoff type; sampling efficiency ~50%. (c) Pressure-difference sampler; the VUV sampler. Hydraulic efficiency, 100%; sampling efficiency, 70%. Suitable for 1–100 mm bedload, storing up to 25 kg. (d) Microphone signal for a 4600 cm^3 s^{-1} flow over a sand-free flume bed (A) and a 4700 cm^3 s^{-1} flow over a sand bed (B), showing passage of a ripple (after Richards and Milne, 1979). (e) Definition diagrams for two methods of bedload measurement based on tracer injection.

sediment by overfilling, vary in their efficiency to retain the bedload as they fill, and cannot provide evidence of variations in the rate of bedload transport during a flood. However, Leopold and Emmett (1976) describe an installation in which eight slots with hydraulically opened gates cover the full stream width and allow measurements of selected parts of the section to be made. Intercepted sediment is weighed and returned to the river downstream by a conveyor whose capacity is 150 kg min^{-1}. A less expensive and potentially more portable system involves a box resting on a pressure pillow in a pit in the bed (Reid *et al*, 1980). The pillow response is adjusted for increasing water-column weight, and a continuous record of bed-load entering the trap is provided; this indicates when transport starts, stops, peaks and oscillates (Figure 4.4a). Direct measurement by portable devices (Hubbell, 1964) such as basket traps in gravel streams and pan or tray samplers in sand-bed streams (e.g. the Polyakoff-type sampler in Figure 4.4b) is problematic. Samplers must be weighted and stable, correctly oriented, and have an entrance which accepts the largest sizes in transport. Installation must avoid 'dredging' bed material and disturbing the transport rate. Hydraulic and sampling efficiencies must be near 100%. The former is the ratio of water discharge through the sampler to the product of inlet area and stream velocity, which is 100% if no acceleration or deceleration occurs. The sampling efficiency is the captured load expressed as a percentage of the true load passing that part of the section where the sampler is installed. This may be measured in laboratory flumes with controlled sediment feed rates, but varies with bedform, location in the section, sediment size and degree of filling, and is therefore generally unknown in the field. A variable, unknown sampling efficiency is, of course, more of a problem than a constant efficiency less than 100%. The best samplers are pressure-difference samplers in which diverging sides, or top and bottom, encourage deposition; the VUV sampler (Figure 4.4c; Novak, 1957) and the Helley–Smith sampler (Helley and Smith, 1971) are the most successful.

Indirect acoustic monitoring involves measuring the noise of interparticle collisions, or impacts on a plate or probe in the bed. Suspended microphones in streamlined housings are most satisfactory as there is no disturbance at the bed. The signal is amplified and transmitted to earphones (Bedeus and Ivicsics, 1963) or recorded, and may be segregated into frequency bands to monitor collisions between different particle-size categories (Tywoniuk and Warnock, 1973). The variation of collision intensity across the section, and fluctuations due to the passage of bedforms (Figure 4.4d), may be identified. However, the signal must be calibrated against known transport rates, and it is unlikely that a simple bedload discharge-signal strength calibration is appropriate (Johnson and Muir, 1969) because some 'noise' is generated by the water turbulence (Figure 4.4d). An approximate average transport rate may be estimated indirectly by measuring the velocity of bedforms (Richardson et al, 1961, Simons et al, 1965). Since the mean depth of mobile bed material is one-half the dune height (h), for a dune velocity $\bar{v}_d$ the bedload transport rate per unit width (g_b) is

$$g_b = \gamma_s(1 - \lambda) \, \bar{v}_d \, \frac{h}{2} \qquad (4.14)$$

where γ_s is the unit weight of sediment and λ the porosity of the bed; the product $\gamma_s(1 - \lambda)$ defines the weight of sediment per unit volume of the bed. The bedform migration may be tracked using sonic depth sounding, or by monitoring the pressure variation as dunes travel over buried pressure transducers.

Various tracer techniques employed in bed material transport studies include painted pebbles, pebbles tagged with metal and traced by metal detector, and pebbles inserted with a radioactive label. Acoustic transponding 'pebbles' detected by side-looking sonar have been used in marine environments (Dyer and Dorey, 1974). These 'pebbles' define the initiation of movement and the direction and distance of transport, but not the total bedload transport rate. The injection of tagged *sand* grains of similar size and density to the natural material does, however, permit an estimation of bedload transport in sand-bed streams to be made. Fluorescent labelling is cheap and effective (Rathbun and Nordin, 1971, Kennedy and Kouba, 1970), and a range of radioactive labels (Table 4.1) are employed with

Table 4.1 Radioactive isotopes used as tracers in studies of bedload movement and transport rates. The half-life must be chosen to balance environmental constraints with the probable rate of tracer dispersion in the river to be studied.

Radioactive isotope	Half-life	Source	Environment used
Sodium-24 (^{24}Na)	15 hours	Crickmore and Lean (1962)	laboratory flume
Bromine-82 (^{82}Br)	36 hours	Crickmore and Lean (1962)	laboratory flume
Gold-198 (^{198}Au)	2.7 days	Crickmore (1967)	flume, river
Iridium-192 (^{192}Ir)	74 days	Hubbell and Sayre (1964)	river
Scandium-46 (^{46}Sc)	84 days	Crickmore (1967)	river
Tantalum-182 (^{182}Ta)	111 days	Courtois and Sauzay (1966)	river

half-lives appropriate to the intensity of bedload movement in the stream being studied. Injected particles mix into the upper mobile bed layer and then travel at speeds comparable to the natural material. The bedload transport may be calculated from their spatial or temporal variation in concentration in the bed (Crickmore, 1967). As suddenly injected material disperses downstream, 'spatial integration' of observed concentrations defines the centroid of the tracer cloud at successive sampling times (Figure 4.4e). The mean particle velocity ($\bar{v}_b$) is the distance moved by the centroid divided by the time period, and the bedload transport rate is the product of bed width (w_b), the depth (d_b) of mobile sand (approximately equal to dune height and measured from the vertical distribution of the tracer in core samples), the velocity, and the weight per volume:

$$G_b = \gamma_s(1 - \lambda)w_b d_b \bar{v}_b. \qquad (4.15)$$

The 'time integration' method measures the varying tracer concentration in bed material passing a section downstream from the injection point. The bedload discharge is defined by the ratio of the input weight of tracer to the integral of the concentration curve. The method is identical to that employed in dilution gauging of stream discharge (Figure 5.2f; Equation 5.4), but is generally inappropriate for slow rates of bedload transport which will vary significantly during the period when the tracer is monitored.

'Steady dilution' techniques are more appropriate. Continually injected tracer added at the rate G_t disperses across the bed and is detected at a downstream section where a constant equilibrium concentration c_b is measured (Figure 4.4e). (Concentrations are measured in weight of tracer per weight of injected or sampled mixture.) By conservation of mass

$$G_t c_t = (G_t + G_b) c_b. \qquad (4.16)$$

If $c_t = 1$ (i.e. all of the injected sample is tracer material) and G_r is small relative to G_b, then

$$G_b \approx G_t / c_b. \qquad (4.17)$$

Bedload transport equations

Bedload transport is unsteady in the short term, with discrete streams of solids wandering laterally across the bed and with temporal 'pulsation' at individual points. However, integration over a suitable time period produces uniform average transport rates across river sections while controlling flow conditions are steady. Bedload transport increases with stream size (width), so this average rate (g_b) is normally expressed as volume, mass or weight (dry or submerged) of sediment transport per unit width per unit time. Bedload transport is continuous under steady flow. Locally, scour may occur at high discharges in constricted sections, but this is abnormal in general since the resulting increase of cross-section area and reduction of velocity would be self-cancelling (Colby, 1964a). Thus, input and output from a short reach are balanced and transport continuity occurs; some input sediment may be deposited, but other particles are entrained to compensate for this.

Bedload transport requires the exceedance of a threshold flow intensity, measured by velocity, shear stress, shear velocity or power. The transport rate increases with excess shear stress (for example) as a function of the difference between the prevailing and threshold shear stresses ($\tau_0 - \tau_{0c}$), or of the ratio τ_0 / τ_{0c} (Kalinske, 1947). Definition of the threshold condition is difficult, as noted in Chapter 3, first because instantaneous shear stresses control initial movement and second because lift forces augment the shear stress. Kalinske (1947) considered instantaneous shear stresses caused by turbulence to be three times the mean bed shear stress, and therefore assumed that initial movement would occur at mean stresses of one-third of those predicted by the principle of moments (Equation 3.71). The threshold is also more difficult to define in heterogeneous bed materials, where interlocking, burial and sheltering all occur. An additional problem is that the cessation of bedload transport occurs at a shear stress significantly less than that at which transport begins (see p. 80). Thus bedload transport will continue at shear stresses *below* the threshold, when a transport equation based on excess shear stress would predict zero transport (Reid *et al*, 1985).

The selection of a measure of flow intensity is often a matter of convenience: mean velocity is easily measured, whereas the energy gradient required to estimate the mean bed shear stress ($\tau_0 = \rho_w gds$) is not. Furthermore, bedform development in sand-bed streams may complicate the use of shear stress as an index of flow strength. The total mean bed shear stress is balanced by the sum of grain and form resistance, of which only the former is translated into shear stress available for sediment transport. Thus, τ_0 correlates poorly with transport rates if bedforms change. In Figure 4.5(a) when a plane bed develops, all of the shear stress is available to transport

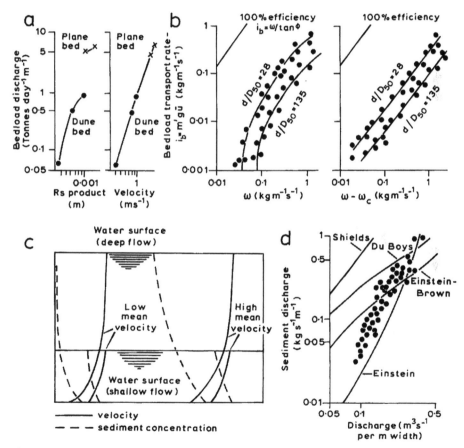

Figure 4.5 (a) Bedload transport in a small river in Mississippi as a function of shear stress and mean velocity, illustrating the effect of bedform changes (after Colby, 1964b). (b) Bedload transport as a function of stream power per unit bed area (ω) and excess power ($\omega - \omega_c$), illustrating the effect of relative depth (after Bagnold, 1977). $D_{50} = 1.1$ mm. (c) Velocity and concentration profiles illustrating the different effect of depth variation on total load transport at high and low velocities (after Colby, 1961). (d) A comparison between computed (curves) and measured (points) values of bedload transport in the Niobrara River (bed material $D_{50} = 0.28$ mm). After Vanoni *et al* (1961).

sediment when the form roughness disappears, and bedload transport increases sharply. The mean flow velocity measures available flow strength more successfully, so this discontinuity is absent (Colby, 1964b). Since power per unit area ($\omega = \tau_0 \bar{v}$) includes the shear stress, it is also discontinuously related to the transport rate. However, it is possible to estimate the proportion of the bed shear stress transmitted to the grains, for example by deriving a grain roughness measure (r') based on bed material dimensions using equations such as (3.21), (3.22) or (3.23), and calculating total roughness (r_0) by inverting the Manning or Darcy–Weisbach equations incorporating measured velocity, depth and slope. The total bed shear may be adjusted to that which is transmitted to the grains by multiplying τ_0 by a function of the ratio r'/r_0. Simple log–log plots of transport rate versus velocity or shear are complicated by bedform changes; Maddock (1969) noted the existence of low, middle and high velocity relationships. The existence of the threshold, however, introduces additional spurious curvature (Colby, 1964b, Bagnold, 1977), which may be removed when plotted against the excess shear or power (Figure 4.5b).

The effect of bedform development in unsteady flow causes hysteresis in the transport rate–discharge relationship (Skibinski, 1967) of sand-bed streams. Small, underdeveloped ripples on the rising stage allow high transport rates because there is little loss of shear stress to form roughness. Large, oversized dunes on the falling stage result in lower transport rates at the same mean bed shear stress.

The overall transport rate is rarely dependent on a single index of the flow. For example, total bed material transport (bedload and suspended bed material) varies with mean velocity but is dependent on depth in a complex way (Colby, 1961, 1964b). At a low mean velocity, bedload transport increases as depth decreases, because the bed velocity is higher in shallow flow. At high velocity, extra depth increases the load because the high suspended bed material concentrations are transmitted upwards into the deeper flow (Figure 4.5c). Transport equations are inevitably of varying success given such complex, nonlinear controls. Predicted rates vary widely in a given river (Figure 4.5d); for example, for the River Tyne, Muir (1970) quotes a ten-fold variation of bedload transport estimated by different methods. Formulae are applicable to specific environments and circumstances, and errors are inevitable when they are used beyond their appropriate range. Theoretical equations can rarely accommodate all the potential variables without becoming practically unworkable. Shen and Hung (1972) have therefore collected a large set of 587 flume and river observations to derive a purely empirical and easily applied formula for total bed material load using regression analysis. Bed material concentration by weight (c_t, ppm) is predicted by

$$c_t = f\left(vs^{0.57}\omega_0^{-0.32}\right) \tag{4.18}$$

which defines an increasing concentration with velocity and energy slope, and decreasing concentration with increasing sediment size measured by the fall velocity. The data are heavily weighted by observations from flume studies with uniform sediment sizes, however, and the relationship is not readily transferable to natural streams. Many of the theoretical and empirical transport rate relations are reducible to a common form in spite of the different initial theoretical considerations (Herbertson, 1969). Thus, three commonly used equations are discussed here to exemplify the general problems noted above and the range of initial approaches.

(a) *The Meyer-Peter and Müller equation*   The classic du Boys equation (1879) predicted bedload transport rates as a function of excess shear stress, and equations of this form have become known as 'tractive force' formulae. Meyer-Peter and Müller (1948) based an empirical relationship of this type on Swiss data ranging over bed material sizes from 0.4 to 30 mm and flow depths from 1 to 120 cm. Their equation is the most popular tractive force equation because of its conceptual simplicity and ease of application. It is

$$K^{3/2}\rho_w gRs - 0.047(\rho_s - \rho_w)gD_{50} = 0.25(\rho_w)^{1/3}g_b^{2/3}. \qquad (4.19)$$

The first term on the left-hand side is simply the mean bed shear stress, τ_0, multiplied by a factor $K^{3/2}$ which defines the effective shear on the grains, τ'_0. K is the ratio n'/n_0 where n' is a Manning coefficient due to grain roughness alone, obtained from the Strickler equation (3.21), and n_0 is the total roughness from observed flow data. This ratio normally lies in the range 0.5–1.0 and reduces the mean bed shear when bedforms are present. The second term on the left-hand side is a threshold bed shear stress defined for hydrodynamically rough beds, as in the Shields entrainment function (3.67). On the right, the bedload submerged weight transport rate is multiplied by $(\rho_w)^{1/3}$ simply to make the left- and right-hand sides of the equation dimensionally equivalent, so that the empirical constants are dimensionless and applicable regardless of the units of measurement. However, using SI units and assuming a water density of 998 kg m^{-3}, the equation can be expressed simply as

$$g_b = 0.253(\tau'_0 - \tau_{0c})^{3/2}. \qquad (4.20)$$

The equation may only be applied to low bedload transport stages in streams with medium sand to gravel beds. The absence from the equation of a relative depth (d/D_{50}) means that it is only applicable to rolling and sliding transport modes and not to intensive saltation or suspended bed material transport, both of which vary with depth of flow at a given mean velocity. Furthermore the Shields-type entrainment function is unreliable, varying with grain exposure or burial (Figure 3.8c), and is only appropriate for rough beds in its original form in Equation (4.19). However, the equation may be applied to smooth beds if the threshold shear stress term is modified appropriately.

(b) *Approaches based on the consideration of stream power expenditure*
The work required to maintain continuous bedload transport is defined by
the force per unit bed area necessary for steady transport, multiplied by the
bedload velocity (work = force × distance, work rate = force × velocity).
The force required to move a solid block on a horizontal surface is $W\mu$,
where W is the weight acting normal to the surface and μ the static friction
coefficient. The force required to maintain steady bedload transport is the

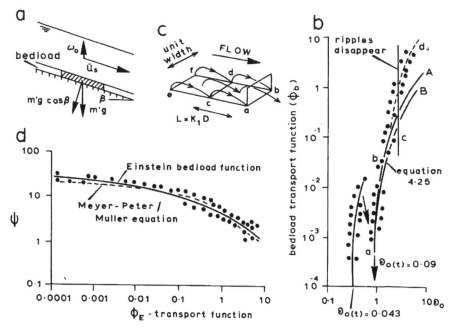

Figure 4.6 (a) Normal force exerted on a sloping bed by steadily moving bedload, and
velocity components acting on suspended particles. (b) The relationship between a
bedload transport parameter ϕ_b and a flow intensity parameter θ_o for bed material of
0.31 mm diameter. See text for explanation (after Bagnold, 1957). (c) Definition
diagram for derivation of the Einstein bedload function, based on grain saltation
jumps. (d) The Einstein bedload function (after Einstein, 1950). The ϕ_E parameter is
a transport function and ψ is a flow intensity function.

submerged weight per unit area of bedload multiplied by a *dynamic* friction
coefficient ($\tan \phi$) approximately equivalent to the static angle of internal
friction for a cohesionless mass of sand grains (Bagnold, 1954). This force
per unit area is $[(\rho_s - \rho_w)/\rho_s]m_b g \tan \phi$, where m_b is the bedload mass per
unit area. On a sloping surface the normal force is actually $m'_b g \cos \beta$ (Figure
4.6a), where $m'_b g$ is the immersed weight, but this slope correction may be
ignored for streams of slope less than 10°. The bedload transport work rate is
therefore

$$i_b \tan \phi = [(\rho_s - \rho_w)/\rho_s]m_b g \tan \phi \bar{u}_b \qquad (4.21)$$

where $\bar{u}_b$ is the mean bedload velocity and i_b is the transport rate, defined by the weight–velocity product, $m'_b g \bar{u}_b$. If the river is envisaged as a sediment transporting machine (Bagnold, 1957, 1960a), the rate at which the flow does work is the available power multiplied by the efficiency with which it is converted to work in moving the bedload. The power expenditure per unit bed area is $\omega = \tau_0 \bar{v}$ (Table 1.3), and thus if the efficiency is e_b we may equate the bedload work rate $i_b \tan \phi$ and the effective power expenditure of the flow, which is also a work rate and is measured in the same units. Thefore

$$[(\rho_s - \rho_w)/\rho_s] \, m_b g \tan \phi \bar{u}_b = \tau_0 \bar{v} e_b. \qquad (4.22)$$

The efficiency e_b is zero at flows below the transport threshold, but increases rapidly when transport begins. It is affected by bedform development, since energy loss due to form roughness reduces the power available for transport. However at high stages, when plane beds develop, a constant efficiency is achieved so that the i_b–ω curve parallels the line of equality representing 100% efficiency (Figure 4.5b). This constant efficiency at high stages is lower for coarser bed material (Leopold and Emmett, 1976).

Bagnold's initial (1957) consideration of stream power and sediment transport emerged from a relationship between dimensionless measures of the bed shear stress (θ_0) and the bedload transport rate (ϕ_b). The shear stress index of flow intensity, θ_0, is defined by

$$\theta_0 = \tau_0/\gamma = \tau_0/(\rho_s - \rho_w)gD \qquad (4.23)$$

where γ is a measure of the bedload weight per unit area in a layer of thickness D (i.e. one grain thickness). This is comparable to the Shields criterion (Equation 3.65). The bedload transport rate is defined as

$$\phi_b = g_b/[\rho_s D \sqrt{(\gamma/\rho_w)}] \qquad (4.24)$$

where g_b is the bedload mass transport rate per unit width. A typical plot of ϕ_b as a function of θ_0 for laboratory flume data (Figure 4.6b) reveals several interesting aspects of the bedload transport process. An initial threshold, $\theta_{0(1)}$, applies to motion on a plane bed. Once a certain intensity of transport occurs, scour in the lee of small-scale primary bed ripples creates larger-scale secondary bedforms (dunes) which dissipate flow energy in form roughness. This causes a discontinuity in the transport rate and an increase in the effective threshold stress, $\theta_{0(t)}$, implied by extrapolation (Figure 4.6b). The main bedload transport curve (A) is fitted by an equation of the form

$$\phi_b = k(\theta_0 - \theta_{0(t)})\theta_0^{1/2}. \qquad (4.25)$$

Since the shear stress is proportional to the square of the velocity (pp. 62–5), the term $\theta_0^{1/2}$ is a measure of velocity, and the bedload transport rate is therefore a function of a power index which is the product of excess, effective shear stress and the prevailing velocity. The factor k may include a variable efficiency factor which results in the curve B predicting lower transport rates for a given dimensionless shear stress θ_0 when adjustment is made for the effects of form roughness. At high shear stresses, a plane bed develops and suspended bed material transport is active; departure from Equation (4.25) then occurs. The general trend a–b–c–d mirrors the low– mid–high velocity regimes of sediment transport identified by Maddock (1969).

A simpler stream power relationship (Bagnold, 1960a, 1966) predicts total load at high transport stages. Total load, i, is the sum of bedload and suspended load, $i_b + i_s$. The former is defined by Equation (4.22), which may be simplified to $i_b = \omega e_b / \tan \phi$. A suspended load immersed weight transport rate i_s is $[(\rho_s - \rho_w)/\rho_w]m_s g \bar{u}_s$, and the appropriate work rate is $i_s \omega_0 / \bar{u}_s$. Here, the ratio of the fall velocity ω_0 to the suspended sediment velocity $\bar{u}_s$ is equivalent to the friction coefficient in the bedload work rate. This is based on the fact that turbulence maintains suspension by providing a vertical velocity component equal to the particle fall velocity, and the rate of lifting work done by the turbulence is proportional to the ratio of the vertical and horizontal velocity components, $\omega_0 / \bar{u}_s$. The suspended sediment work rate is maintained by that part of the power remaining after the utilization of power in the transport of bedload, which is $\omega e_s(1 - e_b)$. This adjustment implies that the model is applicable to suspended bed material rather than wash load, which may actually increase the potential for sandy bedload transport because of its effect on fluid density and viscosity at high concentrations such as occur in desert washes (Gerson, 1977). The suspended load equation is

$$i_s \omega_0 / \bar{u}_s = \omega e_s(1 - e_b). \tag{4.26}$$

Bagnold (1966) defines the efficiencies as $e_b \approx 0.15$ and $e_s \approx 0.015$, with $e_s(1 - e_b) \approx 0.01$. Thus, assuming suspended sediment particle velocities equal to stream velocity, the total load is given by

$$i = i_b + i_s = \omega \left(\frac{e_b}{\tan \phi} + \frac{e_s \bar{u}_s}{\omega_0} (1 - e_b) \right)$$

$$= \omega \left(\frac{e_b}{\tan \phi} + 0.01 \frac{\bar{v}}{\omega_0} \right). \tag{4.27}$$

This provides a satisfactory prediction of total load in the plane-bed regime, but systematically overestimates transport rates at lower stages because no adjustment is made to the shear stress component of stream power to account for energy losses due to form roughness. In addition, problems arise because of the use of mean velocity. Saltating bedload only rises to heights of approximately 10 grain diameters, and is therefore controlled by the bed velocity. For a given mean velocity, the bed velocity, and therefore the transport rate, is higher in shallower flows. Thus the transport rate–power relationship varies with the d/D_{50} ratio (Figure 4.5b; Bagnold, 1973, 1977). Figure 4.5(b) also shows that curvature in the i_b–ω relation is removed by subtraction of a threshold power. However, some curvature or scatter may also exist because the efficiency varies with the bedform regime.

Unit stream power (Yang, 1972) is power per unit volume of water and is obtained by dividing ω by stream depth. It is proportional to the velocity–slope product, $\bar{v}s$ (Table 1.3). Yang (1972) suggested a transport relation of the form

$$\log c_t = a + b \log[(\bar{v}s) - (\bar{v}s)_{\text{crit}}] \qquad (4.28)$$

where c_t is the total sediment concentration in ppm and $(\bar{v}s)_{\text{crit}}$ is the threshold of, or critical, power per unit volume of water required to initiate sediment transport. Such an expression adjusts the power–transport relationship to remove the depth dependency apparent when the power index used is stream power per unit bed area. However, it is not a dimensionless expression, so the constants vary according to the units employed. Yang and Stall (1976) have subsequently generalized the equation, deriving the dimensionless relationship

$$\log c_t = f(\omega_0, D, \nu, \nu_*) \log\left(\frac{\bar{v}s}{\omega_0} - \frac{\bar{v}_{\text{crit}}s}{\omega_0} \right), \qquad (4.29)$$

which has very close affinity to the empirical equation (4.18) developed by Shen and Hung (1972), with the addition of a dimensionless critical unit stream power.

(c) *Einstein's stochastic model of bedload transport* Einstein's (1942, 1950) revolutionary probabilistic approach to bedload transport involved a transport intensity function ϕ_E based on the probability of particle exchange at the bed, and a flow intensity function ψ based on the probability of occurrence of an instantaneous lift forces in turbulent flow capable of entraining the bed particles. This direct appeal to the lift force rather than shear stress is itself unusual, as is the explicitly stochastic rather than deterministic basis of the theory and its consideration of individual particle behaviour rather than bulk bed material movement. Initially applied to uniform bed material (1942), a later generalization to heterogeneous sediment (1950) determined the transport rates for a series of fractional size ranges and then summed these components to obtain the total bedload transport. For simplicity a uniform bed material particle size is assumed here.

Particles are considered to have a uniform saltation jump length L of approximately 100 grain diameters ($L = k_1 D$), independent of both flow and transport intensity. All particles entrained by local eddies from an area of length l and unit width pass through the section AB at the downstream end

(Figure 4.6c), while all particles originating from the equivalent upstream bed area CDEF are deposited in the area ABCD. The number of particles deposited per unit area per unit time (N_d) is the bedload weight transport rate per unit width g_b divided by the unit weight of the sediment γ_s to give a volumetric rate, and then by an individual grain volume (k_2D^3) to give a grain number transport rate. To scale the depositional rate to a unit area, this is divided by the area of bed ABCD ($= k_1D$). Thus

$$N_d = g_b/(\gamma_s k_2 D^3 k_1 D) = g_b/(\gamma_s k_1 k_2 D^4). \tag{4.30}$$

The number of particles eroded per unit area per unit time (N_e) is the unit area divided by the area occupied by a single grain (k_3D^2) to give the potential number of grains, multiplied by the probability of entrainment, p, and divided by the time period during which exchange occurs between bed material and bedload. This gives

$$N_e = p/k_3 D^2 t. \tag{4.31}$$

The time needed for particle exchange is considered to be proportional to the time taken for a grain to fall through a distance equal to its own diameter at its fall velocity, $t = k_4 D/\omega_0$. Thus, using Rubey's (1933a) impact law for the fall velocity of sand-sized grains (Equation 3.64)

$$N_e = \frac{p\omega_0}{k_3 k_4 D^2 D} = \frac{p\sqrt{[\frac{2}{3}Dg(\rho_s - \rho_w)/\rho_w]}}{k_3 k_4 D^3}. \tag{4.32}$$

Under steady, continuous bedload transport, the number of particles deposited per unit area per unit time balances the number eroded, and $N_d = N_e$. Thus the expressions on the right-hand sides Equations (4.30) and (4.32) may be equated and the terms rearranged so that when all constants are collected on the right, we obtain

$$\phi_E = \frac{g_b \sqrt{\rho_w}}{\gamma_s D^{3/2} \sqrt{(\gamma_s - \gamma_w)}} = \left(\frac{\sqrt{(\frac{2}{3})k_1 k_2}}{k_3 k_4}\right) p = f_1(p). \tag{4.33}$$

The transport intensity function ϕ_E is therefore itself a function of the probability of particle exchange, which may also be interpreted as the probability of the instantaneous lift force exceeding the submerged weight of a particle. For any particle a 'factor of safety' against entrainment may be defined as the ratio of the lift force to the submerged weight, F_L/W'. According to Einstein and El-Samni (1949), the lift force F_L is

$$F_L = 0.178 k_3 D^2 \rho_w \bar{v}_b^2/2 = f_2(D^2 \tau_0') = f_2(D^2 \rho_w g R' s). \tag{4.34}$$

The lift force is a function of the mean bed shear stress on a grain because it is dependent on the square of the bed velocity $\bar{v}_b$; the hydraulic radius R' is adjusted to the value expected given the observed slope and mean velocity, and a grain roughness alone. The submerged weight of a grain is

$$W' = g(\rho_s - \rho_w)k_2 D^3. \tag{4.35}$$

The probability of erosion of a grain is therefore dependent on the ratio F_L/W', being controlled particularly by the frequency with which it exceeds unity. Thus, a flow intensity index ψ may be defined as

$$\frac{1}{\psi} = f_3(p) = f_3\left(\frac{\rho_w g D^2 R's}{(\rho_s - \rho_w)gD^3}\right) = f_3\left(\frac{\rho_w R's}{(\rho_s - \rho_w)D}\right) \quad (4.36)$$

which, although derived from entirely different considerations, is essentially similar to those of Shields (Equation 3.65) and Bagnold (Equation 4.23). Although it is difficult to define the probability distribution of the lift force:immersed weight ratio theoretically, Einstein (1942, 1950) was able to determine experimentally the form of the relationship between the transport (ϕ_E) and flow (ψ) intensity indices (Figure 4.6d).

The Einstein bedload function represented in this ϕ_E–ψ relationship is based on a powerful, rational theory, and has been applied with some success to the prediction of bedload transport in sand-bed streams. Total load estimates are made by using the predicted bedload concentration as a reference concentration in fitting the suspended bed material load concentration profile (Equation 4.11). This is a complex procedure, especially for heterogeneous bed material, but Colby and Hembree (1955) have simplified the application and improved predictions by including observed suspended sediment concentrations. A major advantage of the Einstein approach is that no threshold condition is involved (Yalin, 1963), a continuously increasing probability of particle movement with increasing flow intensity being assumed. A satisfactory definition of a threshold shear stress is difficult to achieve. Nevertheless, the rapid increase of the transport intensity function ϕ at low flow intensity ($1/\psi$) may be compared to conditions just above the threshold. A particular limitation, however, concerns the assumed constant saltation step length, which has been shown to increase with stream power (Grigg, 1970) and transport stage (Abbott and Francis, 1977). Furthermore, the probability of particle detachment at a given point is variable, as grains are deposited in the lee of bedforms then buried for a period as the dune migrates forward.

Bedload deposition; the creation of sedimentary structures

As the bedload transport rate varies in the direction of transport, successive locations experience deposition and entrainment. For bedload the range of transport velocities is narrow (Figure 3.8b), and relatively small variations in velocity between sections may cause re-deposition of entrained sediment (Francis, 1973). During deposition, sedimentary structures are created which record the flow and transport direction and are therefore useful for the reconstruction of palaeocurrent strength and direction. Some directional structures are discussed in relation to channel pattern in Chapter 7;

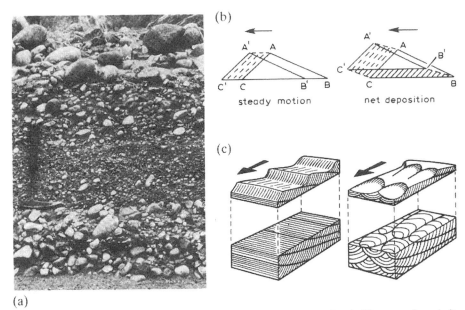

(a)

Plate I (a) Imbricated ellipsoidal pebbles exposed in a stream bank. Flow was from left to right; hammer for scale (after Picard and High, 1973). (b) Migration of a ripple or dune in a unidirectional flow for the conditions of steady motion and net deposition on an aggrading bed. (c) Internal cross-stratification patterns resulting from net deposition by bedforms with straight crests (tabular cross-stratification) and short curved crests (trough cross-stratification). After Allen (1970a).

basic depositional structures resulting from the reduction of flow competence are considered here.

Strong currents are necessary to transport gravel and pebbles, and deposition often occurs because temporary re-orientation of the particle increases its resistance to transport in flows only just capable of maintaining transport under favourable orientations. For example, flat disc-shaped pebbles are commonly deposited when their long axes become oriented transverse to the flow. In addition, the maximum plane of projection (the *ab* plane) often tends to dip upstream when the pebble is deposited in a stable position. This is because in this attitude, the flow cannot undermine the pebble and tip it over. Thus, flat pebbles tend to be deposited in a stacked manner, overlapping one another in an *imbricated* structure (Plate 1a). Once deposited thus, renewed entrainment is considerably inhibited since the required tractive forces are much greater than those expected given the particle size and weight. Those particles temporarily deposited in a non-imbricated structure are more readily eroded and are therefore removed during a subsequent period of high-intensity flow, to be eventually replaced by a pebble which is fortuitously deposited in a manner which resists further entrainment. Thus, the whole deposit becomes imbricated over a period of time.

Sand ripples or dunes migrating downstream leave a record of their passage when the earlier leeside (avalanche face) positions are partially left as inclined foreset layers, or cross-strata (Allen, 1965). In ripples these layers are thin (~1 mm) and are called laminae, whereas the 5–15 mm thick, layers formed by downstream avalanching in larger dune bedforms are cross-beds. If a dune travels parallel to its base and maintains a constant form (Plate 1b) the sediment eroded per unit time on the upstream stoss side equals the deposition in a similar time period on the downstream lee face. Thus, in section,

$$\text{Erosion } ABB'A' = \text{Deposition } ACC'A'. \qquad (4.37)$$

The cross-strata are only preserved if the path of the bedform is inclined upwards relative to its base, which results if the transporting power declines along the flow path, or through time at a single location as the discharge wanes. Under these conditions the bed aggrades, and

$$\text{Net deposition } B'BCC' = \text{Lee deposition } ACC'A$$
$$- \text{Stoss erosion } ABB'A. \qquad (4.38)$$

If the transporting power increases along the flow path, or through time as discharge increases, the dune travels over a scouring bed surface and net erosion occurs, destroying any cross-strata preserved in the bed sediment. Thus extensive preservation of cross-strata demands continuing aggradation, so that successive floods fail to destroy all of the depositional structures created by the falling stage flows of earlier floods.

The detailed form of cross-stratification reflects the three-dimensional shape of the bedforms. Ripples or dunes with long, straight crests generate plane cross-strata in tabular cross-stratification units (Plate 1c). Curved, short dune crests such as those of lunate (Plate 1c) or linguoid forms tend to lay curved cross-strata, filling trough or scoop-shaped erosional hollows in the bed, so that aggraded deposits are characterized by trough cross-stratification.

5

The magnitude and frequency of channel-forming events

Concentrations and transport rates of both suspended sediment and bedload increase with discharge unless exhaustion occurs of the supply of appropriate sediment. The disposition of these components of the solids load defines channel morphology, which is altered by entrainment, transport and deposition during peak flow events of varying magnitude. Sediment transport and adjustment of channel morphology both reflect bed and bank shear stresses or rates of power expenditure, and these measures of flow intensity vary with channel and valley gradients. In a particular reach, these slopes are generally fixed in the short or medium terms and the capacity of the flow to undertake work, in the form of transport or channel creation, reflects the range of event magnitudes imposed by the upstream catchment hydrology.

The significance of high discharges in controlling channel form is suggested by their large sediment loads. Wolman and Miller (1960), however, emphasize that work done over a period by an event associated with a particular magnitude of load also depends on its frequency of occurrence. The magnitude of sediment transported by a given discharge is approximately defined by a power function of excess discharge (or stream power) above the required threshold, in the case of bedload (curve A in Figure 5.1a), while the frequency distribution of discharge events is approximately log-normal (curve B). Thus, the sediment transport over a period by a specific discharge is the product of the magnitude of transport by one such event and its frequency, giving a curve (C) which shows that work done is maximized by events of intermediate magnitude and frequency; extreme events transport large sediment loads, but occur too rarely to be of long-term significance. Bankfull discharge, which fills the channel without overtopping the banks, is an intermediate discharge often considered a critical or dominant channel-forming event in natural rivers. It is therefore conceptually equivalent to the 'dominant discharge' (Inglis, 1949a) of regime canals. This was the single discharge which was representative of the range of flows experienced, and which was thought likely to create the same canal morphology as the varying discharge carried by the canal.

Particularly in semi-arid environments it is an oversimplification to assume that a single event can represent the range of morphologically significant discharges. The timing and succession of events control short-term departures from the regime or equilibrium state, with deposition during low flows and erosion during floods. Wolman and Gerson (1978) argue that the effectiveness of an event reflects the morphological changes it causes through erosion and deposition, as well as the associated sediment transport. In humid environments, extreme events may modify the forms of channels with cohesive banks and increase widths by 20–40%, but subsequent lower discharges allow the forms to 'recover' their equilibrium state in 10–20 years. Recovery periods of 30–100 years are required by the larger changes of semi-arid stream morphology, since the lack of vegetation inhibits accumulation of fines. Forms therefore continue to reflect the effects of the last major flood, as illustrated by the slow recovery of the Cimarron River after widening by the 1914 flood (Schumm and Lichty, 1963). If the recovery period exceeds the mean recurrence interval of the extreme flood responsible, an equilibrium between channel form and a representative channel-forming discharge of intermediate magnitude is inconceivable. Schick (1974) shows that the spotty nature in time and space of arid-region rainfalls, and the short-lived flows which evaporate and percolate, cause successive periods of valley floor aggradation when small floods deposit sediment, and incision and degradation when single 'superfloods' occur. The resulting valley fill sedimentary succession of inset and overlapping deposits (Leopold and Miller, 1954) reflects the sequence of events (Figure 5.1b), and terraces are created by purely local variations in water and sediment discharge. In steep, upland humid catchments flood timing and recovery periods appear equally important. Anderson and Calver (1977) show that channel deepening by the 1952 Lyn flood on Exmoor may survive the mean recurrence interval of the event, and Newson (1980) suggests that different bedload transport rates in major floods on the upper Severn and Wye in Wales arise if the sediment delivered to the streams by slope processes in the first flood is removed by minor events in the recovery period between floods, causing supply-limited conditions in the second. Thus the morphological impact of events is partly a queueing problem, with inter-arrival times between peaks being as significant as peak magnitudes. The cumulative effect of a succession of low magnitude flood peaks may equal that of a single major flood in terms of sediment transport, but not necessarily in relation to morphology. Harvey *et al* (1979) employ a qualitative classification of events based on degrees of bed material movement and morphological change. Moderate events which redistribute bed material occur between 14 and 30 times per year, while major controlling events which change overall channel form occur from 0.5 to 4 times per year. This difference in frequency may reflect the different thresholds of erosion for bed and bank materials (see p. 142) and is of significance in the context of river management.

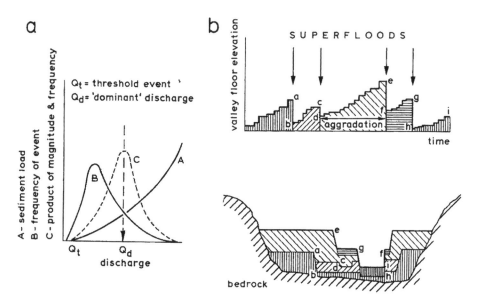

Figure 5.1 (a) 'Dominant discharge' defined by the magnitude and frequency of sediment transport associated with the range of discharges experienced (after Wolman and Miller, 1960). Curve A, sediment load; curve B, frequency of event; curve C, product of magnitude and frequency. (b) Evolution of valley fill by aggradation and degradation caused by minor and major floods in arid environments (after Schick, 1974).

Hey (1976) argues that channel instability occurs if flow regulation increases the frequency of exceedance of threshold bed shear stresses. This may not be the case if a different, higher threshold is important for bank erosion, and the sum of several bed material movements is therefore not morphologically equivalent to one major event.

Measurement of discharge magnitudes

The estimation of representative flood magnitude statistics requires a complete record of the peak flows experienced at a river section. Crest stage gauges provide a simple means of generating these data. Typically, these gauges are inclined tubes fixed at a river bank, with water inlet holes at the base. The tubes contain a floating marker (granulated cork) which records peak stage by adhering to the sides. Peak discharge (Q_p) is obtained from the Manning equation (Dalrymple and Benson, 1967)

$$Q_p = Cd^{2/3} s_w^{1/2} n^{-1} \tag{5.1}$$

using measured channel capacity (C) and an estimated roughness coefficient (n) at the recorded peak stage, and water surface slope (s_w) measured between gauges at each end of a uniform reach. However, the estimation of total sediment yield demands a continuous discharge record, which therefore requires an investment in automated monitoring techniques.

Measurement techniques for natural river sections

To relate hydrological events to river channel forms it is convenient to gauge discharge at a stable natural river section. This is readily achieved by defining a stage–discharge relationship (Figure 5.2a) and recording variations of stage (water level above an arbitrary datum). A range of systems automatically records water-level variations either in analogue form as a pen trace on a calibrated chart, or in digital form on paper or magnetic tape. Stage variation is transmitted to the recording device mechanically, by pressure variation (British Standards Institution, 1971a,b, Buchanan and Somers, 1968), electronically or acoustically. The conventional mechanical method involves a float in a stilling well (Figure 5.2b), but pressure variations may activate a recording device via a diaphragm, the pressure being generated by the height of the water column either above a pressure bulb or by varying outlet pressures in a bubble emission method (Figure 5.2c). The simple stage board may be replaced by a vertical post with electrical contacts at fixed intervals, the rise in water level triggering an electrical signal as each contact is drowned; this may be transmitted telemetrically. Large volumes of redundant data generated during continuous low flow can be eliminated by digital event recorders which are sensitive to the rate of change of water level (Chandler and Patterson, 1973). The energy gradient varies with discharge in some sections (Figure 6.7a), particularly during a rapid hydrograph rise. Discharge therefore varies with slope at a constant stage, and a multivariate slope–stage–discharge relation is required (Corbett *et al*, 1943), slopes being measured using a second stage board. Herschy (1976) describes developments which permit automatic monitoring of flow velocity rather than stage. The time of travel of an ultrasonic signal passing between two submerged transducers on opposite banks along a path oblique to the flow differs according to the direction of travel, being shorter when directed downstream, in porportion to flow velocity. Alternatively, an electromotive force proportional to velocity is induced when the flow cuts the Earth's magnetic field and may be picked up by electrodes in the stream bank. These methods are useful in sections susceptible to

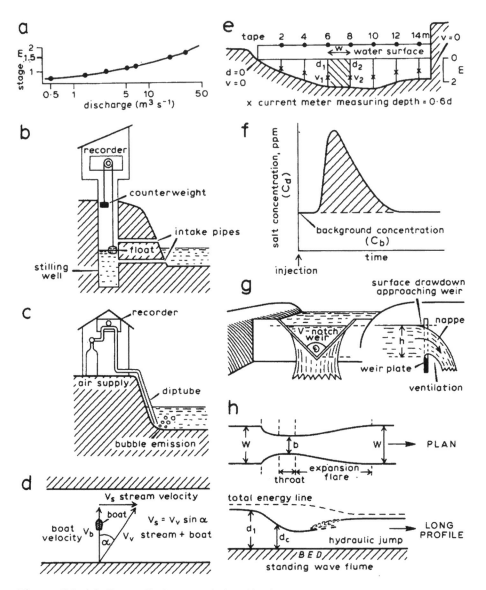

Figure 5.2 (a) Stage–discharge relationship for the River Fowey at Restormel, Cornwall. (b) Mechanical recording of stage variation by float and counterweight. (c) Bubble emission pressure-based monitoring of stage variation. (d) Velocity triangle used in the moving boat method of measuring flow velocity. (e) Mean-section stream gauging method by current meter. (f) Salt wave passing downstream measuring section after gulp injection for dilution gauging method. (g) Thin-plate V-notch weir, with angle of notch $\theta°$ and head h. (h) Plan and long profile of throated flume.

shifting stage relations because of weed growth, but require a velocity–discharge curve which invariably involves more scatter than a depth– or stage–discharge curve, or a measurement of flow cross-section area, before the velocity data can be translated into discharge data.

The initial establishment of a stage–discharge curve requires direct field gauging over a range of discharges, including flood flows, in order that reliance on unsubstantiated extrapolation does not occur. In large rivers, the moving boat method is used to estimate velocity from a velocity triangle (Figure 5.2d), depths being measured by echo-sounder (Herschy, 1976). In bridged or wadeable streams, the normal velocity–area technique is employed (Buchanan and Somers, 1969, British Standards Institution, 1964a). Velocity and depth are measured at equally spaced verticals across the section (Figure 5.2e), and the contributing discharge of each segment, none of which should transmit more than 10% of total flow, is calculated from

$$q_i = w_i \frac{(d_{i1} + d_{i2})}{2} \frac{(\bar{v}_{i1} + \bar{v}_{i2})}{2} \tag{5.2}$$

which is the 'mean-section' method. Total stream discharge, Q, is then

$$Q = \sum q_i. \tag{5.3}$$

Mean velocity in each vertical is measured by current meter by the single point method at 0.6 of the depth (from the surface), or by averaging measurements at 0.2 and 0.8 of the depth. Alternatively, in deep streams an integrated mean velocity can be measured by releasing a bubble or float of known terminal velocity from the bed (Dyer, 1970) and measuring the downstream travel before arrival at the surface. Current meters are inaccurate in rocky, turbulent streams, and here tracer methods are used, including chemical, electrolytic, fluorescent and radioactive tracers (Church, 1975). Although discharge can be measured by constant-rate injection (British Standards Institution, 1964b), the gulp injection method requires less apparatus in the field, and Day (1976) quotes probable errors for this technique in the range of ±3 to ±7%. The sudden injection of a tracer at an upstream site far enough from the measuring section for complete mixing to occur results in the passage of a wave of varying concentrations of tracer through the section (Figure 5.2f). The time to the centroid of this wave permits an accurate estimate of mean velocity which can be used with surveyed flow area to give discharge (Calkins and Dunne, 1970). However, measurement of concentrations of tracer provide a direct discharge estimate without cross-section data. The theory is identical for all tracer types: Ward and Wurzel (1968) describe the use of radioactive isotopes ([198]Au and [51]G), but the ease of field measurement of conductivity, and the cheapness of the tracer, favour common salt as an electrolytic tracer (Aastad and Sognen,

1954). If V litres of salt solution at a concentration C_i are injected (one l per m^{-3} of discharge is recommended) and the variation in concentration downstream is monitored (C_d in Figure 5.2f), the discharge Q is given by

$$Q = \frac{(C_i - C_b)V}{\int\limits_{t=0}^{\infty} (C_d - C_b)\, dt} \tag{5.4}$$

where the integral is the area beneath the plotted wave of varying concentration, adjusted appropriately by the plotting scales. Østrem (1964) and Church (1975) describe a method of field calibration of conductivity measurement which uses relative concentrations and is independent of C_i and C_b, which therefore need not be measured.

Flow measurement at fixed structures

In sand-bed streams, stable stage–discharge curves are difficult to establish (Dawdy, 1961, Simons *et al*, 1973), and in small rivers permanent control structures may be installed. Weirs and flumes provide fixed sections in which a theoretical and stable relation occurs between head and discharge; they therefore require continuous recording of water-level variation. The choice of structure is conditioned by the range of discharges to be measured, the nature of the stream (thin-plate metal weirs are susceptible to damage by debris and to inaccuracy if sediment accumulates behind the weir plate), and the costs. Thin-plate weirs may have rectangular or triangular notches (Figure 5.2g), which must be accurately machined and chamfered if thicker than 2 mm. The nappe, or water overfall, must be 'ventilated' for the standard rating equations to be valid, and thus the water level downstream must be below the notch. For a sharp-crested V-notch weir, the general discharge equation is

$$Q = \frac{8}{15} \sqrt{(2g)} C_D \tan \frac{\theta}{2} h^{5/2} \tag{5.5}$$

where the discharge coefficient C_D varies with head (h) and notch angle (θ), but commonly lies between 0.58 and 0.61 (British Standards Institution, 1965). For a 90° notch, this reduces to

$$Q = 1.40 h^{5/2} \tag{5.6}$$

for average values of C_D. In the discharge equations for rectangular-notch weirs, discharge is a function of notch width, a variable coefficient, and $h^{3/2}$. Broad-crested weirs are robust control structures normally built of concrete, and are therefore more appropriate in larger streams carrying sediment and debris. They are also insensitive to downstream effects, and continue to be effective until the downstream water level rises above the critical depth on the crest. The Crump weir (Crump, 1952) is an example, with an upstream face sloping at 1:2 and a downstream face of 1:5. Combined with a wide-angled triangular cross-section, this forms a flat-V weir which is more robust than thin-plate weirs, but allows accurate gauging of the low discharge confined to the centre of the V. The discharge equations for broad-crested weirs are essentially similar to those for rectangular thin-plate weirs, but with different discharge coefficients. Finally, flumes consist of constrictions in the channel which increase velocity and cause a decrease of water level (Figure 5.2h). For certain combinations of discharge, flume dimensions and downstream water level, the flow will pass through the critical depth in the throat, and a hydraulic jump occurs in the expansion section. Under these conditions, a simplified discharge equation is

$$Q = 1.7bd_1^{3/2}. \tag{5.7}$$

If the critical depth is not reached and no standing wave occurs, the flow is said to be 'drowned' and the flume is a 'Venturi flume', for which a discharge equation is required which includes the ratio of flow areas upstream and in the throat, and the depth difference (British Standards Institution, 1965). Flumes are self-cleaning of sediment and involve a smaller head loss than weirs, but their design and construction are critical. Clearly, however, if discharge data can only be obtained from one of the available types of control structure, they must be related to the natural channel properties of an immediately upstream, undisturbed reach.

Estimation of discharge frequency

'Duration curves' represent cumulative percentage curves of the time each discharge is equalled or exceeded. They define specific flows such as the median discharge ($Q_{50\%}$) or less frequent events including the 1% flow ($Q_{1\%}$, equalled or exceeded on 3.65 days per year). Log-normal discharge distributions are indicated by linear duration curves on log-probability paper (Figure 5.3a). Their slopes reflect flow variability, measured by a 'variability index' which is the standard deviation of logarithms of discharges at intervals of 10% duration between 5 and 95%; values range from 0.14 to 1.17 in sampled American catchments (Lane and Lei, 1950). McGilchrist *et al* (1968) analysed the durations of periods between floods, but the usual basis for relating channel forms to peak events is a flood-frequency analysis, which estimates the magnitudes of events of various return periods (*T*), or probabilities of occurrence. Instantaneous peak discharges are used, but where only daily maximum discharge is available at manually recorded stations, an adjustment can be made (Dury, 1959).

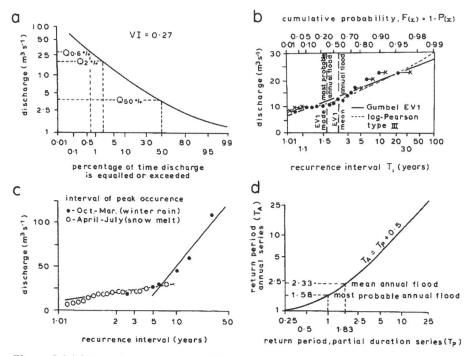

Figure 5.3 (a) Duration curve for the River Fowey at Restormel, 1961–67. Variability index VI = 0.27; $Q_{50\%}$ = 3.68 m^3 s^{-1}; $Q_{2\%}$ = 20.39 m^3 s^{-1}; $Q_{0.6\%}$ = 27.89 m^3 s^{-1}. (b) Flood-frequency analysis for the River Severn at Plynlimon Weir. The method of fitting the Gumbel EV1 distribution is outlined in Table 5.1. Benson (1968) describes a simple method for fitting the log-Pearson type III distribution using mean, variance and skewness of the logarithms of the annual maximum floods. Full circles are return periods by Equation (5.9), crosses by Equation (5.12). (c) Dog-leg flood-frequency curve, South River Fork, Silver Creek near Ice House, California (after Potter, 1958). (d) Relationship between return periods on the annual and partial duration series (after Langbein, 1949).

Flood-frequency analysis

If the annual maximum discharges of N years of record are ranked from highest (rank, $m = 1$) to lowest ($m = N$), the resulting 'annual series' forms $N + 1$ rank classes. The probability of a random event of magnitude x being equal to or greater than an event ranked m is

$$P(x) = m/(N + 1) \qquad (5.8)$$

and the mean return period of this event is

$$T = 1/P(x) = (N + 1)/m. \qquad (5.9)$$

This probability model is distribution-free and based solely on rank orders.

The observed annual maxima may be plotted as a probability distribution, or against these return periods (Figure 5.3b). Discharges of specified return period (Q_T) are estimated by fitting an appropriate theoretical probability distribution, which normally must model the positive skew common in extreme event distributions. This is achieved by natural distributions (exponential, gamma, Gumbel extreme value type 1 (EV1) and Pearson type III), or by transformed distributions (log-normal, log-Gumbel, log-Pearson type III). In the absence of a theoretical justification for a particular model, selection is based largely on convenience, although Chow (1954) suggests that log-normality could reflect numerous random but multiplicative (rather than additive) natural influences on flood hydrology. The favoured Gumbel EV1 distribution (Gumbel, 1958, NERC, 1975), a special case of a general three-parameter log-normal model (Sangal and Biswas, 1970), provides a linear plot on a transformed probability scale (Figure 5.3b). Its distribution function

$$F(x) = 1 - P(x) = \exp\{-\exp[-(x-u)/\alpha]\} \qquad (5.10)$$

is a two-parameter double exponential of fixed skewness with mode u, mean $\mu = u + 0.5772\alpha$ and variance $\sigma^2 = \pi^2\alpha^2/6$, fitted using sample estimates for μ and σ^2 to derive the parameters u and α (Table 5.1). The probability of exceedance of the EV1 mean annual flood is $P(\mu) = 1 - F(\mu) = 0.43$; so its return period is $T = 1/P(\mu) = 2.33$ years, and that for the most probable annual (modal) flood is $T = 1.58$ years. Mean annual flood (Q_{ma}) and $Q_{2.33}$ are only equivalent if the annual flood maxima conform to the EV1 distribution. Simple estimates of Q_{ma} can be obtained by averaging the annual series, or by using

$$Q_{ma} = 1.07Q_{med} \qquad (5.11)$$

if an outlying extreme maximum discharge exceeds three times the median (Q_{med}) (NERC, 1975) and threatens to distort the simple average.

Data plotted on Gumbel paper may display curvature for several reasons. First, a single random sample may appear to deviate from a linear trend (NERC, 1975), thereby encouraging subjective fitting of a curve. This problem arises if an extreme flood occurs within a short record. In these circumstances, the true return period of such an event may be estimated by reference to stratigraphical evidence such as exposed, dateable, buried organic deposits (Costa, 1978). Second, plotting positions defined by Equation (5.9) are distribution-free and strictly are inappropriate for data from a specific probability function. Gringorten's (1963) general plotting rule is

$$\hat{P}(m) = (m-a)/(N+1-2a) \qquad (5.12)$$

with $a = 0.44$ for an EV1 distribution (Figure 5.3b). Third, the data may not fit the EV1 model, and alternatives such as the log-Pearson type III

Table 5.1 Flood-frequency analysis for the River Severn at Plynlimon Weir (Figure 5.3b), including the derivation of exceedance probabilities for the fitted Gumbel EV1 distribution.

Year	Discharge $(m^3 s^{-1})$	Rank	Observed return periods (a)	(b)	Gumbel EV1 $P(x)$
1951	10.4	16	1.31	1.29	0.77
1952	10.9	14	1.50	1.48	0.72
1953	10.6	15	1.40	1.38	0.75
1954	10.1	17	1.24	1.21	0.78
1955	8.5	20	1.05	1.03	0.92
1956	9.6	18	1.17	1.15	0.84
1957	23.7	1	21.00	35.93	0.03
1958	13.1	9	2.33	2.35	0.49
1959	23.1	2	10.50	12.90	0.04
1960	17.0	5	4.20	4.41	0.19
1961	11.4	12	1.75	1.74	0.67
1962	14.8	7	3.00	3.07	0.33
1963	12.3	11	1.91	1.91	0.57
1964	20.7	3	7.00	7.86	0.08
1965	13.4	8	2.63	2.66	0.46
1966	17.5	4	5.25	5.65	0.18
1967	12.4	10	2.10	2.10	0.56
1968	15.8	6	3.50	3.62	0.27
1969	8.8	19	1.11	1.08	0.90
1970	11.3	13	1.62	1.60	0.68

$$\bar{x} = 13.8$$
$$\sigma^2 = 20.7$$

(a): by Equation (5.9); (b): by Equation (5.12).

Derivation of $P(x)$ for Gumbel EV1 distribution

$\sigma^2 = \pi^2\alpha^2/6 \therefore \alpha = 3.54; \mu = u + 0.5772\,\alpha \therefore u = 11.73.$
$F(x) = \exp\{-\exp[-(x-u)/\alpha]\}.$
For $x = 23.7$, $F(x) = \exp(-e^{-3.38}) = 0.967.$
This is the probability of an event being less than x. Therefore $P(x) = 1 - F(x) = 0.033$
$T = 1/P(x) = 29.9.$

(Benson, 1968; Figure 5.3b) or General Extreme Value (NERC, 1975) distributions may be appropriate. A generally applicable model cannot be recommended on the basis of success in fitting empirical data, since the mean absolute, mean percentage or mean squared deviations, and bias in estimating specified events (Q_T), are all equally valid criteria of goodness of fit which favour different models (NERC, 1975). Finally, annual flood maxima may be drawn from two or more event populations; Potter (1958) identified this as a cause of 'dog-leg' plots on extreme-value paper. A probability model for mixed distributions, suggested by Singh and Sinclair (1972), is

$$P(x) = pP_1(x) + (1-p)P_2x \qquad (5.13)$$

where $P_1(x)$ and $P_2(x)$ are two-parameter normal distributions mixed according to the fraction p. The two flood event populations may be rainfall and snowmelt floods (Figure 5.3c), different hydrograph forms generated by summer and winter storm intensities (Harvey, 1971), or a mixture of small floods reflecting varying degrees of antecedent wetness and large events resulting from different storm intensities on fully saturated catchments.

Benson (1960a) showed that a forty-year record is required to estimate mean annual flood within $\pm 10\%$ of the true value with 95% confidence. Even longer series are necessary for a satisfactory estimation of skewness (Matalas and Benson, 1968), the third moment used in fitting three-parameter distributions such as the log-Pearson type III (Figure 5.3b). Such long series, where they occur, may suffer from heterogeneity because of time trends in flood magnitudes, perhaps associated with hydrological changes occasioned by land-use change (Howe *et al*, 1967). An alternative approach is therefore to use the 'partial duration series', consisting of all independent flood peaks above a threshold discharge; the 'annual exceed-ance series' is a special case in which the threshold is fixed so that the number of peaks identified (m) equals the number of years of record (N). The analysis for short periods of record ($N < 10$ years) proceeds as for the annual series, the return periods of ranked discharges being calculated by Equation (5.9) (extended plotting paper is needed because return periods can be less than $T = 1$). The theoretical basis of the model is questionable, in that it requires that events be randomly distributed through time with magnitudes described by an exponential probability distribution. Cunnane (1979) shows that a slight clustering of peaks invalidates the Poisson assumption, and NERC (1975) shows that the implication of an exponential distribution of peak magnitudes above a threshold is that the annual maxima are defined by a double exponential. The Gumbel EV1 distribution is therefore a requirement of the peaks-over-threshold model. This is somewhat restrictive, but alternative probability functions fitted to annual series only diverge markedly in the tails of the distribution and at high return periods in particular. Thus the partial duration series may be used with confidence to estimate the mean annual flood, and other events with $T < 10$ years. As long as at least 1.65 events are included per year, the partial duration series may yield estimates of discharge with lower variance than the annual series (Cunnane, 1973). The mean annual flood maximum (Q_{ma}) has a return period of $T \approx 2.33$ years as an annual maximum, but the event ranks lower ($T = 1.83$) in a partial duration series, which includes several peaks per year, and simply defines the average interval between occurr-ences (Figure 5.3d).

Flood estimation in ungauged catchments

In ungauged catchments, graphical correlation or multiple regression is used to estimate discharges of selected frequency from catchment characteristics, including climate, land use and basin morphology, and particularly the size, slope and network density aspects isolated by principal components analysis (pp. 44–5; Rodda, 1969, Wong, 1963). Several empirical regression models have been developed which predict mean annual flood (Q_{ma}); they are, however, weakened by two general problems. First, the independent variables are usually correlated (large catchments have gentle slopes and low rainfall), and multicolinearity renders the regression coefficients physically meaningless (Nash, 1959). Second, sampling bias is almost inevitable. The nonlinear envelope curve in Figure 5.4(a) reflects the absence of large basins with heavy rainfall and is similar to a discharge–area curve fitted by Alexander (1972). The form of this curve *appears* to reflect flood peak attenuation downstream because of low gradient and floodplain storage, but actually is the result of sample bias. Discharge per unit area (Q_{ma}/A) removes the scaling effect of basin size, and increases with catchment and channel slopes (Armentrout and Bissell, 1970), and rainfall. Reich (1970) finds little evidence that extreme rainfall properties correlate

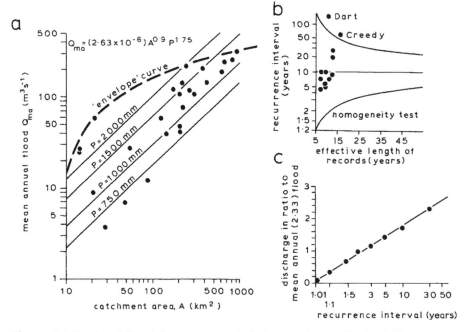

Figure 5.4 Regional flood-frequency analysis for south-west England (Devon and Cornwall). (a) Relationship between mean annual flood and catchment area and rainfall. (b) Homogeneity test: data for the River Dart and Creedy are omitted from subsequent analysis. (c) Median ratios of discharges of various return periods, T, to mean annual flood for stations analysed.

to flood events, and mean annual rainfall surprisingly enters regression equations more frequently (Riggs, 1973, Nash and Shaw, 1966, Figure 5.4a). However, two major studies include a more physically relevant rainfall variable: Benson's (1960b) analysis of 164 basins in New England, and the NERC (1975) study of 482 British basins. The final prediction model of the NERC study is

$$Q_{ma} = 0.0607 A^{0.94} R_{SMD}^{1.03} S_1^{1.23} F_s^{0.27} S_{1085}^{0.16} (1 + L)^{-0.85} \qquad (5.14)$$

where R_{SMD} is 1 day rainfall of 5 year return period after subtraction of soil moisture deficit, S_1 is a soil index, F_s is stream frequency (number per km^2), the slope index S_{1085} is in m m^{-1}, and L is the fraction of catchment area occupied by lake storage. Discharges with other return periods may be estimated by using separate regression models (Benson, 1960b), or by deriving a Gumbel EV1 distribution after predicting the mean and coefficient of variation of annual flood maxima from regressions on basin area and rainfall (Nash and Shaw, 1966). However, standard, regional flood-frequency analyses (Dalrymple, 1960, Dury, 1959, Cole, 1966) predict mean annual flood by a relatively simple regression model (e.g. Figure 5.4a). The hydrological homogeneity of stations with long enough records for a full flood-frequency analysis is then checked as follows. For each station the $Q_{10}{:}Q_{2.33}$ ratio is calculated, and the return period is estimated of the discharge obtained when $Q_{2.33}$ is multiplied by the mean ratio for the set of stations. If the stations have identical $Q_{10}{:}Q_{2.33}$ ratios, this will be 10 years. Non-homogeneous stations are rejected when the calculated return period falls outside the limits plotted on Figure 5.4(b). The homogeneous records are then used to define the median ratios $Q_T{:}Q_{2.33}$, for $T = 1.01, 1.5, 5, 10, \ldots, 2N$, and to develop a curve such as that in Figure 5.4(c). For an ungauged catchment, the mean annual flood is obtained from the regression model; discharges at other frequencies are obtained by multiplying by the relevant ratio.

Bankfull discharge

Ackers and Charlton (1970a) reaffirm the 'regime' concept (Blench, 1969; pp. 18–23) in noting that channel morphology does not adjust with every short-term variation of discharge, but depends on a discharge measure which typifies the range of competent discharges experienced. The flow which just fills the section of an alluvial channel without overtopping the banks has often been accorded this representative status and treated as a 'dominant' or 'formative' event controlling channel form. This has been reinforced by evidence suggesting a uniform frequency of bankfull conditions and a marked process discontinuity associated with overbank flow. However, this evidence must be viewed in relation to the consistency of the methods used to estimate bankfull discharge, which are discussed below.

Identification of bankfull channel properties and discharge

Bankfull discharge is invariably estimated indirectly by identifying the bankfull stage and applying a stage–discharge relationship or by using the slope–area method (Brown, 1971; Equation 5.1). The bankfull cross-section must therefore be defined, and in equilibrium alluvial channels this is achieved by a field survey of the section by levelling and sounding, by horizontal tape and depth staff, or by using an instrument such as the A-frame (Riley, 1969; Figure 5.5a). A morphological definition of bankfull is most successful in straight reaches without point bar development. Wolman (1955) suggested that, by plotting the width:depth ratio against stage for increasing, assumed water levels, a minimum width:depth ratio is apparent at bankfull (Figure 5.5b). Riley (1972) showed that this method is best applied to rectangular sections where depth increases faster than width as stage (discharge) increases (pp. 149–50), and used a bench-index, BI, defined as

$$BI = (w_i - w_{i+1})/(d_i - d_{i+1}) \qquad (5.15)$$

which peaks strongly at bankfull stage (Figure 5.5b). Here, the subscripts i refer to co-ordinates on the perimeter at equal distances, and the index measures local bank slope. A disadvantage of this method, based on field data measured by the A-frame, is that major breaks-of-slope may not be surveyed. Minor benches within the channel cause difficulties for both of these methods, creating a local minimum in the width:depth ratio and a minor bench-index peak. Woodyer (1968) used vegetational and sedimentological evidence to determine that a high bench in incised channels was equivalent to the flood plain level of non-incised channels. The irregularity of bank tops and the flood plain level surface mitigate against reliance on a survey of a single cross-section; rather, an average bankfull cross-section area (C_b) or stage should be defined for a reach whose bankfull slope is s_b, whereupon bankfull discharge can be estimated from

$$Q_b = 4.0 C_b^{1.21} s_b^{0.28}. \qquad (5.16)$$

Williams (1978a) derived this equation from a regression analysis of data from 233 active floodplain and valley flat sites; its similarity to the Manning equation is evident.

Figure 5.5 (a) The A-frame, a simple surveying instrument (after Riley, 1969). (b) Identification of bankfull stage and width by minimum width:depth ratio and maximum bench-index value. (c) The downstream increase in the duration of peak discharges in the White River (after Dury, 1961). (d) Downstream trends of bankfull discharge (full circles) and $Q_{1.5}$ (curve) along the River Wye (after Hey, 1975). (e) Frequency distribution of recurrence intervals of bankfull discharge (after Williams, 1978a). (f) Discrimination between floodplain reaches experiencing scour and deposition during overbank events, based on slope and relative valley widths (after Wolman and Eiler, 1958).

Additional information on bankfull levels is provided by patterns of vegetation and sediments. Speight (1965b) and Nunally (1967) use the channelward limit of trees and tall grasses as an indicator of channel width, and strong tonal variation on colour air photographs caused by variations of vegetation suggests the possibility of a reconnaissance method based on remote sensing. Woodyer (1968) defined distinctive species assemblages

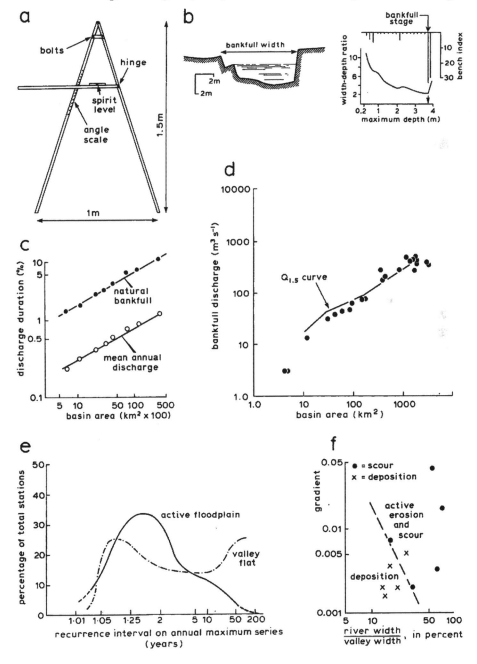

on benches having different inundation frequencies. A sedimentological criterion suggested by Nunally (1967) is the upper limit of continuous sand deposition, bankfull elevation corresponding with the transition between channel and overbank deposits on point bar surfaces. Generally, however, vegetation and sediment patterns are too complex and transitional to provide unequivocal evidence of bankfull stage and usually relate to more extreme events. Sigafoos (1964) identifies flood damage and burial of floodplain trees, but mainly by historical extreme events. Additional evidence of floodwater levels is provided by slackwater deposits. These occur in tributary valleys, and are deposited during extreme events when the mainstream flood flow causes backwater conditions in the tributaries. Again, however, the events responsible usually have return periods of the order of 30–200 years (Patton *et al*, 1979). Furthermore, the rhythmic nature of these deposits is equivocal; each separately identifiable succession of sediments could relate to a single flood, or to surges of flow during a single major event. The only consistent non-morphological evidence of a specific event frequency is the truncated distribution of lichen thalli on boulders and rock-walls, caused by inundation and abrasion by suspended sediment. Gregory (1976) has identified the discharge at the lichen limit as having a recurrence interval of 1.14–1.37 years (annual series). Thus the vegetational and sedimentological criteria are most useful as additional evidence in meandering and incised reaches where the morphological evidence is ambiguous.

The frequency of bankfull discharge

Wolman and Leopold (1957) suggested a common return period for bankfull discharge of 1–2 years, a consistency implying mutual adjustment of flood-plain surface and channel bed elevations. This has fostered assumptions that cross-sections adjust to accommodate a uniform bankfull frequency, with an average return period imposed by the annual hydrological cycle. Dury (1961) showed that the bankfull discharges of the White and Wabash Rivers were a constant fraction of the mean annual flood, and Dury *et al* (1963) estimated the average bankfull return period as $T = 1.58$ years, the most probable annual flood, by estimating bankfull flow from meander wave-length, a dubious procedure given the multivariate control of pattern morphology discussed in Chapter 7. However, more recently Dury (1973) has demonstrated the virtual equivalence of Q_b and $Q_{1.58}$ in American data:

$$Q_b = 0.97Q_{1.58}. \tag{5.17}$$

Similar results emerge from Woodyer's (1968) analysis of the return periods of flows inundating the floodplain of stable Australian rivers, and the high in-channel bench of incised rivers. McGilchrist and Woodyer (1968) estimate a median period between these floods of 335 days. The duration of peak flows tends to vary systematically downstream in a catchment as floods attenuate and overbank storage occurs (Figure 5.5c); a peak event may have constant return period but varying duration. Nixon's (1959) estimate that bankfull discharge is equalled or exceeded 0.6% of the time in British rivers could mean two events per year or a four-day event in two years. Flow duration and flood-frequency concepts are not interchangeable, being based on entirely different types of data (respectively, continuous and discrete).

It is evidently an oversimplification to assume a constant return period for bankfull discharge. Hey (1975) plotted bankfull discharge against basin area, then showed that the downstream trend of $Q_{1.5}$ passed through the scatter of data for the Severn and Tweed rivers. However, on the River Wye, bankfull discharge is less than $Q_{1.5}$ at upstream sites and therefore occurs with greater frequency, while downstream it exceeds $Q_{1.5}$ in magnitude and is less frequent (Figure 5.5d). Hey (1975) suggests that the most probable annual flood, which has a one-year return period on the partial duration series, represents bankfull discharge in gravel-bed streams, while bankfull conditions are more frequent in sand-bed streams. Harvey (1969), however, attributes the systematic variation in bankfull frequency to the hydrological regime. Flashy streams experience more frequent overbank flooding upstream, which could be interpreted as a compensation for a downstream increase in flood duration. Baseflow-dominated streams experience more frequent flooding downstream. Kilpatrick and Barnes (1964) also identify variations in overbank frequency, but relate it to channel slope. Greater entrenchment of steep streams increases their capacity, so that although the return period of bankfull discharge is 1–2 years on a mild slope, it increases on steeper gradients. More locally, considerable variation of floodplain relief (1.7–3.3 m on three Welsh floodplains), reflecting palaeochannel locations, results in complex relationships between degree and timing of inundation and flow magnitude (Lewin and Manton, 1975). Temporal as well as spatial variability in bankfull frequency must also be considered; aggrading reaches experience overbank conditions more frequently until the floodplain elevation is increased. Incision, however, reduces the frequency of overbank flooding. Systematic temporal variation in the frequency of flooding is thus an indication of channel disequilibrium (Blench, 1969, p. 28). Even in relation to equilibrium channels, though, it is not surprising that Williams (1978a) should conclude that the distribution of return periods for bankfull conditions, although centred on a modal value of $Q_{1.58}$ for active floodplain reaches, is too variable for the assumption to be made of a common frequency (Figure 5.5e).

Bankfull discharge and floodplain construction

Mutual adjustment of channel form and floodplain development is reflected in the bankfull capacity of the cross-section and the relationship of floodplain sedimentology and construction to channel pattern, itself closely related to cross-section form (Chapter 7). The classic simplified model of floodplain construction by meandering streams involves lateral deposits of fining-upwards gravels and sands formed by point bar development on the insides of migrating bends, and overbank silt and clay deposition. This justifies Nunally's (1967) sedimentological criterion of bankfull stage. However, Plate 6 (p. 210) shows that the spatial pattern of floodplain sediments is a more complex reflection of palaeochannel deposition, and Wolman and Leopold (1957) note that the simplified model of fining-upwards through the floodplain deposits is far from widespread in reality. Local deposition of gravels by overbank flows is not uncommon (McPherson and Rannie, 1970). The irregular topography of floodplains of upland rivers (Lewin and Manton, 1975) carrying coarse bedload, especially braided streams, may reflect the lack of supply of fine sediments to infill palaeo-channels, which in turn has an important influence on the potential for channel migration in such rivers (Werritty and Ferguson, 1980).

Studies of individual extreme events provide conflicting evidence of their erosional and depositional effects. McPherson and Rannie (1970) and Wolman and Eiler (1958) suggest that floodplain erosion is localized and deposition spotty, averaging 2–3 cm on 25–45% of the surface area and being strongly influenced by pre-existing vegetation. Gupta and Fox (1974) identify erosional effects within the channel – bed widening and removal of sands – but note that re-deposition on channel bars after the flood causes narrowing to the original width, unless back-to-back flood events occur. The 1936 and 1938 Connecticut floods (Jahns, 1947) inundated a 9 m terrace, but caused erosion only on the outer banks of meanders, and resulted in 3.5 cm and 1 cm of silt and sand deposition respectively over areas up to 100 km^2. However, Stewart and LaMarche (1967), reporting a Californian event of a 100 year return period, describe widespread upstream erosion of 14,000 m^3 km^{-1} and downstream deposition of 106,000 m^3 km^{-1} over a shorter total valley length. The discrepancy between these studies reflects local valley conditions, erosion being predominant in steep, relatively narrow reaches, while deposition occurs in wide, gently sloping valleys (Figure 5.5f).

The limited depositional activity of overbank flow on wide floodplains reflects a process discontinuity at bankfull. Wolman and Leopold (1957) describe reductions of suspended sediment concentration at bankfull stage which could reflect exhaustion of supply, but also arise because the rate of increase of velocity relative to depth often declines as bankfull conditions are approached (pp. 150). Furthermore, Barishnikov (1967) shows that sediment transport declines abruptly, by an amount which increases with the roughness of the floodplain surface, when overbank flow interferes with the channel flow. Of course, this does not prove that channels adjust to the process discontinuity, since a similar mutual interference of flow would occur in a concrete channel, which is clearly not self-formed. Nevertheless, the lateral deposits which form the bulk of the floodplain are channel deposits formed at discharges up to the bankfull event. Thus, floodplain construction is predominantly associated with high-frequency events, as long as the sedimentology displays the characteristic point bar fining-upwards. However, Anderson and Calver (1977) describe a boulder deposit by the 1952 Exmoor floods which, after 25 years of infill by fines and vegetation growth, could easily be mistaken for a floodplain, and Baker (1977) argues that in environments dominated by high-intensity rainfall, constructional activity by catastrophic events may be more important. Finally, Tinkler (1971) argues that the dominant discharge controlling active *valley* erosion in Texas is a 20–50 year event, and that floodplain construction fossilizes the valley form because the energy of these events is dissipated on the valley surface. Rather than the floodplain being adjusted to an event of a 1–2 year return period, he argues that the formation of the floodplain itself causes a reduction of the effective dominant event, and that valley filling begins when a threshold sediment yield is reached as valley incision progresses (see pp. 18–19, 55).

Flood frequency, sediment transport and channel form

Magnitude-frequency analysis suggests that most sediment transport is achieved by intermediate events of high frequency (Wolman and Miller, 1960). By combining flow duration curves based on continuous discharge data and sediment rating curves, the sediment transport associated with a range of discharge classes may be calculated, although care is necessary in the selection of class intervals and the choice of simple, multivariate or seasonal rating curves (Walling, 1977). Benson and Thomas (1966) used this method to show that suspended sediment transport is distributed across a range of discharges, with the most effective discharge classes in several streams having mid-points equivalent to an average frequency of 12.4%, ranging from 7.6–18.5% (Figure 5.6a). These discharges are well below bankfull stage. The relative importance of extreme events is, however, dependent on the hydrological regime. In semi-arid environments with variable flow regimes, about 40% of sediment transport is by events of less than a 10-year return period, whereas in humid environments with more consistent flow and lower sediment yield from slopes during extreme events because of the protective effects of vegetation, more than 90% of sediment transport is by frequent events (Neff, 1967). Sediment type also affects the role of discharges of varying frequency. Bedload transport demands the exceedance of a threshold stream power, so the most effective discharge should be more extreme. Following Schaffernak (1950), who introduced the concept of a 'bed generative discharge', Pickup and Warner (1976) calculated the total bedload transport associated with each discharge, using the Meyer-Peter and Müller transport equation. The 'effective discharge' is that carrying most bedload over a period, and it is identified as the peak in the curve of Figure 5.6(b). In the sections investigated, the return period of the effective discharge ranged from 1.15 to 1.4 years, less than the most probable annual flood and the bankfull discharge (whose return period ranged from 4 to 10 years). Andrews (1980) added instantaneous measured suspended load to bedload calculated using the same method to define effective discharges for total load, which range from $Q_{1.18}$ to $Q_{3.26}$, but which in this case were equivalent to bankfull discharge. These results therefore all confirm that relatively frequent discharges dominate the transport of sediment, but disagree over the relationship between the effective and bankfull events.

Channel morphology reflects sediment transport and therefore should be systematically related to runoff frequency. A broad regional variation occurs, with streams in humid environments responding to flood events rather differently from those in semi-arid climates (Leopold and Wolman, 1956). Humid rivers with gravel beds and cohesive banks are stable, whereas the sandy non-cohesive perimeters of semi-arid rivers, which often also lack the anchoring influence of vegetation, are unstable and are associated with marked bed scour and hysteresis in sediment ratings. The influence of sedimentology is also emphasized by Pickup and Warner (1976), whose observations suggest that discharges more frequent than the modal annual flood dominate bedload transport, but that more extreme events control the erosion of cohesive banks. Thus, two groups of events are responsible for creating the channel form: a more extreme group which defines channel capacity; and more frequent events which control bedload movement and bedform construction. Pickup (1976) extends this analysis further. For a given cross-section, the slope and roughness are assumed fixed, and bedload transport is calculated for various widths and discharges (Figure 5.6c). The present bed width of the section is associated with peak bedload transport at a discharge closely approximating the 'effective discharge' defined above. Thus, channel width adjusts to provide the most efficient section for bedload transport under the prevailing conditions of imposed load, slope and roughness, and effective discharge magnitudes.

Figure 5.6 (a) 'Dominant' discharge in the transport of suspended sediment load, Potomac River, Point of Rocks, Maryland (after Benson and Thomas, 1966). (b) 'Effective' discharge in the transport of bedload, Spring Creek (after Pickup and Warner, 1976). (c) Relationship between bedload transport and bed width at various discharges, for Spring Creek (after Pickup, 1976). (d) Relationships between riffle–pool wavelength and channel width, mean annual discharge, and 1.1 year flood for three British rivers (after Harvey, 1975). (e) 'Dominant' discharge for transport of suspended load and bedload.

Most morphological data are less clear-cut in emphasizing the role of a particular flow frequency. Carlston (1965) correlated meander wavelength to mean discharge, mean discharge in the month of maximum discharge, and bankfull discharge. The sub-bankfull events showed the best correlations, and it was concluded that they controlled meander morphology, particularly because bank erosion is often observed to occur on falling stages (p. 164).

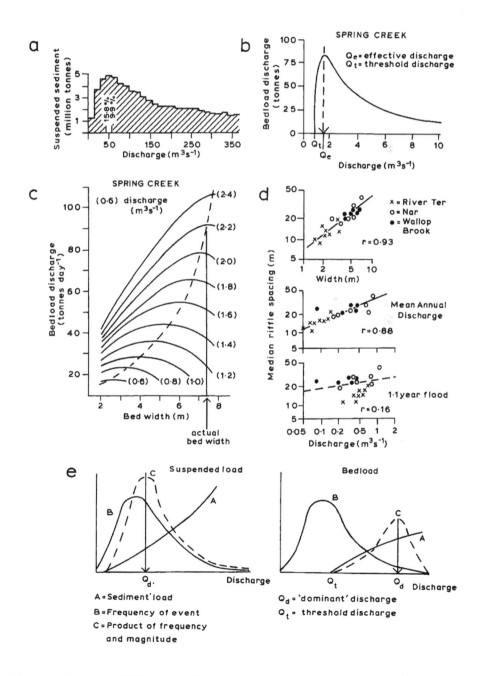

However, falling stage bank erosion usually requires bank saturation at higher discharges, and the correlation alone is a poor guide since it varies with the efficiency of estimation of the various discharge indices used. Ackers and Charlton (1970a) tested Carlston's conclusions in flume experiments which employed varying discharges to model natural hydrographs. For mean discharge (Q_m), Carlston's (1975) relation with wavelength (L) was

$$L = 166Q_m^{0.46}. \tag{5.18}$$

The dominant discharge in Ackers and Charlton's (1970a) experiments is that which creates the same forms as a varying discharge. It is related to wavelength by the expression

$$L = 62Q_d^{0.47}. \tag{5.19}$$

The lower constant implies that the dominant discharge is well above mean flow, and Ackers and Charlton (1970a, p. 250) conclude that 'natural bankfull discharge generates the plan geometry provided the river is not intrenched to such an extent that it does not run bankfull at least annually'. In fact, Harvey (1975) related the wavelength of the riffle–pool sequence which occurs on the bed of gravel-bed rivers to a range of discharge indices and showed that correlations were best with channel width and the subbankfull events (Figure 5.6d), but this is not surprising. Bedload movement and bedforms do appear to be related to frequent flows, but channel capacity and meander morphology reflect slightly more extreme events. Potter *et al* (1968) suggest that events as extreme as the ten-year flood on the upper frequency curve of a dog-leg flood-frequency plot are 'dominant'. They note a discontinuity in the curve of cross-section area as a function of stage, and find that the discharge at this stage is consistently the ten-year event. However, they provide no evidence of the significance of this stage in relation to channel form. More commonly, channel capacity is regarded as being adjusted to events in the range $Q_{1.5}$ to Q_{10}, with the more frequent events 'dominant' in flashy streams and rarer events controlling channel form in baseflow-dominated streams (Harvey, 1969).

However, adjustment to a single specific event does not occur. Gupta (1975) describes channel forms in Jamaica, which is subjected to frequent extreme discharges because of tropical cyclone rainfall. On the north side of the island, seasonal flow is associated with wide, shallow, braided rivers, while on the south side a more uniform discharge regime creates meandering or straight streams. A different response to extreme floods therefore occurs in these contrasting hydrological environments, as a result of different discharge ranges and recovery periods. Pickup and Rieger (1979) conclude that every competent event has some influence on channel form, so that channel form y at time t is a weighted sum of effects of all input discharge events up to and including t, described by a model of the form

$$y(t) = \int_0^x h(u)Q(t-u)\,du \qquad (5.20)$$

where $h(u)$ is an impulse response measuring the effect of discharge on channel form over various time lags (u). The necessity of such a model is greatest in semi-arid streams where adjustments to individual events are so extreme, and recovery periods so long, that streams may be almost permanently out of regime (Stevens *et al*, 1975). Finally, the 'dominant discharge' in any equilibrium channel must reflect the imposed range of discharges and the nature of the sediment load (Figure 5.6e). The effective, or dominant, event is lower in magnitude if the stream carries suspended sediment than if bedload is transported, since threshold conditions must be exceeded in the latter case, and the rate of increase of the sediment transport rate is less (Hey, 1975). Similarly the distribution of event magnitudes will determine the flow carrying most sediment in the long term (Baker, 1977); Figure 5.1(a) must therefore be evaluated in relation to local conditions of sediment character and discharge regime.

6

The morphology of river cross-sections

Hydraulic engineers have long recognized that cross-sections of straight artificial channels must be designed to transmit their sediment load with the required water discharge down the imposed valley gradient, without silting or erosion. Davisian geomorphologists, however, concentrated on long-term long-profile changes in the maintenance of grade until Mackin (1948) drew attention to medium-term cross-section adjustments as one means whereby channels may convey natural discharges at velocities permitting some continuity of sediment transport. Channel width is even considered a primary morphological variable (Leopold and Wolman, 1957, pp. 63–4) to which other aspects of channel form adjust. The strong correlations of meander and riffle–pool wavelengths with width suggest the importance of a width-dependent flow instability controlling these lateral and vertical oscillations (Chapter 7). Rapid cross-section adjustment to equilibrium is apparent in flume experiments conducted by Wolman and Brush (1961). An initial channel was cut by template in a homogeneous sand bed, at an approximately appropriate size for the discharge to be employed. Figure 6.1(a) shows rapid widening early in each run followed by stabilization at an equilibrium width dependent on discharge and sediment size. Aggradation of the channel bed also occurred, producing a wide, shallow, equilibrium channel form.

Normally a natural channel accommodates flows up to approximately the mean annual flood before overbank flooding occurs. If the channel capacity is too small, overbank and bankfull events are more common. Thus the threshold velocity (or stream power) of the perimeter sediment may be more frequently exceeded, and sediment entrainment and perimeter erosion occur. The average velocity over a period will be high in an initially small

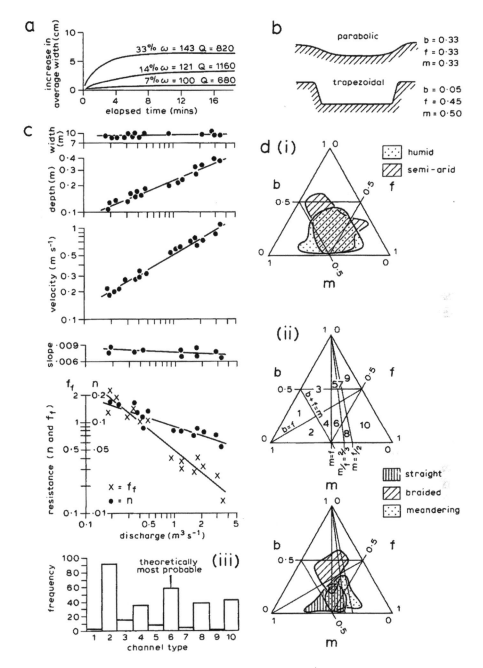

Figure 6.1 (a) Increase in flume channel width through time. Percentages are relative width increase to equilibrium, ω is initial power expenditure per unit bed area (erg cm^{-2} s^{-1}), Q is discharge (cm^3 s^{-1}) (after Wolman and Brush, 1961). (b) Typical channel cross-section shapes and associated at-a-station exponents. (c) An example of at-a-station hydraulic geometry relations (after Knighton, 1975a). (d) Triaxial diagram plots of at-a-station exponents: (i) after Park (1977a) and (ii) after Rhodes (1977); (iii) is a histogram of frequencies of channel types (after Rhodes, 1977).

cross-section, and the excessive momentum of the flow will be transmitted to the banks and bed by strong velocity shear at the perimeter, causing erosion and channel enlargement until the frequency of exceedance of threshold velocities is lessened, the velocities associated with discharges of a given frequency are reduced and the channel is stabilized (Church, 1967). Establishment of an equilibrium section therefore reflects sediment properties as well as discharge magnitudes, since adjustment of the perimeter presupposes sufficient energy for entrainment. The equilibrium section thus results from an interaction between two sets of variables measuring the force applied by the fluid and the resistance to erosion mobilized by the sediment.

Adjustments to stream power: the role of discharge

The discharge regime of the catchment upstream from a river cross-section, often represented by a single 'dominant' discharge index, forms a fundamental independent control of channel cross-section morphology. However, the energy of streamflow reflects the rate of conversion of its potential energy to kinetic energy and work, which depends on stream gradient as well as discharge. In Wolman and Brush's (1961) experiments, the absolute and percentage increase in stream width from that of the initial template-cut channel depended on the initial rate of stream power expenditure per unit bed area, or $\omega = \rho g Q s / w$ erg cm^{-2} s^{-1} (where w is the template width). Figure 6.1(a) shows that the greatest increase occurred in the run with maximum power expenditure rate in the initial channel. When $\omega = 100$ erg cm^{-2} s^{-1}, the equilibrium channel width was only 7% wider than the starting width, and reference to Figure 1.3(a) shows that this stream power value is close to the threshold for the sediment of 0.67 mm median diameter used in these experiments. This suggests that ultimate control of channel morphology in a given sediment reflects the rate of loss of potential energy of the volume of streamflow generated at the frequency of channel-forming events.

In spite of this evidence of the role of stream power, fluvial geomorphology has been dominated by the analysis of adjustments of river morphology to discharge variations. In the context of river cross-sections, this is manifested in the 'hydraulic geometry' concept which is concerned with changes in *flow* geometry at-a-station during changes of streamflow, and adjustments of flow and channel geometry downstream at a constant frequency of flow.

At-a-station hydraulic geometry

At an individual cross-section, changes occur as discharge varies in the water surface width, mean flow depth and velocity, and in other variables such as water surface gradient, Manning's n, the friction factor, and bed shear stress. These changes reflect adjustments of the water prism within the section, but include scour and fill of the channel perimeter when it is formed in non-cohesive sediments. Following Leopold and Maddock (1953a) these adjustments are described by power function relations of the form

$$w = aQ^b; \quad \bar{d} = cQ^f; \quad \bar{v} = kQ^m; \quad s_w = gQ^z;$$
$$n = hQ^{y_1}; \quad f_f = pQ^{y_2}; \quad \tau_0 = tQ^x. \tag{6.1}$$

Since $w\bar{d}\bar{v} = Q$, it follows that

$$b + f + m = 1 \tag{6.2}$$

and

$$ack = 1. \tag{6.3}$$

At-a-station flow geometry reflects the general cross-section shape, which is itself dependent on material properties. In homogeneous non-cohesive sand, a broadly parabolic shape occurs with concave banks steepening to the angle of repose at bankfull water level; adjustment of these mobile sediments occurs in each significant flood. The resulting exponent set approximates to $b = f = m = 0.33$. Examples in cohesionless sediments have been described from semi-arid (Leopold and Miller, 1956), humid tropical (Lewis, 1969) and proglacial (Fahnestock, 1963) environments (Figure 6.1b). Stable, cohesive silty bank sediments sustain rectangular or trapezoidal sections with low width exponents ($b \approx 0.05$; Wolman, 1955) and f and m exponents dependent on the roughness and resistance to scour of the bed material. The velocity exponent is high when roughness decreases rapidly with increasing discharge, which is characteristic of straight cobble-bed reaches with considerable skin resistance (Knighton, 1975a). The high relative roughness at low flow ($D_{50}/\bar{d}$) is rapidly drowned by a rise in water level and not replaced by bedform development and increased form roughness. A typical example is shown in Figure 6.1(c).

Since the three main exponents (b, f and m) sum to unity, the set characterizing a given section can be plotted on a triaxial diagram. Park (1977a) used this approach to seek differences in channel types between environments. The wide distribution of parabolic sections, noted above, and the overlapping distributions of b–f–m sets from humid temperature and semi-arid climates in Figure 6.1(d(i)) suggest that such broad regional controls are relatively unimportant. The bed and bank material properties which determine channel shape are themselves only indirectly related to regional variations in climate, hydrology and weathering regime. Rhodes (1977) divided the triaxial diagram into ten regions delimited by distinct hydraulic criteria, and showed that straight, meandering and braided rivers could be discriminated (Figure 6.1d(ii)). This evidence of a link between cross-section and planform characteristics is supported by investigations by

Knighton (1974, 1975a), and suggests that local hydraulic conditions control the exponent set rather than regional climatic or hydrological factors (see pp. 175–9). The five criteria employed by Rhodes are based on the relative rates of change of width and depth ($b = f$), velocity and depth ($m = f$) and velocity and cross-section area ($m = b + f$), and on the direction of change of the Froude number (which increases if $m > f/2$) and the roughness coefficient (which decreases at constant slope, according to the Manning equation, if $m/f > 2/3$). The frequency distribution of 315 exponent sets in Figure 6.1(d(iii)) shows that the commonest *sampled* channel sections plot in regions 2, 4, 6, 8 and 10 of the triaxial diagram, and therefore experience reductions of flow width:depth ratio as discharge increases. Braided streams are under-represented in the sample, however, and plot in region 3, where increasing width:depth ratio and competence to transport bedload characterize the at-a-station hydraulics (Rhodes, 1977, p. 78).

The exponent set can be predicted using minimum variance theory, which assumes that discharge variations are accommodated by the joint adjustment of the dependent variables within imposed constraints, the most probable channel form being that with minimum total variability (according to Figure 6.1(d(iii)), a Type 6 channel). In the absence of constraints, this occurs when all dependent variables have equal variance. The exponent is a dimensionless measure of variability; for example, in Figure 6.1(c), velocity appears to be more variable than width over the observed discharge range. If a dependent variable Y is related to discharge Q by a power function

$$\log Y = a + b \log Q \tag{6.4}$$

then the exponent, b, is

$$b = r \frac{S_{\log Y}}{S_{\log Q}}. \tag{6.5}$$

Thus, assuming perfect correlation ($r = 1$) and a constant standard deviation (S) of $\log Q$ in a given cross-section,

$$b^2 \propto S_{\log Y}^2 \tag{6.6}$$

(Croxton, 1959, p. 125). Minimum variance thus occurs when the sum of the squares of exponents is minimized (Langbein, 1964a, 1965). To apply this minimization hypothesis, the relevant dependent variables and system constraints must be identified. If only the primary variables (width, depth and velocity) are adjustable within the constraint $w\bar{d}\bar{v} = Q$, $b^2 + f^2 + m^2$ is minimized when $b = f = m = 0.33$. However, Equation (6.1) suggests that at least six dependent variables exist (selecting one only of the friction measures), although some are derived, secondary variables, and it is not clear whether all should be included and given equal weight (Williams, 1978b). As an example, consider a straight, uniform reach where slope is constant as discharge increases ($z = 0$) and the banks are vertical in cohesive sediments ($b = 0$). In addition to these constraints are those provided by the definitions of the derived variables. As $f_f = 8g\bar{d}s/\bar{v}^2$, it follows that

$$y_2 = f + z - 2m \qquad (6.7)$$

when the equation is re-expressed in terms of the exponents, and as $\tau_0 = \rho g \bar{d}s$,

$$x = f + z. \qquad (6.8)$$

Since $b = 0$ and $z = 0$, all exponents of the selected dependent variables can be defined in terms of the depth exponent f:

depth exponent	$= f$
velocity exponent, m	$= 1 - f$
friction factor exponent, y_2	$= f - 2m = f - 2(1-f) = 3f - 2$
and shear stress exponent, x	$= f$

The minimum variance solution requires that

$$f^2 + (1-f)^2 + (3f-2)^2 + f^2 \rightarrow \text{minimum}$$

or

$$12f^2 - 14f + 5 \rightarrow \text{minimum}. \qquad (6.9)$$

By differentiating with respect to f and setting equal to zero,

$$24f - 14 = 0$$

and therefore minimum total variance occurs when $f = 0.58$, $m = 0.42$, $y_2 = -0.26$ and $x = 0.58$, which defines a Type 6 cross-section in Figure 6.1(d(iii)).

Minimum variance theory predicts adjustments of the flow geometry within a stable section, assumed invariant during a change of discharge; predicted exponents define the general trends of depth and velocity irrespective of scour and fill, which are assumed to balance. Thus the shape of the section within which the flow adjusts is predetermined (in the example above it is rectangular). Langbein (1964a) used a theoretically defined stable cross-section with sloping banks, for which a plot of width against depth for successively increasing assumed water levels gave $\omega \propto d^{0.55}$. Thus an additional constraint, $b = 0.55f$, could be used, and the predicted exponents ($b = 0.25$, $f = 0.45$ and $m = 0.30$) agreed with the *mean* exponents identified by Leopold and Maddock (1953a) for river sections in the Midwest USA ($b = 0.26$, $f = 0.40$, $m = 0.34$). However, only slight differences occur in the predicted exponents for various combinations of dependent variables and constraints (cf. Langbein and Leopold, 1964, p. 790), and since the exponents b, f and m lie between 0 and 1 and sum to unity (and are therefore not independent) it is impossible to compare observed and predicted values statistically. Furthermore, minimum variance theory only predicts the most probable exponent set under given constraints, and individual cross-sections may deviate from these 'modal' values. Ideally, the exponents should be related to independent sedimentological constraints in individual sections. Knighton (1974) has identified a chain of causation linking bank silt–clay content directly to bank angle, which is then inversely related to the width exponent. Williams (1978b) also recognizes that the width exponent reflects bank angle and channel shape, which themselves depend on sediment properties, but in an as yet undefined way. He shows that a survey of an individual cross-section permits the definition of a water width–area relationship for various assumed water levels; this takes the form $w \propto C^t$. The width exponent can be predicted from the regression relationship

$$b = 0.8t \qquad (r^2 = 0.86) \tag{6.10}$$

and the depth exponent from

$$f = 0.6 - 0.58t - 0.0018D_{50} \quad (r^2 = 0.66). \tag{6.11}$$

The inverse relation between f and median bed material size for a given width exponent reflects the inhibited scour of coarse bed material, as well as the lower rate of depth increase but faster velocity increase in cobble-bed streams where skin resistance is drowned out by a rise in water level. However, the relation may be nonlinear over a wide range of bed material sizes (Figure 6.2a), with silty beds being more resistant to scour than sand beds, as predicted by the Hjulström curve (Figure 3.8b).

Most natural river cross-sections display more complex behaviour than conventional hydraulic geometry suggests. Lewis (1966) noted that flow at low discharge is often confined to a secondary channel in the stream bed, and the width–discharge curve displays a break-of-slope when its capacity is exceeded. Richards (1976a) extends this to the consideration of the whole cross-section: breaks of bank slope caused by bank slumps, in-channel benches (Woodyer, 1968), bank foot sediment accumulation and bank top curvature may all be translated into complexities in the width–discharge curve, and thence to the depth and velocity relationships. There is no *a priori* reason why power functions should best describe variations of flow geometry with discharge, and roughness changes in particular cause departures from these simple curves. More complex variations may be described by polynomials, the optimal degree of polynomial being determined by testing the significance of improved variance explanation by successively higher-order curves (Chayes, 1970). The exponents in the power function relationships are analogous to the first derivatives of more complex curves (see Figure 6.2b), and therefore

$$\frac{d(\log w)}{d(\log Q)} + \frac{d(\log d)}{d(\log Q)} + \frac{d(\log v)}{d(\log Q)} = 1.0 \tag{6.12}$$

is equivalent to Equation (6.2).

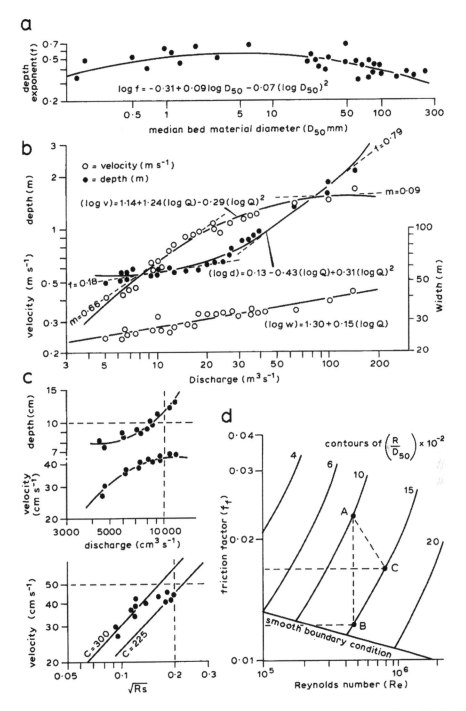

Figure 6.2 (a) Nonlinear relationship between depth exponent and median bed material diameter. (b) Flow geometry variations at Draycott on the River Derwent, England. (c) Depth, velocity and roughness variations reflecting the effect of increasing form roughness at a cross-section in a sand-bed flume. (d) The relationship of the friction factor to relative smoothness and Reynolds number for a flat bed experiencing sediment transport (after Lovera and Kennedy, 1969).

The data plotted in Figure 6.2(b) are best fitted by the curves

$$\left.\begin{array}{l} \log w = b_1 + b_2 (\log Q) \\ \log d = f_1 + f_2 (\log Q) + f_3 (\log Q)^2 \\ \log v = m_1 + m_2 (\log Q) + m_3 (\log Q)^2 \end{array}\right\} \quad (6.13)$$

and since

$$\frac{d(\log d)}{d(\log Q)} = f_2 + 2f_3 (\log Q) \quad (6.14)$$

it follows that

$$b_2 + f_2 + m_2 + 2(f_3 + m_3) (\log Q) = 1 \quad (6.15)$$

Thus, the instantaneous gradients of the three curves sum to one in the manner of conventional exponents, with increasing slope of the depth curve balancing decreasing slope of the velocity curve (assuming a constant width exponent). Richards (1973) noted a consistent tendency for this form of curvature, which arises because of consistent behaviour of channel roughness (Knighton, 1979). In sand-bed streams below a Froude number of 1, the smooth bed gives way to ripples and dunes with increasing discharge and sediment transport rate, and although the hydraulic radius increases to make the section more efficient and encourage higher velocities, this is partly offset by the greater form resistance of dunes, which can increase roughness six-fold (Raudkivi, 1967). Figure 6.2(c) shows the effect of dune forms in a flume cross-section. At low discharges the Chezy C (inversely related to the friction factor) is 300, but at high discharges it is 225, and the effect of increased form resistance is reflected in a reduced rate of increase of velocity. When upper-regime flow is achieved, a further discontinuity occurs in the depth–discharge curve as plane beds replace dunes, and the Chezy C increases again to a constant value dependent only on particle size (Dawdy, 1961). Short-term disequilibrium between stream power and bedform type during changes in streamflow (Allen, 1976a,b) results in different roughness values at comparable discharges on rising and and falling hydrograph limbs. This creates hysteresis in stage–discharge curves (Figure 1.2d), making stream gauging difficult in sand-bed rivers; Simons *et al* (1973) have, however, outlined a simple empirical method for predicting stage variation resulting from these bedform changes. In gravel-bed rivers (such as that whose hydraulic geometry is shown in Figure 6.2b) bedform changes are not significant, but the friction factor may be a function of 'relative smoothness' (R/D_{50}) and Reynolds number (Re), both of which increase with discharge. Figure 6.2(d) shows that the rate of decline of the friction factor with increasing relative smoothness is much less when the Reynolds number increases (AC) than when it is constant (AB). Thus, over a flat bed with no form roughness, the rate of decline of the friction factor with discharge may decrease at high discharges, with the result that the rate of increase of velocity also lessens.

Downstream hydraulic geometry

Channel cross-section morphology adjusts to accommodate the downstream changes of stream discharge and sediment load. The concept of 'downstream hydraulic geometry' uses the power function relationships of Equation (6.1) to describe trends in the dependent variables with discharge (Leopold and Maddock, 1953a) and encompasses two forms of analysis. The first considers downstream changes of flow geometry within the varying overall cross-section and at a constant flow frequency, so that between-section comparisons of flow width, depth, velocity, etc., are made at a standard flow duration or return period. An inherent limitation of this approach geomorphologically is that the selected flow frequency may be competent at some sections but not at others, and is therefore of variable significance. The second approach represents the geomorphological analogue of 'regime theory', an empirical method of determining stable artificial channel dimensions of sediment-bearing canals carrying a dominant discharge (pp. 286–94). Downstream adjustment of the bankfull channel cross-section is analysed by relating its width, depth, capacity and bank top (bankfull water surface) slope to a bankfull discharge estimate which represents the equivalent of the dominant discharge. Velocity and roughness are then derived from $\bar{v} = Q_b(wd)^{-1}$ and a friction relationship. Bankfull discharge must be obtained indirectly at ungauged sections where no stage–discharge curve exists. It may be calculated using the Manning equation (Brown, 1971), although this involves circular analysis if it is then correlated with the channel properties used in its estimation. Alternatively, a regression relationship between a peak flow index and physical and climatic properties of hydrologically comparable gauged catchments may be used (pp. 134–5; Benson, 1960b). Such regression models can often only be defined for standard peak flow indices ($Q_{2.33}$ or $Q_{1.58}$) rather than for bankfull discharge. To use one it is necessary to accept the assumption that, for example, the $Q_{2.33}:Q_b$ ratio is constant, or that these events are equivalent (Brush, 1961). Basin area or total upstream channel length may be used as surrogates for discharge in perennial rivers (Hack, 1957, 1965). However, the exponent of, for instance, a channel width–basin area power function cannot be compared with a width–discharge exponent because it reflects the discharge–area relationship, which is itself dependent on area-related variations in catchment slope and rainfall.

The complete hydraulic geometry of a river is encapsulated by a series of quadrilaterals, one for each dependent variable. Two sides of the quadri-

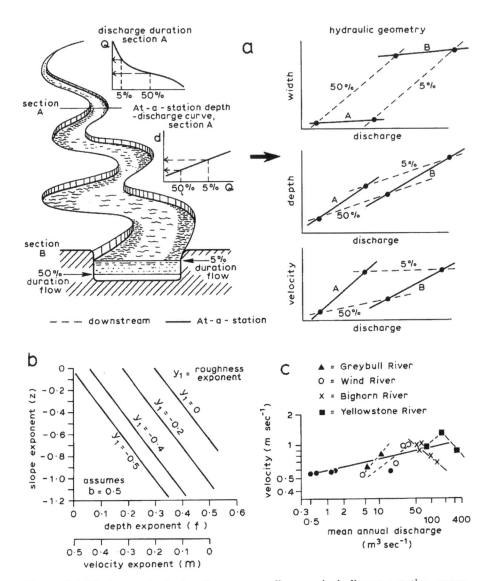

Figure 6.3 (a) Composite hydraulic geometry diagrams including at-a-station curves for upstream and downstream sections and downstream curves at 5% and 50% duration flows: the left-hand diagram shows the identification of depths at section A at these flow frequencies. (b) Interrelationship between depth, velocity, slope and roughness downstream exponents according to the Manning equation (assuming $w \propto Q^{0.5}$) (after Leopold and Miller, 1956). (c) Downstream trends in velocity for rivers in Wyoming and Montana, each represented by a different symbol (after Mackin, 1963). ● Single observations on small tributary streams.

lateral are defined by at-a-station power functions, and two sides by downstream functions. The at-a-station relationships are for an upstream and a downstream section, while the downstream trends are for one high-frequency flow (such as the mean annual discharge) and one low-frequency event (such as $Q_{2\%}$ or the mean annual flood, Q_{ma}). Figure 6.3(a) illustrates the derivation and nature of the hydraulic geometry diagram (Leopold and Maddock, 1953a, p. 19), which should be viewed as a simplified approximation to general trends because of the complex at-a-station behaviour noted above, and the multivariate controls of variation in channel form in the downstream direction.

Significant differences exist in the downstream hydraulic geometry of perennial and ephemeral rivers. Average width (b), depth (f) and velocity (m) exponents for downstream trends in Midwest American rivers were defined as $b = 0.5$, $f = 0.4$ and $m = 0.1$ at mean annual discharge (Leopold and Maddock, 1953a), although Carlston (1969) amended these to $b = 0.46$, $f = 0.38$ and $m = 0.16$ after a regression analysis. The more rapid downstream increase of width compared to that of depth implies an adjustment of width:depth ratio, and hence a scale-related change in channel shape, with large rivers being slightly wider relative to their depth than small streams. In ephemeral semi-arid streams, Leopold and Miller (1956) defined the average downstream exponent set $b = 0.5$, $f = 0.3$ and $m = 0.2$ at an unknown flow frequency for individual flash-flood events travelling downstream and diminishing in discharge as evaporation and seepage losses occurred. Their use of power functions implies that width, depth and velocity increase downstream to the point of maximum discharge and then decrease according to the same relationship.

The apparently consistent form of the width–discharge relationship has encouraged the assumption of a square-root 'law' ($b = 0.5$), particularly in regime theory (Lacey, 1930). If such a relationship is assumed, it follows that from Equation (6.2), $m = 0.5 - f$. The Manning equation can therefore be expressed in exponent terms as

$$0.5 - f = \frac{2}{3}f + \frac{1}{2}z - y_1 \qquad (6.16)$$

which can be simplified to

$$z = 1 - 3.33f + 2y_1 \qquad (6.17)$$

and plotted as a trivariate diagram showing the interrelationships between the exponents (Figure 6.3b). Downstream roughness and slope exponents are partly independent factors related to sedimentological trends and to the inherited long-profile characteristics; thus the velocity and depth exponents respond to their influence. If roughness is constant downstream ($y_1 = 0$), perhaps because gravel skin resistance gives way to sand form roughness, and if an (average) downstream slope exponent of $z = -0.67$ occurs, the velocity exponent will be zero. Consistency of the width exponent must be assumed for this diagram to be valid, and Park (1976) provides some evidence in support of this. Channel capacity–basin area power functions for several Devon catchments have exponents inversely related to basin relief ratios (Schumm, 1956), implying a less rapid downstream increase of channel size in steep catchments. Capacity is the product of width and mean depth, and significantly the width–area exponent is independent of the relief ratio, with all of the variation in the capacity–area trend reflecting differences in the depth–area exponent. However, the square-root relationship cannot be regarded as a law. Carlston (1969) notes that width increases faster downstream in small streams while depth adjustment predominates in large rivers, and Henderson (1961) argues that a square-root relationship only arises in rivers with a certain slope–discharge relation (or long-profile configuration). In fact, Henderson (1961) developed the multivariate relationship

$$p = 1571Q_{bs}^{1.17}s_b^{1.17}D_{50}^{-1.5} \qquad (6.18)$$

in which the wetted perimeter p ($\approx$ width) is shown to increase with stream power, represented by the discharge–slope product, rather than being related to discharge alone, and to decrease with median bed material size. This is one of the few studies which reiterate the power:resistance interaction shown to be a determinant of channel form by the experiments of Wolman and Brush (1961), who also emphasize that erosive power is dependent on slope and suggest that the square-root relationship only arises if bank materials are constant.

Leopold (1953) identified a tendency for velocity to increase downstream, contradicting Davisian assumptions. Although long-profile gradient normally decreases downstream, velocity can increase because of increasing cross-section efficiency (hydraulic radius) and decreasing skin resistance as cobbles are successively replaced by gravel, sand and silt. However, the varying competence of low-magnitude sub-bankfull flows needs to be considered. At a flow such as the mean annual discharge, the downstream sand bed may only experience limited transport, with lower-regime plane-bed or ripple bedform phases and little form roughness. Thus the downstream decrease of roughness is well defined, and velocity increases (Figure 6.3b). At a more extreme flow, the sand bed is live, dune bedforms develop and form roughness is increased. The downstream decrease of roughness is less marked therefore, and velocity is approximately constant. Table 6.1 shows two sets of exponents which illustrate that velocity is nearly constant downstream at bankfull stage, whereas it increases at sub-bankfull flows. Both examples are for trapezoidal channels with cohesive banks and coarse bed material; different patterns of behaviour may occur in streams with different sedimentological trends. Downstream velocity trends vary along a given river at different flow frequencies and between rivers according to their bed sediments, gradient and channel roughness. Thus, Mackin (1963) suggested that different subsystems in a major catchment may display downstream increases, decreases or constancy of velocity (Figure 6.3c) depending on these influences and Carlston (1969) argues that the modal downstream velocity exponent is zero, with a range either side of this. In ephemeral streams, a more rapid downstream increase occurs in both velocity and width:depth ratio, which appears to represent an adjustment of flow and channel geometry optimizing conditions for the continuity of sandy bedload transport in streams characterized by a less rapid downstream decrease of long-profile gradient (pp. 225–7). Such a systematic response of the exponent set to physical controls cannot be accommodated by minimum variance theory, which predicts a single, modal downstream hydraulic geometry which roughly fits

Table 6.1 Downstream velocity exponents in two rivers at different flow frequencies, showing less rapid increase at bankfull or near-bankfull discharges, and, in one case, the less rapid downstream decrease of roughness.

Source	Exponent	Discharge frequencies				River
		$Q_{50\%}$	$Q_{15\%}$	$Q_{2\%}$	Q_b	
Knighton (1974)	m†	0.38	0.23	0.08	–	Bollin–Dean, Cheshire
Wolman (1955)	m†	0.32	0.32	0.17	0.05	Brandywine Creek, Pennsylvania
Wolman (1955)	y_1‡	−0.40	−0.51	−0.32	−0.28	Brandywine Creek, Pennsylvania

†velocity; ‡Manning's roughness coefficient.

perennial, but not ephemeral, streams: $b = 0.55$, $f = 0.36$, $m = 0.09$, $z = -0.74$, $y_1 = -0.22$ (Leopold and Langbein, 1962). However, the application of the theory is questionable, as it assumes slope and roughness to be dependent variables when in reality they are partially independent. T. R. Smith (1974) attempts an alternative prediction based on a simple sediment transport law and the continuity equation. Although its physical basis is laudable, its simplifying assumptions (e.g. linear increase of discharge with basin area) are dubious and its results ($b = 0.6$, $f = 0.3$, $m = 0.1$) unrealistic, particularly the high width exponent.

The downstream hydraulic geometry concept has several limitations. First, downstream trends are complicated by changing magnitude–frequency properties, with upstream sections adjusted to more frequent events than downstream (Harvey, 1969). Discontinuous trends may arise, therefore, if channel or flow geometry is related to a constant flow frequency. Figure 6.4(a) shows a composite sub-bankfull width-discharge trend for both headwater and mainstream elements of an Amazonian basin.

To provide the maximum level of explanation of the width variance, it is necessary to fit the two separate regression relationships shown (Thornes, 1970). The discrepancy between the headwater and main-stream trends may reflect changing flow frequency relations, as well as increased sand supply to the smaller streams following deforestation in headwater catchments. Second, considerable systematic variation occurs in channel geometry at the local scale, because of the interrelationships between cross-section properties and plan-form and long-profile characteristics (pp. 175–9).

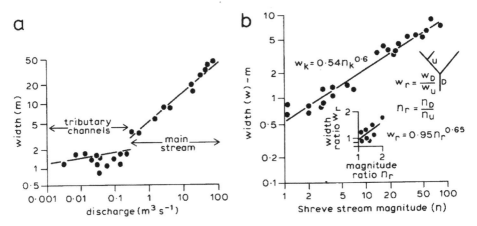

Figure 6.4 (a) Downstream width–discharge relationship in the Xingua–Araguaia region, showing the difference between headwater tributary and main-stream trends (after Thornes, 1970). (b) Relation between channel width and Shreve magnitude of a stream link; inset shows width and magnitude ratios at tributary junctions.

Thus, scatter about the general downstream trend may result from an inconsistent choice of measurement section with respect to the meander pattern and riffle–pool features (Wolman, 1955, p. 39). It may therefore be difficult to distinguish random scatter reflecting sections possibly in disequilibrium from systematic effects related to local variability. Third, the multivariate control of channel geometry must be emphasized. Even at-a-station flow geometry exhibits this, in spite of the constancy of most independent variables at a single location. Knighton (1975a, p. 196) quotes a minimum r^2 of 96% in multiple regressions predicting section mean velocity from discharge and a resistance coefficient. Downstream, bed and bank materials, valley gradient, roughness and riparian vegetation all vary, and adjustments of channel geometry must be interpreted in relation to these multivariate controls. Mackin's (1963) re-evaluation of downstream velocity trends identified distinct relations for different river sub-systems (Figure 6.3c), reflecting the effects of unmeasured independent variables. These additional controls are considered in the following section.

Finally, it is interesting to consider an alternative model to conventional downstream hydraulic geometry, with its assumption of smooth, continuous adjustments of width, depth, capacity, etc., as functions of discharge or basin area. Major adjustments occur at tributary junctions, and downstream trends of discharge, area and channel geometry are step functions (pp. 32–7) reflecting drainage network organization. Within a single Shreve stream link, channel width (for example) may vary stochastically about a constant mean value, so that successive deviations (w_j) from the mean, measured at equally spaced distances along the link, are correlated and related to one another by a first-order autoregression of the form

$$w_j = \phi_1 w_{j-1} + \epsilon_j \qquad (6.19)$$

where ϕ_1 is a coefficient between +1 and −1 and the ϵ_j are random, independent error terms. Adjustments of mean width then occur at tributary junctions. Miller (1958) described these adjustments, and those of depth, slope and bed material size, using expressions of the form

$$w_a = k_1(w_b + w_c) \qquad (6.20)$$

where the main stream below the junction is defined by subscript a and the two tributaries by b and c. When $k_1 = 1$ tributary widths are summed, and when $k_1 = 0.5$ they are averaged. Normally, k_1 averages 0.67, but only when tributaries are of comparable size. Richards (1980) suggests a more general model for adjustments occurring at junctions of tributaries of dissimilar size, where the ratio of downstream (D) to main upstream (U) tributary widths (w_R) is a function of their link magnitude ratio (n_R), as shown in Figure 6.4(b). This is defined by

$$w_R = c_1 n_R^{k_2}. \qquad (6.21)$$

The magnitude ratio tends to unity as the discrepancy in tributary magnitudes increases, and so does the width ratio. Thus the constant $c_1 = 1$, and if all adjustments at junctions in a homogeneous area are fitted by an equation with a constant value of k_2, then Equation (6.21) is equivalent to

$$w_k = w_1 n_k^{k_2} \qquad (6.22)$$

after successive upstream substitution of the junction ratio relationship. Thus the width of link k is a function of link magnitude and mean exterior link width, w_1 (Figure 6.4b). Stream width is therefore a link-associated variable adjusting downstream in discrete steps particularly where large sub-networks join. This alternative model, which combines step changes at junctions and stochastic variation within links, may also be applicable to other aspects of channel geometry.

'Resistance': the role of perimeter sediment

No general formulae exist which relate sediment properties to relative erosional resistance, and thence to equilibrium channel shape. Consequently, channel stability must often be assessed by flow-based indices rather than by bed and bank material resistance to erosion. Blench (1969) used the bed (F_b) and side (F_s) factors

$$F_b = v^2/d \quad \text{and} \quad F_s = v^3/w \tag{6.23}$$

to define the range of flow conditions between silting (low velocity) and eroding (high-velocity) flows for which channel stability may be expected, and suggested that a stable width could be defined by

$$w = \sqrt{(F_b/F_s)}Q^{0.5}. \tag{6.24}$$

Although useful in indicating the range of 'safe' designs capable of maintaining sediment transport, such equations fail to provide additional physical relationships; substitution of Equations (6.23) in (6.24) yields $wdv = Q$, the basic continuity equation for steady flow. Furthermore, the bed and side factors fail to correlate with perimeter sediment properties (Kellerhals, 1967) or to improve the explanation of variation in width or depth when added to discharge in multiple regressions (Richards, 1977b). The lack of a general model of the sedimentological influence over channel form reflects several difficulties.

First, there is a range of channel types with varying perimeter sediment characteristics, including those with entirely non-cohesive perimeters, cohesive banks and non-cohesive sand beds, cohesive banks and cobble beds, and cohesive bed and banks. Erodibility of the non-cohesive sediments (sands and coarser) is governed by particle size (weight) and shape, but for cohesive sediments (silt and clay) electrochemical interparticle bonds which relate to mineralogy, the presence of an electrolyte and the ionic charge on the particles (Masch, 1968) are more important. Second, bank material varies considerably along a reach and vertically within a bank. Although bank retreat might be considered a function of the weakest component, this depends on its position. A basal sand layer commonly found in the floodplains of meandering rivers (Klimek, 1974, Thorne and Lewin, 1979) controls the retreat of an overlying silty unit; this would not be the case if the sand overlay the silt. In addition, sedimentological differences occur between opposite banks in a section, as on a meander bend with a sandy point bar on the inner bank. Channel migration is controlled by the nature of the eroding bank (i.e. that adjacent to the thalweg), and not necessarily by some weighted average of sediment properties in the whole perimeter (Schumm, 1960a). Third, a given material varies in resistance with the growth of bank vegetation, the antecedent moisture conditions prior to a potential erosion event, and the existence of a protective basal accumulation of the results of the previous erosion event (Klimek, 1974); the sequence of floods is thus particularly important. Finally, bank retreat results from particle or aggregate detachment and large-scale bank failure, two distinct but often interrelated modes of erosion controlled by different material properties.

Bank erosion rates and processes

Bank stability is an important control of equilibrium channel form, and so the rates and processes of bank erosion must be considered. Bank retreat is measurable over different time scales (Hooke, 1980): by field measurement using erosion pins (Wolman, 1959) over 1–10 years; by comparison of variously dated maps and air photographs of appropriate scale (Lewin *et al*, 1977) over 10–200 years; and by using sedimentological and palaeobotanical evidence over longer periods when bank erosion must be distinguished from palaeochannel abandonment. Average bank erosion rates (E_b, m yr^{-1}) reflect river size, and Hooke (1980) has summarized data from several studies to show a general relationship with basin area (A, km^2) similar to that based on field data from south-east Devon,

$$E_b = 0.025A^{0.45}. \tag{6.25}$$

In south-east Devon, average rates of 0.08–1.18 m yr^{-1} were measured, implying that complete floodplain reworking by sampled rivers could occur in periods ranging from 600 to 7000 years. If width is also a power function of basin area with a similar exponent, average bank retreat is a constant proportion of channel width, while deviations from this quantity reflect additional controls. Erosion rates are highest in asymmetric sections where the flow is diverted against one bank (Knighton, 1973), and measured rates may partly overestimate the true average by concentrating on studies of outer banks in meander bends. The ease of the removal of debris results in a faster migration of lower banks (Klimek, 1974). However, random variability is considerable. Twidale (1964) reports the fastest observed retreat on the steepest bank monitored, but also the longest period of stability. Such short-term temporal variation makes it difficult to compare field and map-based measurements. Spatial variation is also marked (Figure 6.5a); erosion-pin studies invariably employ a pattern of pin sites along a bank and at various heights. On a bend, the point of maximum erosion may migrate around the curve as the flow adjusts to the changing bend shape, and in one vertical there may be temporary burial of basal pins. Bank material properties influence retreat rates in general and in detail. Hooke (1980) found the bank silt–clay percentage to be the most important independent control of erosion rates after river size. Within a composite bank, however, Thorne and Lewin (1979) quote erosion rates of 0.35–0.60 m yr^{-1} in a basal gravel unit and 0.03 m yr^{-1} in the overlying silty clay, a variation which markedly influences the general process of retreat (Figure 6.5b).

Some bank erosion is non-fluvial in origin, and Twidale (1964) discusses a considerable variety of such processes. Rainwash, with a strong winter maximum, was measured on an abandoned river bank by Thorne and Lewin (1979) and could account for much of the retreat between major collapses in the silty upper unit of active banks in the same area. Frost action has been emphasized frequently (Wolman, 1959), the role of needle ice development (Plate 2) being foremost. Hill (1973) shows that its effect is dominant in a susceptible fissile till, whereas a massive till is eroded more by fluid action. The effect of streamflow is complex. Knighton (1973) suggests that 70% of erosion is achieved by 'high' discharges, but this includes a range of flows up to bankfull. The timing of a flood may be more important than its magnitude, since less stress is needed for erosion of wet cohesive banks in winter. Wolman (1959) also emphasizes the seasonality of erosion, little occurring even during extreme events in summer when banks are dry. This leads to an important conclusion, namely that width:depth ratios are higher in a given cohesive sediment if the stream is baseflow-dominated and therefore maintains a high moisture content in the banks. Flashy hydrological regimes result in more stable banks and narrower, deeper streams (all else being equal).

Fluvial bank erosion is also varied in nature. Direct fluid shear, and rainsplash and wash, detach individual particles of sand and gravel or crumb-sized aggregates from silty banks, and in the latter case the aggregate stability may control erodibility. Mass failure of the bank is more dependent on bulk mechanical properties of the sediment, and includes rotational failures and tension or beam failure after undercutting (Figure 6.5c). Rotational failures occur when the bank is at limiting equilibrium and the restraining forces (cohesion and friction) just balance the sliding forces (the component of overlying sediment weight) acting along the failure plane. The factor of safety (F) is therefore unity, that is

$$F = \frac{\text{Restraining force}}{\text{Sliding force}} = \frac{Slr}{Wz} = 1 \qquad (6.26)$$

(Figure 6.5c), where S is a measure of the material strength (or 'shear resistance') and W the overburden weight. Failure may occur as a result of increased weight because of waterlogging, decreased frictional shear strength caused by increased pore water pressure, and increased height because of basal scour. Failure often occurs on the falling stage of floods (Twidale, 1964) when the buttress of basal debris has been scoured and the banks are thoroughly wetted. In large rivers, significant rotational failures occur which may be preserved in the geological succession in the deposits of low-gradient deltaic streams (Laury, 1971). Assuming cohesive material, Schofield and Wroth (1968) provide an approximate model of critical stability conditions of a river bank, in which the maximum loading differential sustained along a cylindrical failure surface is a function of effective

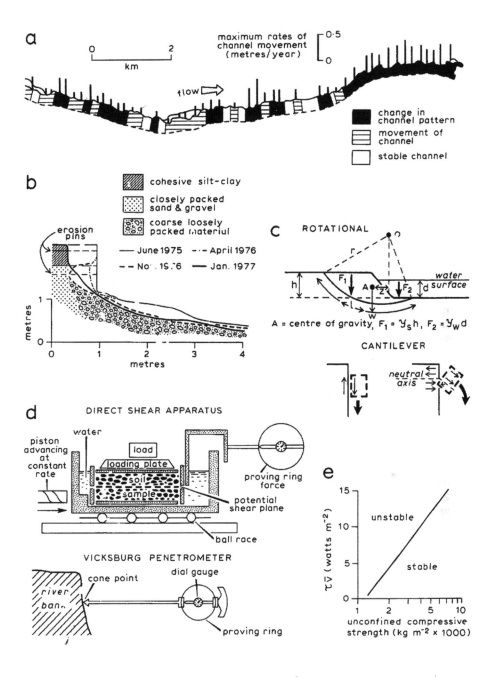

Figure 6.5 (a) Variations of channel migration and bank erosion along a reach of the River Yarty, south-east Devon (after Hooke, 1980). (b) Formation of a cantilever overhang by fluvial undercutting of a composite river bank (after Thorne and Lewin, 1979). (c) Rotational and cantilever failures, including definitions of quantities in Equations (6.26) and (6.27). (d) Direct shear apparatus and the Vicksburg penetrometer. In each case shear resistance is measured by deformation of a proving ring. (e) Unstable and stable streams discriminated by power per unit bed area ($\tau \bar{v}$) and unconfined compressive strength of bank material (after Flaxman, 1963).

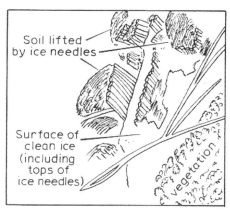

Soil lifted by ice needles

Surface of clean ice (including tops of ice needles)

vegetation

Plate 2 Needle ice development on a cohesive river bank, lifting particles and aggregates from the surface to be washed into the stream on melting.

cohesion, c':

$$\gamma_s h - \gamma_w d = 5.53 c' \qquad (6.27)$$

where γ_s and γ_w are sediment and water unit weights respectively. For $c' = 3$ t m^{-2} and $\gamma_s = 1.6$ t m^{-3}, the maximum stable bank height h of a dry channel ($d = 0$) is 10.4 m, whereas when full ($d = h$) it is 27.7 m. The alternative mass failure process, tension or beam cantilever failure, is a smaller-scale event controlled by rapid undercutting of basal gravel, the strength imparted by the root system, and the fissures between soil peds (structural blocks) along which failure ultimately occurs (Thorne and Lewin, 1979).

Channel shape and perimeter sediments

Absolute channel dimensions (width, depth) depend on the discharge magnitude imposed by the upstream catchment, and in particular on the energy expended by the streamflow. Channel shape is partly scale-dependent because the width:depth ratio increases slightly downstream as width increases faster than depth. However, the main control of channel shape is the perimeter sedimentology, although this has proved difficult to quantify. Wolman and Brush (1961) illustrate this in the simplest case of the cross-section with homogeneous perimeter sediment, where channel shape reflects the bank stability. In a non-cohesive 2 mm sand, only one depth was stable for a given stream power, this being the depth at which the flow was just competent. In a slightly cohesive 0.67 mm sand, however, a range of depths was stable even when significant bed material transport occurred. In sections whose bed and banks are composed of different material, even greater confusion is evident. Kellerhals (1967) argues that bed material has no influence on width, Henderson (1961) that streams are wider if fine sandy bed material is present (Equation 6.18), and Schumm (1960a) that coarse material inhibits scour and encourages widening. An analysis of empirical data often generates conflicting results, each of which depends on the characteristics of a particular set of data. Generally, large rivers are wide, deep, gently sloping with fine bed material, and because of this covariation simple correlations are difficult to interpret. The true relationship of width to bed material size is only revealed by a partial correlation in which discharge, slope, depth and bank material properties are all fixed. Furthermore, there are conflicting physical influences: cobbles are more resistant to erosion than sand and may encourage bank erosion instead of bed scour, but they may also dissipate the flow energy through greater roughness, and thus limit the capacity for bank erosion. Bed material thus seems to have no obvious effect on channel width, but should affect depth. Kellerhals (1967) produced the following relationship for streams with non-cohesive bed material:

$$d = 0.42 Q_d^{0.4} D_{90}^{-0.12} \qquad (6.28)$$

which suggests that scour in finer (sandy) bed materials increases the channel depth. This equation includes a dominant discharge index (Q_d) which ranges from the 3- to 5-year floods for the cross-sections investigated. One successful approach to the development of a multivariate hydraulic geometry has involved relationships between dimensionless channel properties (e.g. w/D_{50}, d/D_{50}) and dimensionless discharge (Parker, 1979). Andrews (1984) carefully sampled mobile, gravel-bed, single-thread channels in Colorado and identified a set of equations based on these dimensionless variables for bankfull channel dimensions. British data for comparable streams appear to obey similar relationships, indicating the generality of hydraulic geometry defined in this way, by preliminary classification of river type and standardization of geometric properties using grain size statistics.

Bank material characteristics represent a more important sedimentological control, but different properties control stability in relation to the various bank erosion processes. Direct particle removal by fluid shear reflects particle size in the sand–gravel–cobble spectrum, but aggregate size in the silt–clay range. Sprinkler experiments and soil erosion studies suggest that the proportion of water-stable aggregates in the soil controls its erodibility, and that this is increased by base-rich bedrock which provides the flocculating bivalent cations Ca^{2+} and Mg^{2+}, a high organic matter content, and high percentages of silt and clay, especially non-swelling clay such as kaolinite rather than montmorillonite (André and Anderson, 1961, Bryan, 1977). Aggregate stability controls soil resistance to rainsplash, but may be a less relevant index of resistance to fluid shear, particularly since entrainment of small clay aggregates requires a lower shear stress than individual clay particles. Mass bank failure reflects the bulk mechanical properties of the sediment, whose shear resistance (S in Equation 6.26) is a function of cohesion (c') and friction, measured by the angle of internal friction ϕ'. Total frictional strength is dependent on overburden normal stress σ', so the shear resistance is defined by the Coulomb equation (Whalley, 1976)

$$S' = c' + \sigma' \tan \phi'. \qquad (6.29)$$

The primes indicate that all parameters are measured with respect to the effective normal stress, σ', which is

$$\sigma' = \sigma - u. \qquad (6.30)$$

The pore pressure u is usually positive (saturated pore spaces) in the case of river bank failure, when the pore water carries part of the overburden stress σ, which therefore reduces the frictional strength. The parameters c' and ϕ' require laboratory testing, for example by direct shear apparatus (Figure 6.5d), and removal from the field environment often results in sample disturbance. McQueen (1961) suggests that material strength is highly dependent on variable properties such as density, moisture content and loading, and that these are difficult to replicate in laboratory tests. Variations of moisture content alone can cause order of magnitude changes in strength estimates (Partheniades and Paaswell, 1970). However, Flaxman (1963) succeeded in discriminating stable and unstable streams by comparing bankfull power expenditure per unit bed area with the unconfined compressive strength of bank material (Figure 6.5e), this laboratory test providing a measure of strength which is related to fixed material properties such as plasticity and percentage of fines ($<5\ \mu$m). A field estimate of bank strength may be obtained using the Vicksburg penetrometer, shown in Figure 6.5(d) (Chorley, 1959, 1964), which actually measures compaction rather than shear resistance. Skempton and Bishop (1950) note that *in situ* penetrometer strength estimates are very sensitive to moisture content variation, however, and Park (1978) showed that bank strength thus measured has no simple correlation with a fixed material property, although it is strongly dependent ($R^2 = 67\%$) on the combination of moisture content and percentage of silt and clay.

It would generally appear that bank retreat and the processes of bank erosion are strongly affected by the percentage of fines, especially of the silt and clay particles which provide the cohesion which is the dominant source of shear resistance when frictional resistance is minimized by the effect of high pore water pressure. Schumm (1960a) therefore adopted a weighted index (M) of the silt–clay content of the channel perimeter to account for the different percentages in the bed (S_b) and banks (S_c)

$$M = \frac{[(S_b.w) + 2(S_c.d)]}{w + 2d} \tag{6.31}$$

and showed that, for stable channels without coarse gravelly beds, this index correlated with channel form ratio, F (Figure 6.6a), in the relationship

$$F = 255M^{-1.08}. \tag{6.32}$$

Channels with silty banks are narrow and deep in cross-section, while those with sandy, erodible banks are wide and shallow. The silt–clay index is a surrogate for the complex array of influences of bank stability discussed above. Schumm (1960a) suggested that stable channels would plot close to the fitted regression, while aggrading and degrading channels would be respectively wider and shallower (above the curve) and narrower and deeper (below). Width and depth both increase downstream with increasing discharge, but if M increases downstream, depth will increase faster than width as the channel shape adjusts to the changing perimeter sediment. Melton (1961) criticized the F–M correlation on the grounds that the weighted index M includes width and depth in its derivation, which produces a ratio correlation effect (Benson, 1965). However, Schumm (1961a) defended the relationship in principle, and proceeded (Schumm, 1961b) to identify separate multivariate expressions predicting width and depth as a function of discharge and the weighted silt–clay percentage. Ferguson (1973c) has suggested a model relating width more logically to bank silt–clay content (B), and Miller and Onesti (1979) reiterate this, adding a multi-variate expression for channel depth in which bed material size is a significant third control, deeper channels having sandy beds as suggested by Equation (6.28). The role of bank material resistance, as measured by the bank silt–clay percentage, is summarized by the following regressions based on data from Schumm (1960a, 1968a):

$$w = 25.5 Q_{ma}^{0.58} B^{-0.6} \qquad (6.33)$$

and

$$d = 0.03 Q_{ma}^{0.35} B^{0.6}. \qquad (6.34)$$

Thus, at a given discharge, streams with silty banks are narrow and deep, while those with less silty (and therefore sandy) banks are wide and shallow (Figure 6.6b). By inference from $Q = wdv$, these equations imply

$$v = 1.23 Q_{ma}^{0.07} \qquad (6.35)$$

in which velocity at mean annual flood increases slightly downstream, but is independent of the bank material. In addition,

$$F = 800 Q_{ma}^{0.15} B^{-1.20} \qquad (6.36)$$

which suggests that a slight increase in form ratio occurs downstream with discharge, but that the major control of channel shape is the sediment-ological one. Both width and depth are therefore seen to reflect a discharge: sediment interaction, the control being multivariate rather than bivariate as implied by the original concept of downstream hydraulic geometry.

Channel shape and sediment transport

The status of perimeter sediment as a control of channel shape varies between rivers. In non-migrating, inactive streams with a low total sediment load, the perimeter sedimentology is an inherited independent influence reflecting the depositional activity of palaeochannels. In migrating, active rivers constantly reworking the upper surface of the valley fill and carrying relatively large sediment loads, perimeter sediments reflect the type of sediment load presently carried. Thus, streams carrying suspended sedi-ment have high perimeter silt–clay contents, while bedload streams have

Figure 6.6 (a) Relationship between channel width:depth ratio (F) and weighted mean perimeter silt–clay percentage (M). After Schumm, 1960a. (b) Multivariate relationship between F, bank silt–clay percentage and mean annual flood Q_{ma} (note the much narrower range of *bank* silt–clay percentages). (c) Relation of suspended sediment load to channel width, depth and velocity for channels at a constant discharge of ~14 m^3 s^{-1} (after Leopold and Maddock, 1953a). (d) Relation of competence to width, depth and velocity at constant discharge of ~11 m^3 s^{-1} (after Wilcock, 1971). (e) Variations of depth, velocity, suspended sediment load and stream bed elevation during a flood hydrograph at Bluff, Utah, on the San Juan River (after Leopold and Maddock, 1953a).

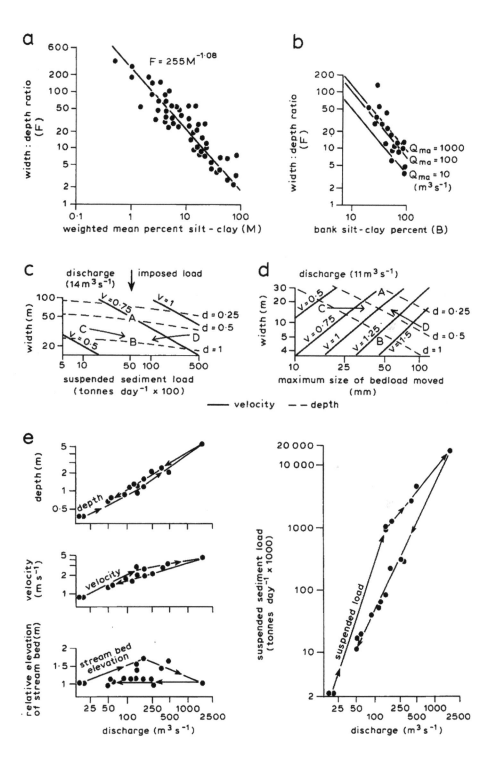

Table 6.2 Classification of types of alluvial channel based on the nature of sediment load (after Schumm, 1968a). F = width:depth ratio, P = sinuosity, M = perimeter silt–clay percentage.

Channel type	Perimeter sediment, M (%)	Bedload as percentage of total load	Channel behaviour		
			Stable	Depositing	Eroding
Suspended load	>20	<3	$F < 10$	deposition on banks	erosion of bed
			$P > 2$ gentle slope	narrowing	deepening
Mixed load	5–20	3–11	$10 < F < 40$	deposition on banks,	erosion of bed, then banks
			$1.3 < P < 2.0$ moderate slope	then bed	
Bedload	<5	>11	$F > 40$	deposition on bed	erosion of banks
			$P < 1.3$ steep slope	bar formation and shallowing	widening

high sand contents. Schumm (1968a, 1971) justified this argument by showing that M is inversely correlated with the bedload, expressed as a percentage of total load, and based a classification of alluvial channels on their prevailing load characteristics (Table 6.2). The morphology of equilibrium channels thus reflects their load, which is reflected in turn by their bed and bank sediments. The adjustments made by disequilibrium aggrading (excess load) or degrading (load deficit) rivers also vary according to their transport regime (Schumm, 1960b), and occur in such a way that bank stability is maintained and optimum conditions develop for the transportation of the type of load imposed from upstream. The qualitative illustration of this is based on an empirical diagram derived by Leopold and Maddock (1953a) who obtained width, depth, velocity and suspended sediment load data for several streams at a fixed discharge (Figure 6.6c). A given sediment load can be carried by this discharge in a range of channel forms: in a wide, shallow stream by rapid flow (A); or in a narrow, deep stream by slow flow (B). For the stability of the banks, the latter is preferable. If a channel is relatively wide and deep (C) the imposed load is in excess of that which can be carried at the rather slower velocity required by $Q = wdv$. Table 6.2 shows that deposition on the banks will narrow the stream and increase velocity until it can carry the imposed load. If it is narrow and shallow (D), its high velocity will render it load-deficient, but it will erode its bed until, again, its channel form is appropriate for the maintenance of transport continuity. Wilcock (1971) has derived a similar diagram to illustrate the varying conditions of channel form within which bedload transport may occur (Figure 6.6d). A given bedload competence, and hence transport rate, can be maintained by a specific discharge either at a slow velocity in a wide, shallow channel (A) or at a rapid velocity in a narrow, deep one (B); relatively slow velocity is efficient in a shallow channel because of the high bed shear stress created by the close spacing of isovels near the bed. If coarse material arrives at a wide, deep section (C) it is deposited on the bed (shallowing the section; Table 6.2). A narrow, shallow channel (D) experiences rapid velocities and is therefore overcompetent if fine bedload is introduced, and it erodes its banks. The adjustment again permits continuity of transport and, by reducing velocity, improves bank stability. These examples show that rivers can readily adjust their cross-sections to maintain a condition of sediment transport continuity; major changes of long-profile gradient by degradation or aggradation are not essential. Andrews (1979a) illustrates this with data from the East Fork River, Wyoming, which maintains bedload transport imposed on it by its tributary, Muddy Creek, entirely by adjusting its cross-section, particularly increasing velocity so that bedload transport can occur across the full bed width.

A general theory which attempts to explain the equilibrium geometry of the cross-sections of straight rivers, in relation to bank stability and sediment transport, has been developed by Parker (1978). This notes that the lateral slope of a river bank inevitably drives sediment towards the channel centre, so that bank erosion is a necessary consequence of sediment transport. In a stable channel this must be balanced by deposition, which Parker (1978) suggests is provided by the outward diffusion of suspended sediment towards the bank. Bedload is thus derived from bank erosion and moves to the channel centre, while sediment in suspension moves to the banks. In the absence of suspended sediment, bank angles will be shallow to inhibit bedload transport to the centre, and wide shallow sections will result. This model has been tested empirically, with some success, by Pizzuto (1984).

These are medium-term adjustments of channel form to the prevailing load in streams carrying significant quantities of sediment. Leopold and Maddock (1953a,b) have also interpreted short-term adjustments of flow as a reflection of load introduced to a section during a flood hydrograph, echoing Mackin's suggestion that sediment load is both a cause and an effect of velocity (Mackin, 1948, p. 471). As a flood wave passes through a section of the San Juan River (Figure 6.6e), rapid increases of velocity and suspended load occur on the rising limb. This is compatible with the finding that the sediment rating curve exponent is higher when velocity increases faster than depth (i.e. when $m/f > 1$). However, Leopold and Maddock (1953a,b) argue that the increasing velocity does not *cause* the higher suspended load, because this would necessitate scour of the bed, which actually experiences fill as the suspended load increases. Thus, they suggest that the sediment introduced to the reach cannot be carried through it, and that its deposition on the bed reduces the rate of increase of depth and increases that of velocity so that further sediment input can continue to be carried through the reach. They also argue that the high concentration of suspended sediment damps turbulence (Vanoni and Nomicos, 1960; pp. 66–71), thereby reducing roughness and encouraging high velocities. However, large-scale scour and fill of the bed are unusual because negative feedback renders them self-cancelling; scour, for example, enlarges the cross-section and reduces velocity, which initiates renewed fill. Thus, Colby (1964a) suggests that flood discharges are mainly accommodated by changes in the water surface elevation rather than by scour and fill, except in confined sections. Furthermore, development of dune bedforms may make it difficult to measure mean bed elevation during a flood, and an alternative interpretation of the hysteresis of velocity in Figure 6.6(e) could be that dune development lags behind the discharge variation. On the rising limb, dunes are smaller and underdeveloped, and present less form resistance, than at comparable discharges on the falling limb when they are large and overdeveloped. Furthermore, in small streams the energy slope will be greater on the rising limb as the flood wave arrives than after the peak has passed, and this also contributes to the hysteresis of velocity. There may be an alternative interpretation of the trends indicated in Figure 6.6(e) therefore, with a wave of bedload passing through the reach, affecting the flow geometry and then the suspended load, and a complex interaction between the components of load and the flow characteristics.

Riparian vegetation and channel form

Bank vegetation exerts a strong control over bank stability and therefore has some influence on channel form. This is particularly evident in small channels, along which Zimmerman et al (1967) have identified fluctuations in width as streams pass through meadowland and woodland. Dense, fine networks of grass roots stabilize banks even against extreme flood events, and channels are narrow. In forests, larger tree roots are less capable of binding bank material, and streams are wider and shallower, although also more variable in form. Individual tree roots and tree throws may dominate local variation in form in headwater streams. Keller and Swanson (1979) have illustrated this effect in the western USA Douglas fir and Californian redwood forests, where large fallen trees control the distribution of loci for sedimentation and are responsible for the formation of 50–90% of pools and result in 60% of the total fall in stream elevation by creating 'organic' riffles (Keller and Tally, 1979). These unsystematic effects are pronounced in low-order streams which are small in relation to tree size and which cannot move fallen trees. Evaluation of the systematic effect of bank vegetation on bank stability and erosion is more difficult since quantification of vegetation presents problems. However, Smith (1976) found that the bank erosion rate, measured using conventional erosion pins or soil blocks in an 'erosion box' at a constant flow velocity, decreases significantly as the percentage by weight of vegetation roots in a sample increases. In root-free silt, an erosion rate of 160 cm h^{-1} occurs, whereas with 16–18% of roots and 5 cm of root 'rip-rap' the rate is 0.02 cm h^{-1}. A limitation of this approach, however, is that comparable percentages may arise for samples containing grass or tree roots, although their resistance effect differs markedly. Finally, vegetation within the channel strongly affects roughness and, in causing energy loss, detracts from the availability of energy for sediment erosion and transport. This is also difficult to evaluate. Powell (1978), for example, shows that winter values of Manning's n in one channel average 0.03, but reach 0.4 to 0.5 in summer as a result, apparently, of reed growth. However, low summer flows are much more shallow, so part of the seasonal roughness difference is a reflection of greater relative roughness; high values of n for low winter discharges demonstrate this.

Local variation

Random variations in hydraulic geometry arise because of the shifting nature of steady state alluvial channels. Bank collapse after undercutting during channel migration can result in short-term variability of at-a-station flow geometry (Figure 1.7d), even though medium-term equilibrium of cross-section morphology is maintained. However, systematic temporal and spatial variations are apparent even in equilibrium channels. Temporally, the seasonal growth of channel vegetation increases flow resistance (Powell, 1978) to create non-stationarity in the response of depth and velocity to discharge changes. Systematic local spatial variation associated with longi-tudinal and transverse oscillations in the river channel, represented by the riffle–pool sequence and the meander pattern, may introduce considerable scatter to downstream hydraulic geometry trends unless filtered deliberately by a systematic sampling of measurement sections (Wolman, 1955, p. 40, Brush, 1961). Local variability may be extreme: Richards (1977b) notes that depth differences between adjacent riffles and pools on the River Fowey are equivalent to those arising downstream at riffle sites as a result of a 350% increase in the contributing catchment area. Both at-a-station and downstream variations in flow and channel geometry reflect local site influences, but the systematic differences between riffles and pools, for example, cannot readily be rationalized by incorporating an additional continuous numerical variable in a multiple regression equation. Instead, different hydraulic geometry relationships may be required for each type of section, including riffles and pools in straight and meandering reaches (e.g. Figure 6.7c). Ultimately, a family of multivariate regression models may be required to define the full spatial variation of channel geometry. This would include relationships similar to Equations 6.33–6.36, in which the channel geometry variables are defined as functions of discharge and sediment properties, but with separate expressions for each discrete class of river section.

The riffle–pool systematic effect

Energy conditions in gradually varied flow in a straight channel with non-uniform bed slope are defined by the Bernoulli equation (Equation 3.48). In Chapter 3, this is used to establish an energy balance equation between two sections separated by an incremental distance. If discharge, channel width, bed slope, roughness and the flow depth of the downstream section are known, this nonlinear equation can be solved numerically for an unknown upstream depth, which becomes the known downstream depth of the next incremental reach (Prasad, 1970, Humpidge and Moss, 1971, Fread and Harbaugh, 1971). Thus a long river reach may be subdivided, and flow depths and velocities calculated from a starting depth. Flow geometry variations through a riffle–pool sequence may therefore be mathematically modelled (Richards, 1978a), starting from a downstream control depth at a riffle site. Such a simulation indicates that pool depth is controlled by the altitude of the downstream riffle, but is independent of channel width and roughness, whereas flow velocity in the pool is dependent on channel width. At-a-station hydraulic geometry in the pool is predicted by modelling over a range of discharges and appropriate starting depths in the riffle; the depth exponent is less than that in the riffle while the velocity exponent is greater. Water surface slope and energy gradient increase in the pool and decrease in the riffle as discharge increases, and thus the pool experiences a faster increase in bed shear stress.

These model results assume an invariant channel perimeter and adjustment of the flow within a 'rigid' container. They are, however, confirmed by empirical data collected at discharges up to approximately mean annual discharge on the River Fowey, a narrow range of flows within which power function relationships can be accepted. Figure 6.7(a) shows that riffle and pool sections can be discriminated on the basis of exponents describing rates of change of depth, velocity, shear stress and gradient with discharge. Pools have higher velocity, shear stress, and slope exponents and lower depth exponents. This is important in the context of longitudinal variation of shear stress and sediment transport, and therefore for the formation and maintenance of the riffle–pool sequence and the meander pattern (see pp. 184–91). Furthermore, it lends support to the view that water surface slope, flow depth and velocity become approximately equalized through a reach at higher discharges (Langbein and Leopold, 1966), although it must be emphasized that depth–discharge power functions may 'converge' while the absolute difference in depth between adjacent pool and riffle sections actually increases. Exponents in riffle and pool sections strongly reflect not

only the overall bed topography, but also the bed material size. Figure
6.7(b) shows that although riffle–pool differences are consistently maintained, depth and velocity exponents relate to bed material size, with the
former decreasing and the latter increasing as material coarsens. This
accords with Williams' (1978b) results, summarized in Equation (6.11), and
reflects the effect of a stable, rough bed of coarse material in causing a lower
rate of depth increase with discharge.

The riffle–pool sequence affects overall channel form as well as the flow
behaviour represented by the at-a-station exponents. The advantage of the
discriminant functions based on the exponent values (Figure 6.7a) is that
these criteria are dimensionless. The v^2/d ratio (Wolman, 1955) is dependent

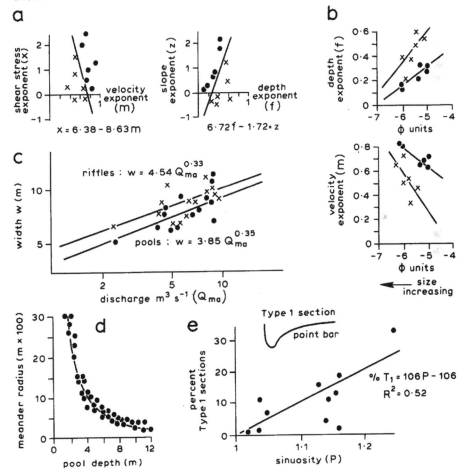

Figure 6.7 (a) Discrimination between riffle and pool sections based on at-a-station
exponents: pools are represented by dots, riffles by crosses. (b) Relationships of depth
and velocity exponents for riffles and pools to bed material size. (c) Downstream
width-discharge trends for riffle and pool sections, River Fowey, Cornwall. (d)
Relationship of pool depth to meander bend radius (after Konditerova and Popov,
1966). (e) Percentage frequency of highly asymmetric cross-sections in reaches of
various sinuosity (after Milne, 1979).

on flow stage and is less discriminatory at bankfull stage when velocity and depth are less differentiated between riffles and pools. Mid-channel riffle bars in straight gravel-bed reaches of the River Fowey divert the flow to undercut the banks and widen the channel, so at bankfull stage riffles are defined by the discriminant function

$$w > 8.51\bar{d}^{0.83}. \tag{6.37}$$

Figure 6.7(c) shows the downstream variation of these channel widths related to mean annual flood, with the 'outer limit' riffle and pool regressions. Riffles are about 12% wider than pools on average, and since riffle water surface slopes remain steeper even at bankfull stage in straight reaches, the joint variation of width and slope tends to minimize variation of the slope:width ratio, and hence of power expenditure per unit bed area (Richards, 1978b).

Channel cross-sections in meander bends

Knighton (1974) showed that width exponents are low in channels with steep, cohesive banks, whereas the gentle bank slopes of point bars in meander bends and gravel bars in braided rivers result in more rapid increases of width with discharge. Migrating rivers constructing bar features can therefore be discriminated from stable, straight streams by their high b exponent. Furthermore, they are associated with low rates of increase of velocity (m exponents). This is partly because, as bar surfaces are gradually inundated as the water level rises, new marginal zones of shallow depth and high flow resistance are added to the flow cross-section (Lewis, 1969, Knighton, 1974, p. 1074), and partly because, when roughness is primarily a function of curvature resistance in a bend, there is less of a tendency for it to decrease as discharge increases than when skin resistance dominates (Knighton, 1975a). At low discharges, Richards (1976b) reports water surface slopes and friction factors approximately equal in straight and curved pools, but twice as large in straight riffles as in the riffles in meander bend inflections. The overdeepened pool in a meander bend creates a backwater curve which 'drowns' the skin resistance of the upstream riffle (Ippen and Drinker, 1962). This means that the low-flow water surface slope tends to be less strongly differentiated between the riffle and pool in a bend, and together with a more rapid increase in slope in curved pools and less rapid decrease in 'curved' riffles, water surface slope becomes equalized throughout a curved reach at lower (sub-bankfull) discharges than in straight reaches (Leopold et al, 1964, p. 206).

A particularly prominent characteristic of meandering channels is the development of an asymmetric cross-section at the bend apex, where the pool is scoured to a depth which increases as bend radius decreases (Figure 6.7d). Konditerova and Popov (1966) demonstrate that relative pool depth

(d_{max}/w) increases with reach sinuosity and relative radius of bend curvature (r_c/w) in stable rivers, but that this relationship breaks down in 'underfit' streams where meander bends may have multiple pools. Tight bends have excessively deep pools and asymmetric sections at the apex, and such bend and cross-section characteristics are common in sinuous reaches. Milne (1979) classified nearly 600 sections from 11 representative reaches of upland rivers in Britain into 11 types, of which type 1 is strongly asymmetric with well pronounced point bar development on one bank. Figure 6.7(e) shows that the percentage of type 1 sections in a reach is significantly correlated to the reach sinuosity. Using the asymmetry index A_s (Table 1.2), type 1 sections have sub-area ratios in excess of 2 and often greater than 5. Cross-section morphology can, however, also affect bend shape. Chitale (1970) suggests that outer bank erosion is concentrated at the apex if the cross-section is narrow and deep, which results in marked amplitude development. In wide, shallow cross-sections erosion is concentrated downstream from the apex, resulting in asymmetric bends and downvalley migration. A complex interaction therefore occurs between cross-section forms and channel pattern, which is the second dimension within which river morphology is adjustable.

7

River channel pattern:
processes, forms and sedimentology

The channel pattern, or plan form, of a reach of an alluvial river reflects the hydrodynamics of flow within the channel and the associated processes of sediment transfer and energy dissipation. Adjustments of equilibrium channel pattern may occur over time scales of 10 to 100+ years, depending on the mobility of the channel-forming sediments, and may reflect the direct or indirect effects of man (e.g. channelization or land-use changes) on the controlling water and sediment discharges that determine channel processes. River management thus demands an understanding of the freedom of the river to modify its pattern. However, most approaches to channel design (Chapter 10) have assumed straight channels, or adopted simple rule-of-thumb generalizations such as the wavelength–width relationship of Figure 1.1(b). Progress beyond this is inhibited by the physically indeterminate mutual adjustment of several degrees of freedom (as defined in Chapter 1). Restricting attention to single-thread channels alone, it is evident that adjustment of channel gradient on a given valley slope demands a change of sinuosity, because $P = s_v/s_c$. However, a given sinuosity can be created by an infinite range of channel alignments of varying meander wavelength and regularity. Furthermore, meander development results in cross-section asymmetry, so at least one extra variable, in addition to width and depth, is needed to define cross-section shape. Channel patterns form a continuum in response to varying energy conditions, ranging from single-thread (straight, sinuous or meandering) to multi-thread (braided) forms. Whilst the sedimentology of these channel types has been well researched to provide the basis for the palaeohydraulic interpretation of sediments, the natural reluctance of the engineer to build structures on unstable braided streams has inhibited the development of our understanding of the behaviour of this type of stream.

The continuum of river channel patterns

Qualitative definitions of pattern types include distributary, reticulate and anabranching patterns (Figure 7.1a), for which specific local explanations are often appropriate. For example, Riley and Taylor (1978) explain the distributary pattern of the upper Darling system in Australia by reference to its Quaternary palaeohydrology, the development of alluvial fans formed by coarse load deposition, and discharge losses downstream through seepage and evapotranspiration. Anabranching patterns are particularly difficult to interpret. Avulsion during floods may create anabranches which eventually carry most of the flow, although slow depositional processes delay complete abandonment of the original channel. The anabranching pattern may therefore represent temporal disequilibrium between the effects of rapid erosion and slow deposition, possibly over considerable time scales. Schumm (1977, pp. 297–303), for example, demonstrates long-term oscillations between channels in the Mississippi alluvial plain, where upstream aggradation occurs as the delta builds seawards, lowering the channel gradient and now encouraging the flow to switch to the alternative, steeper Atchafalaya channel which carries almost 40% of

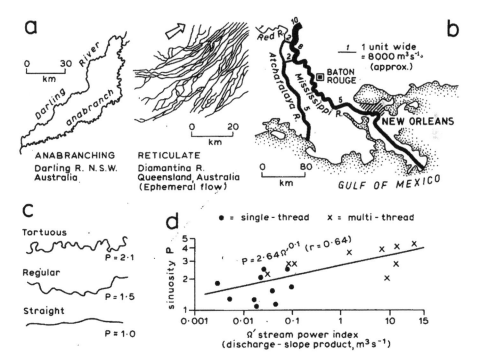

Figure 7.1 (a) Two types of qualitatively defined channel patterns (after Dury, 1969). (b) The lower Mississippi River, showing the potential avulsion down the anabranch of the Atchafalaya channel. (c) Classification of single-thread channels based on sinuosity (after Schumm, 1963). (d) Relationship between total sinuosity and a power per unit length index, applied to single-thread and multi-thread channels.

the discharge and requires regulation to prevent major avulsion (Figure 7.1b).

Single-thread (straight or meandering) and multi-thread (braided) streams may, however, be related to a single general model in which each results from a particular 'force:resistance' balance. Straight streams occur in low-energy environments, braided streams in high-energy environments (i.e. large, variable discharges carrying heavy sediment loads on steep slopes), with meanders in intermediate conditions. It is thus illogical to distinguish meandering from straight streams at an arbitrary sinuosity ($P = 1.5$; Leopold and Wolman, 1957) or classify streams by sinuosity as in Figure 7.1(c) (Schumm, 1963). A preferable approach is to define pattern morphology by a common index such as 'total sinuosity' (Figure 1.4d and Table 1.2), which is the ratio of total active channel length to valley length. The continuum of channel patterns may be identified by relating total sinuosity to an index of stream power, which measures the rate of potential energy expenditure by the stream. Figure 7.1(d) uses the index Ω', defined as the discharge–channel slope product and therefore a measure of power expenditure per unit length of channel. Ideally, total sinuosity should be related to maximum potential power expenditure, controlled by the valley gradient (Table 1.3), but published data on river characteristics rarely include this. Scatter in the plot reflects the variation in sediment 'resistance' as well as the practical problem of obtaining discharge measurements of comparable magnitude–frequency properties in lowland single-thread and upland multi-thread streams. A zone occurs where tortuous, high-sinuosity meanders and low-intensity braids overlap, which perhaps indicates the need to pass a certain threshold of power expenditure before adjustment of pattern type occurs (see pp. 213–17).

The creation of a particular channel pattern is thus dependent on the total energy available. In a given sediment, higher rates of total, or potential, power expenditure on steep valley surfaces result in greater total sinuosity, which increases bed area by lengthening the channel and reducing slope, or by increasing channel width, so that the excess stream energy is dissipated in overcoming extra frictional resistance. Attention is diverted from the continuum of patterns by the qualitative and quantitative classification of pattern types, which in any case are often dependent on flow stage. For example, straight channels with mid-channel gravel bars appear braided at low flow, and meandering streams in upland environments often have extensive unvegetated point bars across which chutes are formed to create pseudo-braided patterns at high stages (Plate 3). However, it is convenient to use the traditional straight–meandering–braided classification in order to highlight the hydrodynamic and sediment transport processes involved in pattern formation, as well as the sedimentological features associated with different types of current pattern. An example of a less generalized classi-fication is that of Kondrat'yev and Popov (1967), which identifies a more

Plate 3 A sinuous upland river (the River Elan, Wales) showing extensive low elevation point bar development with potential flood chutes across the bar surface.

complete continuum of pattern types including two categories of straight stream distinguished by different bedform characteristics; restricted, free and incomplete meandering; and braided channels.

'Straight' alluvial river channels

The rarity of uniform, straight natural channels longer than about ten channel widths, thought by Langbein and Leopold (1966) to indicate that meandering is a more probable state, reflects the variability of bank materials and the influence of bank vegetation and random bank collapses. Sinuosity can also be an ambiguous criterion of departure from straightness. The reach of the River Elan, Wales, in Figure 7.2(a) is underfit (see pp. 270–2), with a meander wavelength much larger than the normal 10–14 times stream width, but with a succession of riffles and pools around each bend. The bends occur because the stream is confined between valley bluffs which divert it back and forth across the valley floor. By a conventional operational definition, the sinuosity is $P = 2.2$, but it is approximately $P = 1$ if the 'straight' line is defined as the sum of all riffle–riffle distances (the logic of which is shown on pp. 194–204). Thus a conventional sinuosity exceeding ~1.1 may simply indicate that a low-power stream is being diverted from a uniform flow direction by sedimentological and topographical constraints which it is incompetent to modify.

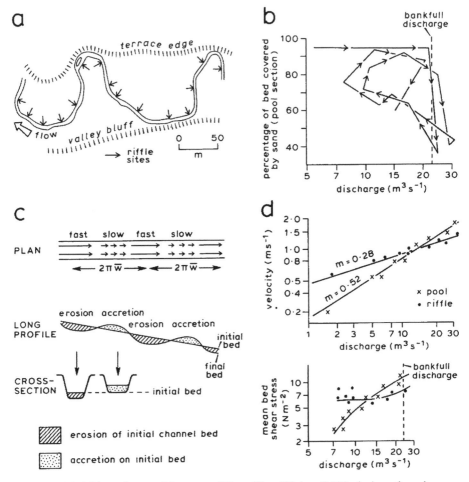

Figure 7.2 (a) Plan of part of the upper River Elan, Wales. (b) Variation of sand cover of a pool bed during three flood peaks (after Lisle, 1979). (c) Erosion and accretion of stream bed reflecting alternate zones of fast and slow flow. (d) Velocity and shear stress 'reversals' at two sites (after Andrews, 1970b, and Lisle, 1979).

The oscillatory behaviour of flow in a straight channel suggests a close relationship with meandering. Oscillation or perturbation of flow is evident in most fluids passing a boundary at which resistance to motion occurs, which may be a density interface as in the Gulf Stream. Gorycki (1973) demonstrated the role of hydraulic drag in meander development by showing that filaments of water flowing down inclined hydrophobic surfaces exhibit alternating deeps and shallows and hence relatively slow and fast flow. At steep inclinations filaments became unstable and meandered, which suggests that vertical oscillation develops into lateral oscillation above a threshold gradient, and which echoes the evidence described above of a continuum of energy conditions controlling different pattern types.

'Vertical' oscillations of flow

The local variation in hydraulic geometry described in Chapter 6 reflects the longitudinal non-uniformity of flow apparent even in roughly straight streams, particularly those with heterogeneous bed material organized into riffles and pools (Plate 4). Riffles are topographic high points in an undulating bed long profile. They are spaced about 5–7 channel widths apart and are composed of coarser sediment such as pebbles (Leopold *et al*, 1964, pp. 206–8). Pools are low points with sandy beds, although scour at high

Plate 4 A riffle–pool sequence in a straight river reach, River Fowey, Cornwall. The riffle is a medial bar, and the channel is wider at the riffle because of lateral under-cutting of the banks by flow diverted around the riffle.

discharges may expose coarse lag sediment, which is then covered again by sand during periods of low flow (Hack, 1957, Lisle, 1979; Figure 7.2b). At low flow, the bed topography influences the flow geometry: riffles have rapid, shallow flow with a steep water surface gradient, and act as a broad-crested weir (Richards, 1978a) to dam the flow back in the upstream pool, where it is deep and slow-flowing with a gentle surface slope. This oscillation in the vertical plane is explained by a combination of turbulence-induced velocity pulses and 'velocity reversal' between adjacent riffles and pools.

Yalin (1971) demonstrated theoretically that turbulence generated at the boundary of a straight, uniform channel would produce large-scale roller eddies associated with alternate acceleration and deceleration of flow. The scale of the eddies depends on channel size, and spacing between successive fast or slow zones averages $2\pi\bar{w}$. Longitudinal velocity variation would be stochastic, but at each section, j, the velocity, v_j, would relate to velocities at upstream sections as well as to a random disturbance, ϵ_j. Thus, along the channel the variation in velocity is described by a second-order auto-regression

$$v_j = \phi_1 v_{j-1} + \phi_2 v_{j-2} + \epsilon_j. \qquad (7.1)$$

(An autoregression is a lagged regression of a series of measurements on itself; second-order autoregression implies dependence back over two sampling intervals.) The coefficients ϕ_1 and ϕ_2 are analogous to partial regression coefficients in multiple regression, and for certain combinations of these parameters this autoregression describes a pseudo-oscillatory, damped cyclic variation of velocity. This alternate fast and slow flow may interact with mobile-bed sediment in an initially uniform channel (Figure 7.2c) to cause erosion and accretion, assuming the threshold velocity is exceeded. Scour enlarges the cross-section where velocity is high; this reduces the velocity. Accretion reduces the cross-section area and increases the velocity. Thus, a uniform channel with varying velocity is transformed to a channel whose bed undulates and in which longitudinal velocity variation is minimized by adjustments of the cross-section.

Riffles are evidently zones of accretion and in this model are associated with slower-than-average flow velocity; this appears to conflict with the observed faster flow at riffles. However, this observation refers to low discharges when the flow is controlled by the gross bed topography which is itself created at high, competent discharges. A velocity or bed shear stress 'reversal' occurs between adjacent pools and riffles over a range of discharges (Keller, 1971, Andrews, 1979b, Lisle, 1979). As Figure 7.2(d) shows, at low flow the riffle experiences higher velocity and shear stress, but as discharge increases, convergence and cross-over occurs so that at higher, competent, channel-forming discharges the pool experiences hydraulic conditions favouring scour and erosion. The bed topography formed at higher discharges forces the low flows, which are of low competence and capacity to transport sediment, to conform to the pattern of shallows and deeps so produced. This 'reversal' also explains sediment sorting in the riffle-pool sequence. Peak flow velocities may be competent to transport coarse sediment throughout a reach, but a large pebble scoured from the pool will enter a zone of slower flow in the downstream riffle, where deposition may occur. The riffle thus becomes an accumulation of coarse sediment. At low flow, the overall competence of a reach may permit sand transport, but even this may be deposited as a deltaic wedge at the entry to a pool, after being carried over the riffle by its faster flow. The sediment sorting between pools and riffles thus depends on recent flow events and sediment supply. At high discharges little difference in bed material size is evident, but at low flow it is exaggerated by superficial sand infill in the pool as long as an appropriate supply of sand-size sediment exists (see Figure 7.2b).

Alternative theories of the origin of riffles and pools provide less comprehensive explanations of their geometry and of the processes of initiation, maintenance and sedimentological differentiation. For example, Yang (1971b) invokes his law of minimum rate of potential energy expenditure, arguing that energy is dissipated less rapidly over a reach consisting of segments with different energy gradients, than over a uniformly sloping reach. Excessively rapid energy loss in the uniform reach results in erosion and compensatory accretion which disrupt the uniformity of bed gradient; this is similar to Yalin's model described above. Yang, however (p. 1571), argues that the tendency to minimize the rate of energy loss 'should form a pool–riffle sequence with a higher average velocity and a steeper water surface slope at the riffle than at the pool', which then results in a concentration of coarse sediment on the riffle bed. This ignores the fact that at competent discharges which form the bed topography the pool experiences faster velocities. Langbein and Leopold (1968) suggest a kinematic wave model for gravel riffle and sand-dune forms, based on the mutual interference of particles during transport. A flow–concentration curve (Figure 7.3a) indicates that the transport rate is zero when the linear concentration of particles is either zero, or very large in a 'bumper-to-bumper' jam (the model derives from the theory of traffic flow by Lighthill and Whitham, 1955). The maximum transport rate (given by mean concentration multiplied by mean particle velocity) is at some intermediate concentration. The kinematic wave is a group of particles. For example, beads in a narrow flume may travel in groups whose speed is less than that of individual beads, which detach themselves from the downstream end of a group and catch up the next, so that the group appears to move back upstream (inset, Figure 7.3a). The average speed of a particle is given by the gradient of OC (where C defines the flow rate and concentration of a particular transport system), whereas the wave velocity is given by the tangent AB at C. (In Figure 7.3a its negative slope implies a wave travelling against the flow of particles.) This model does not explain the sediment sorting or riffle–pool wavelength, although recent development (Naden, 1981) of a stochastic version, based on probabilities of particle movement or deposition and applicable to sediment-size mixtures, is capable of predicting the distribution of sizes of gravel bedforms as measured by numbers of particles. However, riffles are large, three-dimensional bedforms which are often remarkably static (even though particles move through them), as shown by Dury's (1970) resurvey of the Hawkesbury River after 100 years. This static nature of the waveform demands a horizontal tangent at the peak of a flow–concentration curve, which implies a condition of maximum transport rate. This condition seems improbable for many riffles in stable, gravel-bed streams.

The morphology of riffle–pool sequences generally agrees with Yalin's

(1971) theory. Wavelengths average 5–7 times the mean width (Keller and Melhorn, 1973), and this range includes the theoretical $2\pi w$ period of velocity pulsation. Furthermore, if a regression line is used to generalize the trend in bed elevation along a reach, measured at equally spaced points, the residuals from the trend are often fitted by a second-order autoregression.

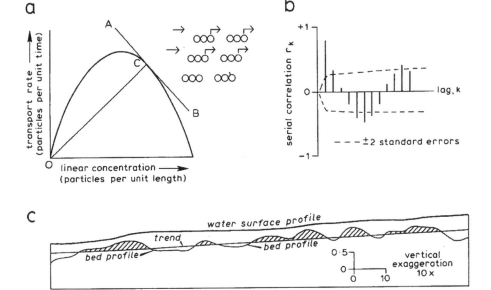

Figure 7.3 (a) Relation of transport rate to linear concentration of bed particles. The inset shows a kinematic wave moving upstream, as predicted by the tangent at C. (b) Correlogram of bed elevation residuals from the downstream trend of bed elevation in the reach shown in (c). (c) Bed profile and low-flow water surface profile of a reach of the River Fowey, Cornwall. Riffles are positive residuals from trend.

The pseudo-cyclic nature of the bed oscillation is shown by serial correlation, in which the series of residuals is correlated with itself after lagging by one, two, . . ., etc. sampling intervals. Serial correlations are plotted against the lag to form a correlogram, and for riffle–pool bed profiles this is commonly pseudo-cyclic (Richards, 1976b). Figure 7.3(b) shows such a correlogram for height deviations at 2 m intervals from the trend fitted to the bed profile of the reach in Figure 7.3(c), where riffles and pools are zones of positive and negative residuals. By defining the height residuals as z_j, this series is described by the stochastic model

$$z_j = 1.23 z_{j-1} - 0.49 z_{j-2} + \epsilon_j. \tag{7.2}$$

Table 7.1 lists the parameters ϕ_1, ϕ_2, the variance of the random error terms ϵ_j and the explained variance of the fitted equation (r^2) for several bed profiles, including that of the River Elan reach in Figure 7.2(a). Pseudo-oscillatory variation of bed elevation is implied when

$$\phi_1^2 + 4\phi_2 < 0. \tag{7.3}$$

In some streams it is possible to identify a statistical model describing the joint variation of bed elevation and width noted on pp. 177–8 (Anderson and Richards, 1979), which occurs if low-sinuosity streams have sufficient energy to erode their banks at riffle sites. However, coarse sediment inputs from bluffs, terraces and gullies often distort the riffle–pool sequence by supplying sediment the stream cannot sort, and this may encourage locally wide, shallow pool cross-sections (Milne, 1979).

Table 7.1 Parameters of second-order autoregressive models fitted to bed elevation residuals (all in metres) from bed profile trends of several river reaches.

River	ϕ_1	ϕ_2	σ_e^2	$r^2\,(\%)$
Fowey	1.23	−0.49	0.00	82
Fowey	1.00	−0.39	0.00	69
Mole	0.77	−0.21	0.04	41
Popo Agie	1.03	−0.45	0.04	62
Middle	1.37	−0.63	0.01	81
Pole Creek	0.86	−0.20	0.06	54
Bronte Creek	1.24	−0.36	0.01	98
Hoaroak Water	0.97	−0.29	0.01	60
Elan	1.16	−0.33	0.01	77

Secondary circulation

Longitudinal velocity oscillation in uniform channels is related to large-scale circulatory motion superimposed on the main flow. A simple example of such secondary circulation consists of a pair of spiral currents rising at the banks and plunging in mid-channel (Figure 7.4a). Secondary circulation is caused by (i) the creation of turbulence by boundary roughness, (ii) density differences caused by variations of temperature or suspended sediment concentration (Vanoni, 1946), and (iii) the interaction of the flow with dead spaces in the corners of rectangular or trapezoidal sections (semi-circular sections generate less large-scale turbulence, but rarely exist in natural sediments). The number of cells of secondary circulation reflects the channel width:depth ratio (Figure 7.4b).

Leliavsky (1955) and Keller and Melhorn (1973) emphasize that different forms of secondary circulation occur in riffles and pools, and detailed current meter measurements in meander bends confirm this (Hey and Thorne, 1975). In pools, flow is 'convergent' at the surface, and so floating debris collects in the shallow hollow at mid-channel where the flow plunges.

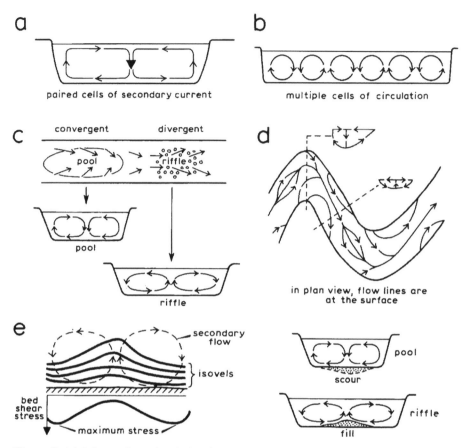

Figure 7.4 (a) Secondary circulation in channel with relatively low width:depth ratio. (b) Secondary circulation in wide, shallow channel. (c) Convergent and divergent surface flow in pools and riffles. (d) Surface flow vectors in a meander bend, showing convergent flow in the pool at the bend and divergent flow at the riffle in the inflection (after Hey and Thorne, 1975). (e) Effect of secondary circulation on bed shear stress, and interaction with mobile bed sediment to create scour (pool) and fill (riffle bar).

In riffles, surface 'divergent' flow occurs, with a slight surface super-elevation where the mid-channel current rises (Figure 7.4c). In a meander Hey and Thorne's results (Figure 7.4d) show this basic pattern except that the surface convergent flow in the pool, at the bend apex, involves asymmetric cells with one compressed against the outer bank. A prerequisite for meandering is the development of this asymmetry, leading to bank erosion and channel migration on alternate banks at pool locations. This is discussed further on pp. 200–2.

Secondary circulation is superimposed on the downstream flow and causes longitudinal and transverse variations of bed shear stress (Bathurst, 1979, Bathurst *et al*, 1979). Where circulation currents ascend, the faster water in the vertical profile is 'lifted' from the bed; velocity shear at the bed (the velocity gradient) is reduced, and therefore so is the bed shear stress (pp. 68–70). Thus, when flow converges *at the bed* (the riffle section), this favours deposition and a central bar may develop (Figure 7.4e). Conversely, descending flow increases the bed shear stress and encourages scour, as in the pool. Clearly, therefore, the non-uniformity of the flow contributes significantly to creating the geometry of 'straight' streams, in which riffles and pools reflect accretion and scour associated with secondary circulation and pulsation of velocity. What distinguishes 'straight' from meandering streams is their limited capacity for perimeter erosion, which only occurs where diversion around medial riffle bars causes bank undermining, or where plunging flow causes bed scour. There is insufficient excess energy, beyond that needed to transport supplied load and overcome friction, to change the flow direction markedly by selective bank erosion and channel migration. In fact, deviation from a uniform flow direction is mainly the result of external environmental influences such as the valley topography.

Sedimentology, sedimentary structures and flow direction

Equilibrium patterns reflect a force:resistance ratio, so low sinuosity reflects not only low stream power, but also coarse, relatively immobile sediments (pebble and cobble bed material). Bluck (1976) suggests a general downstream trend of changing bedforms and channel patterns as bed material sizes decline. Upstream, coarse sediments in medial bars characterize low-sinuosity streams. Medial bars (Figure 7.5a) are similar to braid bars (Leopold and Wolman, 1957), spool bars (Krigström, 1962) and longitudinal bars (Williams and Rust, 1969). Most bars have two components, namely, a bar platform of coarser sediment representing a basal element of roughly constant form and composition, and the overlying finer sediments known as 'bar supraplatform deposits', which are more variable and subjected to removal and replacement during floods (Bluck, 1976). Riffles are usually the downstream faces of the basal platform, in medial or lateral bars, and are therefore confined to the low-flow channel and formed of a plinth of coarse sediment in a morphologically constant unit. The directional properties of sedimentary structures in a medial bar may indicate the existence of currents converging to the channel centre (at the *bed*, cf. Figure 7.4c). The downstream face is steep, with large-scale gravel cross-strata dipping into the channel bed. This gravel plinth is a high stage deposit; the supraplatform sediments only exist if there is a sand supply and form the bar tail drapes over the downstream face. These sands have foreset beds dipping into the pool and may form deltaic structures. These sandy sediments are deposited on the falling stages of floods and at low flow. The pattern of flow is controlled by the gross bed topography at these lower discharges, and often tends to involve radial divergence of currents over the downstream

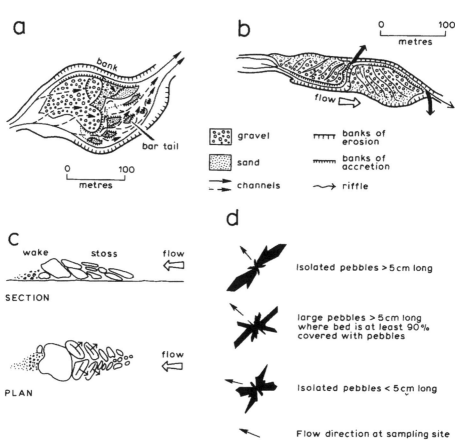

Figure 7.5 (a) Sedimentology and morphology of a medial bar (after Bluck, 1976).
(b) Alternating orientation of riffle faces of lateral bar platforms (after Bluck,
1976). (c) Pebble cluster in plan and section. (d) Long axis orientations in three
samples of 50–80 pebbles from different environments of the River Donjek, Alaska
(after Rust, 1972).

face of the riffle. Thus the dip directions of the foreset beds of the deltaic
accumulations at the head of a pool will be very variable.

Current directions influence sedimentary structures at different scales.
First, the bar shape depends on channel orientation and flow patterns within
the channel. The long axes of successive medial bars deviate little from mean
channel direction, but orientations of the riffle faces of successive lateral
bars are on alternate sides of the mean direction (Figure 7.5b). Second,
cross-strata of bedform structures vary in orientation, according to sediment size and location. Bar platform gravels have a smaller variability of
cross-strata orientations than bar tail sands, being high stage deposits.
Third, imbrication occurs particularly in discoid pebbles, which are stacked
with the plane of their *a* and *b* axes dipping upstream (Plate 1a). The dip
direction of this maximum projection plane indicates the flow direction at

the time of deposition, and since gravel deposition is at high flow, it is probably a dominant current direction. This structure results because particles are deposited when in the position of maximum resistance to movement; downstream dips are unlikely because the flow would tip the pebbles over. Fourth, pebble clusters develop when large pebbles obstruct the movement of smaller pebbles and gravel, which accumulate on the upstream or 'stoss' side. Finer particles are deposited during falling stages of floods in the lee of the obstacle clast to form a wake, so that the entire cluster is eventually about twice as long as it is wide. The particles are strongly imbricated and resist entrainment, and thus the feature is preserved (Brayshaw, 1984). About 90% of clusters have the largest boulder downstream from a recognizable stoss accumulation, and about 80% have wake long axes within ±10° of current direction (Dal Cin, 1968). Individual clasts in the wake have dip orientations which may depart markedly from that of the current (Teisseyre, 1977), and thus care is needed in interpreting individual clast dips (Figure 7.5c). This also applies to the final directional indicator – the individual clast itself. Axis orientations vary with clast shape and behaviour during transport and deposition. Rods may roll with long axes perpendicular to the flow, blades slide with long axes parallel to the flow and discs slide or flip over, with either long or intermediate axes parallel to the flow. However, if a pebble is just transportable when oriented such that resistance to movement is minimized, it will remain deposited if it falls to the bed in an orientation demanding a higher stress to move it (i.e. a blade at right-angles to the flow). Thus axis orientation is an unreliable guide to current direction, and the *ab* plane dip direction is preferable. Rust (1972), for example, illustrates that pebble long axis orientations vary with pebble size and emplacement site (Figure 7.5d): large, isolated pebbles are more consistently oriented than small ones, partly because smaller pebbles tend to be more sphere-shaped (Figure 1.5c) and therefore have less directional preference.

Meandering river channels

Sinuous rivers are common because frequent topographical and sedimentological constraints disturb the directional uniformity of low-power streams. However, a distinction between straight and meandering streams emphasized in Chapter 6 is that the latter are actively migratory as a result of selective bank erosion and point bar development. They therefore result from a higher power:resistance ratio, and their sinuosity is the product of active, inherent processes rather than a passive response to external influences, although their degree of morphological regularity reflects these environmental controls. Zeller's (1967) summary demonstrates the ubiquity of meanders, all obeying a general wavelength–channel width relationship, in a variety of materials (Figure 1.1b). Free surface underground streams in limestone conduits meander when conditions allow the hydrodynamic processes to dominate the environmental influences of joint control; moderate structural dips and sufficient relief provide the necessary hydraulic gradient, whereas steep dips disrupt the bend sequences (Deike and White, 1969, Ongley, 1968). Thus, sufficient energy for bank erosion, and intensified (skew-induced) secondary circulation to make it selective, appear to be necessary for active meandering, while sediment transport may not be (given its absence in the geometrically similar bends of supraglacial streams and the Gulf Stream; Leopold *et al*, 1964, p. 302).

Randomly distributed environmental disturbances distort meander patterns and therefore seem unlikely to account for regular successions of bends. Nevertheless, local bank projections (collapse blocks, boulders, logs) may initiate skew-induced asymmetric secondary circulation which amplifies to create a bend. This then transfers the spiral current downstream from the bend exit (Prus-Chacinski, 1954), resulting in damped meander oscillation depending for its extent on the available stream energy. This is apparent in laboratory studies (Friedkin, 1945) when the flow is introduced at an angle to the initial straight channel direction. Langbein and Leopold (1966) suggest that meanders are the most probable realizations of random walks of a specified sinuosity between two points, which implies that regularity can develop *on average* from random influences. Transverse oscillation of flow is a necessary prerequisite for meandering (Anderson, 1967, Fujiyoshi, 1950), whatever the reason for its initiation. For example, the Coriolis force may impose transverse currents on the flow of high-latitude rivers (Neu, 1967), although globally the effects of vortex generation at rough channel boundaries (Einstein and Shen, 1964) are of much greater general significance. Amongst the numerous explanations of meandering are many which refer to various types of turbulent oscillation. Hjulström's (1949) theory based on gravity waves (seiches) is typical, but its emphasis on the normally weak relationship between meander wavelength and meander belt width illustrates that theories must be evaluated in the context of observation of meander form.

Meander morphology

Traditional meander scale and shape indices are the axial wavelength and radius of curvature (Figure 1.4b; Fergusson, 1863, Leopold and Wolman, 1957, 1960). Axial wavelength (L) is difficult to measure in irregular bends, and radius of curvature (r_c) varies if a bend is not an arc of a circle, so that a careful operational definition must precede measurement of a mean (Figure 1.4c) or minimum (apex) radius. Nevertheless important relationships are based on these indices. Dury (1956) relates wavelength to bankfull discharge as

$$L = 54.3Q_b^{0.5}. \qquad (7.4)$$

Although Ackers and Charlton (1970a) confirm the role of bankfull discharge, a stronger correlation exists with channel width, for which various relationships exist including Zeller's (1967; Figure 1.1b) and those of Leopold and Wolman (1957, 1960):

$$L = 7.32\bar{w}^{1.1} \tag{7.5}$$

$$L = 12.13\bar{w}^{1.09}. \tag{7.6}$$

As exponents do not differ significantly from unity, the linear regression

$$L = 12.34\bar{w} \tag{7.7}$$

is quite satisfactory. Here, the constant is very close to 4π (12.57), which suggests a close link with the riffle–pool wavelength ($= 2\pi\bar{w}$), because in a regular meander bend there are two riffle–pool cycles to one bend with pools at the apices and riffles in the inflections.

The strong correlation with width reflects the role of secondary circulation controlled by channel size. As shown in Chapter 6, channel width is caused by an interaction of discharge and bank material properties, and a stream in silty sediment is narrower and deeper than one in sand, assuming equal discharges. Discharge and sediments jointly control width, which then controls the wavelength of secondary flow (Yalin, 1971) and hence that of the riffle–pool sequence and meander bends (Figure 7.6a). Width, and thus wavelengths, are indeterminate given discharge alone; hence the poorer correlations with discharge. By using various discharge indices, Schumm (1968a) demonstrated that multivariate relationships best explain variation in meander wavelength:

$$L = 1935 Q_m^{0.34} M^{-0.74} \quad (r^2 = 0.89) \tag{7.8}$$

$$L = 618 Q_b^{0.43} M^{-0.74} \quad (r^2 = 0.88) \tag{7.9}$$

$$L = 395 Q_{ma}^{0.48} M^{-0.74} \quad (r^2 = 0.86). \tag{7.10}$$

All show the expected direct relation with discharge and inverse relation with the weighted silt–clay index (M), paralleling their correlations with width. The highest percentage of the variance of meander wavelength is accounted for using mean annual discharge (Q_m), echoing Carlston's (1965) findings that sub-bankfull flows relate closest with meander wavelength. This may have physical significance, but could mean that bankfull (Q_b) and mean annual (Q_{ma}) floods have greater estimation errors. Chitale (1970) provides some evidence that axial wavelength also varies with slope and relative roughness. Meander scale and shape are related since

$$L = 4.59 r_c^{0.98} \tag{7.11}$$

(Leopold and Wolman, 1960), and therefore $r_c \approx 2$–3 $\bar{w}$. Bagnold (1960b) showed that resistance and energy loss due to bend curvature are minimized

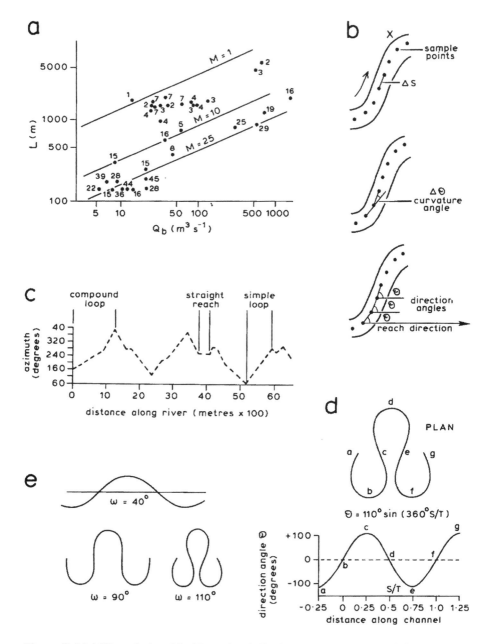

Figure 7.6 (a) The relationship (Equation 7.9) of meander wavelength (L) to bankfull discharge (Q_b) and weighted perimeter silt–clay percentage (M). Numbers are measured silt–clay percentages (after Schumm, 1968a). (b) Sampling for series analysis: direction (θ) and curvature ($\Delta\theta$) series. (c) Bends of the White River as circular arcs forming simple and compound loops identified in azimuth–distance plots (after Brice, 1973). (d) A sine-generated curve with maximum direction angle $\omega = 110°$ (after Langbein and Leopold, 1966). (e) Sine-generated curves with various maximum direction angles (after Langbein and Leopold, 1966).

when $r_c/\bar{w} \approx 2$–3, and so bends appear to adopt as efficient a shape as possible (see p. 203).

Several model bend shapes have been proposed. Sine curves are generally inappropriate, since they cannot model those acute bends in which the channel is locally directed in the up-valley direction (cf. Figure 7.6d). However, bends can be represented as a series of straight segments (Figure 7.6b) by defining equally spaced sample points and by identifying the azimuth, direction angle with respect to overall reach direction (θ), or direction change angle (curvature, $\Delta\theta$) of successive segments (ΔS). By plotting segment azimuth against channel distance, Brice (1973) showed that straight reaches (constant azimuth) could be distinguished from meanders which are simple or compound circular arcs, the latter having two or more sections with curvature directed to the same side of the river (Figure 7.6c). A single circular arc has a linear azimuth–distance plot (constant azimuth change between segments). Langbein and Leopold (1966), however, favour a model in which segment direction angle (θ) is a sine function of distance (S) measured as a proportion of total bend or reach length (T). Thus

$$\theta = \omega \sin\left(\frac{S}{T} 2\pi \right) \tag{7.12}$$

where ω is the maximum angular deviation at bend inflection points. An example of such a sine function with $\omega = 110°$ is shown in Figure 7.6(d) together with its 'sine-generated' meander bend (the bend is not a sine curve, but is generated from a sinusoidal variation of direction). Figure 7.6(e) compares bends with different values of ω. In such bends, curvature increases towards the apex; the curvature is $d\theta/dS$, and therefore a cosine function of distance. Ferguson (1973a) discusses other models and means of comparing them with natural bends. The 'Fargue spiral' model, for example, has a linear increase in curvature to a maximum at the apex, and it is interesting because its development followed the observation that curvature was linearly related to depth around stable bends. If the maximum direction angle, ω, is defined for a natural bend, it may be compared with the value for theoretical bends of the same sinuosity (Figure 7.7a). For circular and sine-generated curves, ω increases with sinuosity P to give outer limits between which natural bends can be plotted. Departure from the theoretical curves probably reflects bend asymmetry caused by downvalley migration, as bank erosion is concentrated downstream from the bend apex. Kondrat'yev (1968) accommodated this tendency by including a skewness (asymmetry) parameter in a meander bend model in which curvature was described by a sine function of distance along the channel centre line. Indeed, Carson and Lapointe (1983) argue that asymmetry is inherent, with a majority of bends exhibiting a downvalley convexity of plan, similar to that in Figure 7.8(d), which they refer to as 'delayed inflection asymmetry'. They conclude that symmetrical models such as the sine–generated curve should be abandoned because of this prevalence of asymmetry.

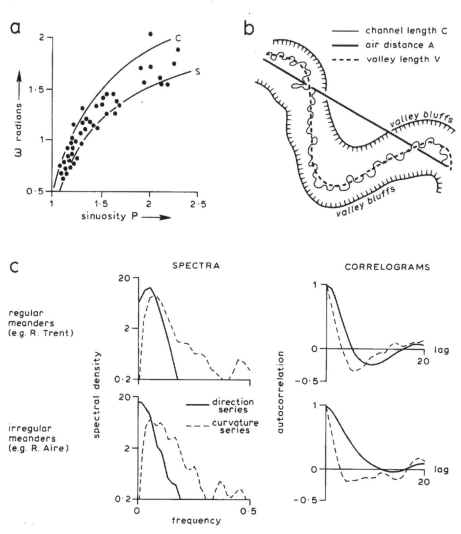

Figure 7.7 (a) The relationships between ω and sinuosity P for circular (C) and sine-generated (S) bends as criteria for natural bend shapes (after Ferguson, 1973a). (b) Hydraulic and topographic sinuosity. (c) Spectra and correlograms for regular and irregular meander patterns (after Ferguson, 1976). Results are for both curvature and direction series.

Meander patterns are usually complex successions of irregular and compound bends, for which traditional indices and theoretical models are less appropriate. This complexity partly reflects topographical effects. Table 1.2 indicates that the sinuosity of a regular bend is the arc:axial wavelength ratio, and this sinuosity is the result of hydrodynamic processes. Figure 7.7(b) shows a sinuous channel following a meandering valley axis; the total sinuosity (TS) is channel length divided by air distance (C/A), and the valley sinuosity (VS) is valley length divided by air distance (V/A). Mueller (1968) suggests that two indices define the components of sinuosity:

$$\text{Hydraulic sinuosity (HSI)} = \frac{100(\text{TS} - \text{VS})}{(\text{TS} - 1)} \tag{7.13}$$

and

$$\text{Topographic sinuosity (TSI)} = \frac{100(\text{VS} - 1)}{(\text{TS} - 1)} \tag{7.14}$$

of which the first defines the percentage of sinuosity attributable to hydraulic meanders. For a stream such as that in Figure 7.2(a), however, it is necessary to discriminate between an actively meandering state and passive sinuosity as low-power streams are diverted by environmental controls. This may be achieved by dividing channel length by the sum of all straight-line riffle–riffle distances (an operational definition more objective than the 'valley axis' of Mueller). The resulting sinuosity would be high for a bend with active outer bank migration and inner bank point bar accretion.

An alternative approach to meander morphology is based on series analysis. The series of direction angles (θ) representing a reach (as defined in Figure 7.6b) retains information about valley direction changes, but the curvature series ($\Delta\theta$) is independent of topographical effects. Both direction and curvature angles may be defined as positive and negative according to the local channel curvature, so that in Figure 7.6(b) the curvature angle is near zero in the bend inflection, becomes positive in the bend towards X, zero at the next inflection, and negative in the subsequent bend. The oscillating series may be analysed by serial correlation methods (in the distance domain) or by spectral analysis (in the frequency domain). The former shows the correlation between successive angle measurements, displayed in the correlogram, while the latter assesses the relative importance in the series of waveforms of different frequency (or wavelength). Since the spectral density function is a Fourier transform of the correlogram, the two provide essentially the same information. Speight (1965a) first used spectral analysis to define hydraulic and topographic scales of meandering, identified as peaks in meander spectra. Chang and Toebes (1970) and Thakur and Scheidegger (1970) subsequently refined the technique and interpretation. Ferguson (1975, 1976) showed that both techniques can be used to define a range of types of sinuosity, from regular to irregular bend sequences (Figure 7.7c). Regular meandering is associated with a well-defined consistent wavelength, shown by a sharp peak in the spectrum and a pseudo-cyclic correlogram; a series of sine-generated curves would produce a 'spike' in the spectrum and a sinusoidal correlogram. The hydrodynamic influences here dominate environmental influences in powerful streams or homogeneous sediments. Irregular meanders occur in low-power streams with dominating environmental influence by heterogeneous valley fill sediments. Ferguson's 'disturbed periodic' meander sequence model (1976) incorporates scale and irregularity parameters in a mixed deterministic–stochastic framework. A regular pattern is dominated by deterministic repetition of bend form, as in a natural stream where hydrodynamic influences prevail, whereas irregularity reflects stochastic environmental disturbances. For all types of meander pattern, but especially those with compound and irregular bends, objective average wavelength estimates may be obtained from the spectrum and correlogram (Ferguson, 1975), and these estimates correlate with channel width. They are arc wavelengths, however, since the meander pattern is being quantified along the channel, not along the bend axis (Figure 7.6b), and a different constant is therefore necessary in the wavelength–width relationship.

Meander processes and development

Skew-induced secondary flow is a necessary precursor of meandering. Once initiated, perhaps because of high shear stress and vortex generation adjacent to a rough bank (Einstein and Shen, 1964), it may be amplified by a positive-feedback process. Leliavsky (1955) showed that the centrifugal force acting on water flowing around a bend causes·water surface super-elevation of magnitude

$$z = \bar{v}^2 w / g r_c \tag{7.15}$$

on the outside of the bend. The resulting transverse water surface gradient 'drives' a transverse current which plunges near the outer bank and crosses to the inner bank at the bed. Cause and effect are easily confused here, super-elevation being characteristic of a pre-existing bend. However, even in a straight reach, deflection of the current to one bank produces super-elevation and initiates skew-induced secondary flow, whereupon the channel is potentially unstable if bed material movement occurs. Callander (1969) employed hydrodynamic stability analysis to show that an initial perturbation of flow (Figure 7.8a) leading to lateral 'vibration' is unstable because sediment transport at the bed intensifies the perturbation by moving bed material to the inner bank (an incipient point bar). This produces a transverse bed slope, accentuates the thalweg oscillation, encourages scour hole formation at the outer bank, and eventually results in bank retreat and true meandering. Parker (1976) noted that secondary flow alone cannot result in meander formation; a second process is necessary to enable the instability to be effective. In alluvial rivers, this is bed material transport, but in supraglacial streams, it is differential melting and freezing.

Bank erosion is necessary before a meander can develop, but not excessive bank retreat which creates a wide, shallow cross-section in which multiple circulation cells occur (Figure 7.4b). Parker's stability analysis (1976) suggests that straight streams have small form ratios ($F < 10$), while meanders have values in the range $10 < F < 100$ (approximately), and wider, shallower channels experience mid-channel bar development and braiding.

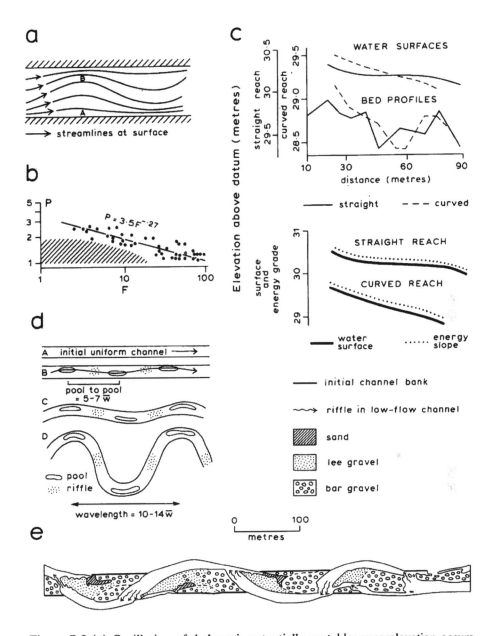

Figure 7.8 (a) Oscillation of thalweg is potentially unstable; superelevation occurs at B, and intensified skew-induced secondary flow causes deposition at A. (b) Relationship between sinuosity (P) and width:depth ratio (F) (after Schumm, 1963). The curve is a limit and the hatched area represents the probable plotting area for narrow, deep, 'straight' channels. (c) Water surface and energy gradients in straight and curved reaches, Baldwin Creek, Wyoming (after Leopold *et al*, 1964). (d) Transformation of a straight to a meandering stream (after Dury, 1969). (e) River Ystwyth, June 1970, after development of lateral bars and incipient meandering after artificial straightening (after Lewin, 1976).

Schumm (1963) identified an interrelationship between sinuosity (P), form ratio (F) and weighted mean channel silt–clay percentage (M), where

$$P = 0.94M^{0.25} \tag{7.16}$$

and

$$P = 3.5F^{-0.27}. \tag{7.17}$$

Sinuous streams thus have narrow, deep sections with high silt–clay contents in their perimeter sediments. Such streams favour meandering by maintaining a cross-section form which permits continued maintenance of the double cell secondary circulation. Schumm (1963) measured braided streams as having low sinuosity by treating the whole active channel zone as a single river when measuring channel length. Thus both relationships may represent limiting envelope curves. Figure 7.8(b) shows the relationship between stream sinuosity and form ratio for meandering streams and for braided streams when the latter are defined as having low sinuosities. Narrow, deep low-power straight streams with single-thread channels should plot on this diagram in the shaded region. These rivers would also tend to plot beneath the curve defined by Equation 7.16 on a diagram relating sinuosity to the channel perimeter silt and clay content.

Bank erosion and channel migration demand available stream energy, which is largely dissipated in overcoming various scales of frictional resistance, including skin resistance, form roughness, and macroscale roughness due to channel pattern (bends, bars and islands). In a straight reach, pools are hydraulically less rough than riffles, with deeper flow and higher velocities at peak discharges, since their lower relative roughness compensates for shallower energy gradients (Figure 7.8c). Higher velocity, shear stress (Figure 7.2d) and power expenditure, lower frictional losses, and flow directed against one bank, result in bank retreat on alternate sides in successive pools if the local power expenditure rate exceeds the erosion threshold. An excess of initial energy is therefore essential before active bank erosion and bend formation occur, and this is most likely on steeper valley surfaces. Meandering may thus be interpreted as a mechanism for reduction of excess gradient (Shulits, 1955). Rapid power expenditure per unit length (Table 1.3) on a steep valley gradient is associated with excessive bed material transport and bank erosion, which, by creating meander bends, reduces channel gradient and lessens the rate of power loss. This is evident in supraglacial streams, where maximum sinuosity develops in those channels with the highest initial stream power (Ferguson, 1973b), because excessive fluid shear at the outer bank in asymmetric secondary flow cells locally raises the water temperature to encourage melting, while refreezing occurs on the inner bank. The bends created during meander development add a new source of roughness due to curvature in the pool, which helps to reduce the excess of initial energy. Meander amplitude growth continues until the extra curvature resistance itself dissipates the excess energy, and further morphological change is inhibited. Bend shape exerts a strong control over flow resistance, which is excessive in tight bends of low r_c/w ratio because flow separation occurs at the inner bank, leading to the onset of 'spill resistance'. Bagnold (1960b) considers that bends tend to r_c/w values of 2, which gives minimum curvature resistance. However, Davies and Sutherland (1980) suggest a more logical interpretation, namely that bends (which develop to offset excess energy expenditure) tend towards a local *maximum* of curvature resistance at $r_c/w \approx 2.5$ to dissipate energy most efficiently.

Extra energy loss due to curvature in the pool tends to create a uniform distribution of frictional resistance through a curved reach, which consequently experiences a smoother water surface and energy gradient (Figure 7.8c). Total energy loss is not necessarily increased, however, because the rate of loss per unit length is reduced in the longer meandering channel and because a backwater curve upstream from the deep pool drowns skin resistance in the riffle (Ippen and Drinker, 1962). Meandering minimizes the variance of energy loss and associated variables (shear stress, friction factor), and the sine-generated curve minimizes the energy expenditure per unit mass of water since it reduces slope (Yang, 1971c). However, teleological and anthropomorphic arguments that rivers tend to minimize power loss, or its variance, cannot replace an explanation based on the physical mechanisms of meander development, in relation to both river management and analysis of form and sedimentology.

Although mutual adjustment occurs of plan form, cross-section, bed forms and slope, explanations of meandering commonly invoke consideration of the transformation from initially straight channels, largely for simplicity. Figure 7.8(d) shows Dury's (1969) model, which mirrors the explanation of the preceding paragraphs. Tinkler (1970) outlined an alternative transformation in which riffles became locations of bend formation, but this is denied by the clear sedimentological differences between riffles and the point bars which occur adjacent to bends. Keller (1972) suggested a five-stage transformation whose final phases added to the Dury model the creation of secondary pools in the lengthened meandering stream to maintain a riffle–pool spacing along the channel of 5–7$\bar{w}$. This was based on a spatial study from which inference of temporal change was made. There is little justification for making this ergodic assumption, however, particularly since the streams in question were not of a high *active* sinuosity. In fact, Gorycki's (1973) study suggests that riffles and pools are elongated in meanders, and analysis of Leopold and Wolman's (1960) arc wavelength data yields

$$\lambda = 17.2\bar{w} \qquad (7.18)$$

the higher constant reflecting bend sinuosity (Table 1.2) and matching the wavelength:width ratio obtained by doubling the riffle–riffle distance measured along the channel (Richards, 1976b).

Direct observation of the transformation of initially straight laboratory or natural channels illustrates the meandering process more clearly. Friedkin's (1945) flume studies showed that alternate scour holes and transverse bars develop, followed by bank erosion; in the non-cohesive sands used in flume experiments, channels remain wide and shallow, and only become truly meandering if suspended sediment (clay) is added (Schumm and Khan, 1972) to stabilize the banks. Artificially straightened natural streams rapidly redevelop meanders given high power:resistance ratios, with meander amplitude growth outstripping wavelength changes (Noble and Palmquist, 1968). Lewin's (1976) study of the re-adjustment of the straightened River Ystwyth shows that lateral (transverse) bars form at high stages (bar platform deposits) and accelerate bank erosion by modifying flow patterns (Figure 7.8e), with diagonal riffle faces terminating each bar platform. Point bar sedimentation (supraplatform deposits) then causes dominance of one sinuous channel and alternate bank attachment of the bars, although chute channels remain, possibly because the lack of supply of sand prevents full point bar growth. Again, bank erosion follows bar formation and the development of lateral oscillation of flow.

Meander bends continue to migrate even when their curvature resistance dissipates excess energy and equilibrium forms have developed. Outer bank erosion continues since the current is directed at the bank, and perfect equilibrium never exists in a single bend where erosional and inner bank depositional processes are temporally out of phase (the latter being slower). Energy losses are distributed over a whole reach, in which one bend grows in amplitude to compensate for cut-off in another bend, thereby maintaining total sinuosity and gradient (Figure 1.7a). This approximate reach equilibrium recalls the 'chain model' of meanders (Thakur and Scheidegger, 1970), which compares a meander sequence with a chain whose ends are fixed, but which can lie in an infinite range of sinuous paths as the chain is shaken. Cut bank migration results from direct shear, undercutting and collapse, and non-fluvial processes of bank erosion discussed in Chapter 6, and is balanced by point bar growth. Asymmetry of the bend arises from the downstream displacement of maximum velocity with respect to the centre line, and is reflected in a downstream lag between bed profile variations and channel direction. For the Popo Agie River (Figure 7.9a) this is defined by

$$|\Delta\theta_j| = -6z_{j-1} + \epsilon_j \qquad (7.19)$$

where $|\Delta\theta_j|$ is absolute curvature. This is inversely related to bed elevation residuals from a linear trend (z_j) because maximum curvature occurs with minimum bed elevation. Hickin (1974) identified the importance of this lag in actively migrating Canadian rivers, where scroll bar deposition gives the growing point bar a ridge and swale topography. Figure 7.9(b) shows the

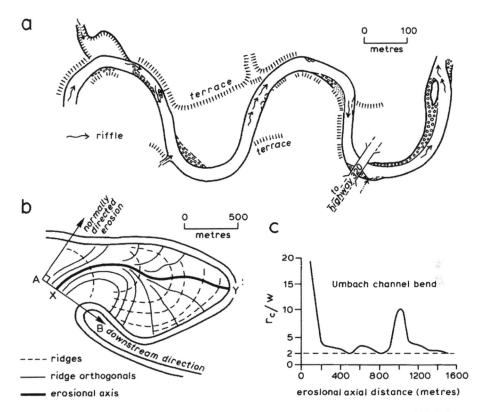

Figure 7.9 (a) Popo Agie River, a regular meander pattern with $\omega \approx 80°$ (after Leopold and Wolman, 1957). (b) Ridge and swale pattern on the Umbach bend, River Beatton, British Columbia. Fine continuous curves are ridge orthogonals; XY is the 'erosional axis' (longest orthogonal) and AB is the original channel orientation. Angle of XY with respect to AB defines relative importance of downstream and normally directed erosion (after Hickin, 1974). (c) Minimum radius of curvature:width (r_c/w) ratio for ridges of Umbach bend, related to distance along erosional axis (after Hickin, 1974).

pattern of ridges together with the orthogonals which indicate the direction of channel migration. The longest of these, the 'erosional axis', identifies the line of maximum migration, and its orientation indicates the relative importance of lateral and downstream migration. For a given bend it can be shown that, during migration, bend shape maintains a value of $r_c/w \approx 2$–3 (Figure 7.9c), and dendrochronological studies of trees on ridges and swales (Hickin and Nanson, 1975) show that migration rate is related to the r_c/w ratio. This echoes the results of Soviet investigations (Kondrat'yev, 1968) which characterize bend development by the 'angle of turn', α, that is, the change in channel direction between successive inflection points; bend migration tends to be fastest in bends with $\alpha \approx 200°$, which consequently have strongly asymmetric velocity fields at the apex.

The sedimentology of meandering streams

The predominant sedimentological unit in meandering rivers is the point bar, which differs in form, sediment sizes and structures both between and along streams as sediment supply and bend sinuosity vary, but which always retains the influence of transverse secondary currents. The general distribution of point bar facies in a gravel-bed river (e.g. the Endrick, in Scotland; Bluck, 1971) and its relation to current patterns are shown in Figure 7.10(a). At low flow, a pool current operates, with a transverse component rising across the convex inner bank. At high stages, the maximum current shifts to the inside of the bend and the bar head current dominates. The bar itself consists of two elements: the basal platform of gravel, which varies little around the curve, is subject to modification only in extreme events and is laterally continuous with the riffle; and the supraplatform deposits. These are falling-stage deposits of the bar head current and are more variable in space and time. Sediment size declines from bar head (gravel) to bar tail (sand) at a rate depending on the prevailing gravel:sand supply ratio, and structures change in successive floods. McGowen and Garner (1970) demonstrate a similar pattern of deposition in a sandier river (Figure 7.10b) where lower and upper bar surfaces occur separated by a break of slope (Bluck's 'inner accretionary bank'). In this case, sands on the lower surface are transported over a wider range of low flows and are dominated by cross-stratification. Hickin's (1969, 1972) studies of point dunes illustrate clearly the sedimentological distinction between active upper sands and stable platform deposits. Point dunes are actively migrating bedforms at the apex of the convex bank at high flow. Their transit across stable sediment is marked by a discontinuity in the trends of sediment size and sorting with depth beneath the dune surface (Figure 7.10c). A supply of mobile supraplatform sand appears to be necessary before lateral (transverse) bars (cf. Figure 7.8e) can become loci for point bar development.

Secondary flow produces a transverse, inward-flowing current at the bed, making an angle $\delta°$ with the main flow direction. Bed velocity, shear stress and power expenditure per unit bed area decrease down-current and thus up the point bar surface. Figure 7.10(d) shows both the skin-friction lines along which the shear stress resultant operates and the lines of equal power. Individual bed material particles migrate to that point where the transverse upslope component of the fluid force and the transverse downslope component of the particle weight are balanced, and then move parallel to the main flow. Since stream power declines inward, sediment sorting places larger particles near the thalweg where the higher fluid force balances their greater immersed weight, while smaller particles migrate to the upper bar. Since sand bedforms reflect stream power and particle size (Figure 1.3a; pp. 84–8), dunes occur in coarse sand where stream power is high and ripples develop in the fine sand higher up. Upper-regime plane beds may occur between the dune region and the channel lag deposits in the thalweg (Bridge and Jarvis, 1976). The point bar sediment size distribution has been

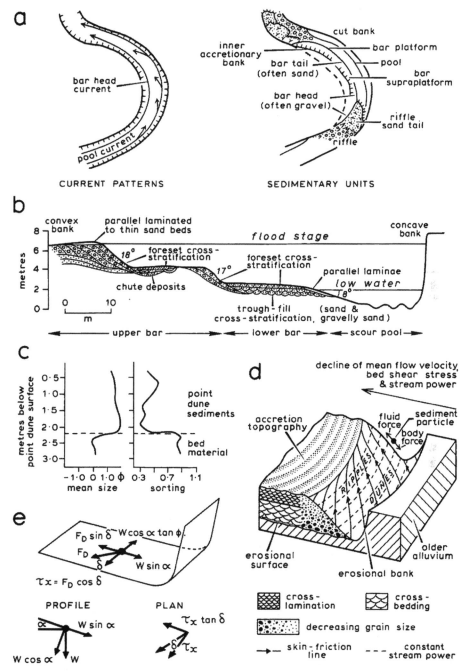

Figure 7.10 (a) Current pattern and depositional units in a meander bend (after Bluck, 1971). (b) Point bar structure, including chute deposits (after McGowen and Garner, 1970). (c) Vertical size and sorting variations in point dune (after Hickin, 1969). (d) Flow pattern, shear and power variation, force balance on particles, and sediment size sorting in a meander bend (after Allen, 1970a). (e) Definition diagram illustrating force balance on particles on point bar surface; F_D is drag force on a particle along skin friction line, τ_x is mean longitudinal shear stress.

theoretically modelled for given bend size and shape (wavelength, radius of curvature, channel width) and flow geometry (Bridge, 1976, 1977). For a circular bend, the transverse point bar surface is defined by

$$\tan \alpha = \frac{11d \tan \phi}{r} \qquad (7.20)$$

where α is the local surface slope, d the water depth and r the radius of curvature at that point, and $\tan \phi$ is the dynamic friction coefficient of the sediment. This equation is derived by balancing the forces acting on grains in both the longitudinal and transverse directions (Engelund, 1974) as in the definition diagram, Figure 7.10(e):

$$\text{longitudinally,} \quad F_D \cos \delta = W \cos \alpha \tan \phi \qquad (7.21)$$

and

$$\text{transversely,} \quad F_D \sin \delta = W \sin \alpha \qquad (7.22)$$

so that

$$\frac{W \cos \alpha \tan \phi}{\cos \delta} = \frac{W \sin \alpha}{\sin \delta} \qquad (7.23)$$

and therefore

$$\tan \delta = \frac{\tan \alpha}{\tan \phi}. \qquad (7.24)$$

Rozovskii (1961) developed an expression for δ in terms of d and r:

$$\tan \delta = \frac{11d}{r}, \qquad (7.25)$$

and by combining Equations (7.24) and (7.25) we obtain (7.20). With the transverse bed slope defined, equilibrium particle size is obtained by balancing the drag force on a particle by its immersed weight. The drag force is the upslope component of the mean bed shear stress, τ_x, multiplied by the projected area of the particle; since τ_x acts longitudinally, the upslope component is $\tau_x \tan \delta$. The downslope component of the immersed particle weight is the product of its volume, the density excess in water, the gravitational acceleration (which together give the immersed weight), and the local bed slope. The balance equation is therefore

$$\pi \left(\frac{D}{2} \right)^2 \tau_x \tan \delta = \frac{4}{3} \pi \left(\frac{D}{2} \right)^3 (\rho_s - \rho_w) g \sin \alpha. \qquad (7.26)$$

After substituting Equation (7.24) into (7.26) and rearranging, an expression for equilibrium particle size D at any point can be obtained:

$$D = \frac{3\tau_x}{2g(\rho_s - \rho_w) \cos \alpha \tan \phi} \qquad (7.27)$$

Here, particle sizes are seen to increase with mean shear stress and point bar transverse gradient. Since point bars are normally convex in cross-profile, particle size increases as the slope steepens towards the thalweg. This model predicts the broad pattern of grain size variation in a bend, but ignores the detailed effects of inertia and secondary circulation induced by bedforms (Dietrich *et al*, 1979). The former is important in that it delays the crossing of the shear stress maximum to the outer bank, displacing pool scour and bank erosion downstream and encouraging bend asymmetry. The bedform effect results in complex transport paths as grains zig-zag between troughs and crests, reversing their transverse direction as they are being spatially sorted.

The particle size segregation in river bends creates a characteristic stratigraphy in point bar deposits (Plate 5). As bends migrate, compensatory lateral accretion occurs on the inner bank, the sediment sometimes being derived from the upstream cut bank and carried within the appropriate secondary flow cell (Figure 7.4d). There is little evidence of material crossing the channel from outer to inner bank. Rapid migration results in scroll bar deposition, and the spacing between successive scroll ridges reflects the speed of migration (Hickin and Nanson, 1975). The stratigraphy of lateral accretion deposits displays the fining upwards of point bar surface sediments in the following sequence: channel lag gravels on an erosional surface representing maximum scour depth; coarse sand with large-scale cross-stratification recording dune bedforms; progressively finer sand with ripple cross-laminae; and silt and clay with vegetation remains reflecting vertical accretion by

Plate 5   A gravel-dominated channel-fill sedimentary unit, with slightly imbricated basal gravel, trough-bedded sands, and silty floodplain deposits arranged in a fining-upwards succession (after Picard and High, 1973).

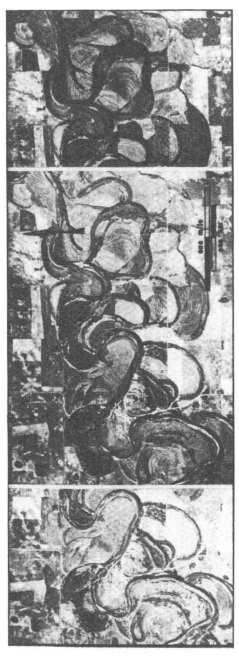

Plate 6 Stereo-pairs illustrating the alluvial deposits of the Notikewin River, Alberta. Ridge-and-swale microtopography, produced by scroll bar deposition on active point bar margins, is evident, as are abandoned channel traces (ox-bows) at various levels above the present river and various stages of infill (after Mollard, undated).

overbank flow. As meanders sweep back and forth during valley aggradation a succession of complete or partial 'fining-upwards cycles' is deposited, with this sedimentary sequence identified as the most probable by transition probability analysis (Allen, 1970b). However, Jackson (1976) suggests that the classic fining-upwards sequence requires fully developed flow with equilibrium between flow velocity, bed material sizes and bedforms. This may not occur in tight bends ($r_c/w < 1$) and only occurs in the latter half of bends with $r_c/w > 5$. In addition, stratigraphical details reflect both the prevailing balance between gravel and sand supply and the more local downstream sorting within the bend from bar head to tail. Other complexities in meander sedimentation include chute deposits (Figure 7.10b) comprised of gravel, sands and mudballs deposited during the falling stage as the bar head current wanes. Furthermore, in some rivers benches form against the *concave* cut-bank of sharply curved bends upstream from the point of maximum curvature, and fine sand and mud accretion at this site of expanded flow may contribute significantly to floodplain sedimentation (Page and Nanson, 1982). More generally, the floodplain deposits of a meandering stream include cut-off meander bends and avulsion channels through levees. Plate 6 illustrates the variability of surface alluvium, which may include lateral and vertical accretion, abandoned channel fill and backwater deposits. Fisk's (1944) study of the Mississippi valley is a classic interpretation of the complex sedimentology of valley fill created by active meandering.

Braided (multi-thread) channels

Braided channel patterns reflect particular environmental conditions, and are no longer considered necessarily to represent disequilibrium in aggrading systems. Generally, braiding is favoured by high-energy fluvial environments with steep valley gradients, large and variable discharges (e.g. proglacial streams), dominant bedload transport, and non-cohesive banks lacking stabilization by vegetation. The resultant wide, shallow cross-sections develop secondary circulation with multiple cells (Figure 7.4b), and bar formation occurs where flow converges at the bed. Braids may develop locally where chance bar emplacement or channel widening occur. Cause and effect is difficult to determine, but flume experiments by Leopold and Wolman (1957, p. 46) suggest that a bar of coarse sand diverts flow to cause bank erosion, and positive feedback then accentuates bar development and widening (Figure 7.11a). Transport occurs over the bar surface, which develops a tail of fine sand, while incision in the lateral channels lowers the water surface to expose the bar which then becomes dissected. The complex of islands is stabilized by vegetation in natural streams and experiences further high-stage sedimentation.

Distributary channels formed by braiding are less hydraulically efficient than parent single channels, and so the braid represents a major modification of flow patterns and energy losses. Table 7.2 shows that combined distributary widths are greater and depths less than in the single channel,

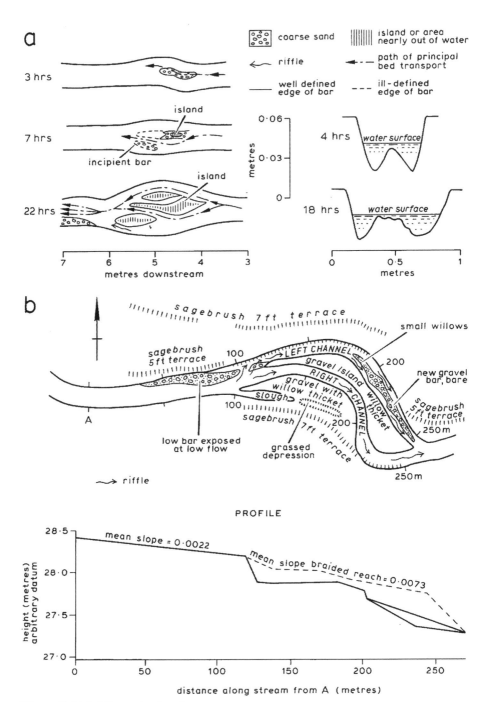

Figure 7.11 (a) Stages in the plan and cross-section development of a braided reach in a laboratory flume (after Leopold and Wolman, 1957). (b) Plan and profile of Horse Creek, near Daniel, Wyoming, showing local steepening of gradient in a braided reach (after Leopold and Wolman, 1957).

with increased friction compensating for steeper gradient to maintain comparable velocities. The adjustment of width is the inverse of the behaviour at tributary junctions noted by Miller (1958): if width varies downstream as the square root of discharge, then the sum of the widths of two equal distributaries is 1.41 times the width of the undivided parent channel. Maintenance of the stream power necessary for sediment transport demands increased slope in the divided reach (Figure 7.11b), and mutual adjustment of several hydraulic variables occurs in order to permit transport continuity despite lower water discharges in each channel. Although braided channel forms reflect physical transport processes, complex braided reaches may appear topologically random (Krumbein and Orme, 1972). Each link can be

Table 7.2 Ratio of hydraulic properties of divided and undivided reaches of braided streams (Leopold and Wolman, 1957).

Variable	Green River	New Fork River (reach A)	New Fork River (reach B)	Flume at Cal Tech				
				A	B	C	D	Mean
Cross-section area	1.30	1.03	1.60	0.94	1.08	0.78	1.07	1.11
Width	1.56	1.83	2.00	1.05	1.34	1.48	1.70	1.57
Depth	0.88	0.56	0.79	0.90	0.80	0.52	0.63	0.73
Velocity	0.77	0.97	–	1.06	0.93	1.27	0.93	0.99
Slope	5.70	2.30	1.40	1.30	1.40	1.90	1.70	2.24
Friction factor, f_f	10.50	1.30	–	1.10	1.30	0.63	1.25	2.68

defined according to its starting and terminating node (fork F, or junction J), and under the assumption of random channel link combination, the probabilities of link types are FF = 2/9, FJ = 4/9, JF = 1/9 and JJ = 2/9; natural braid forms display relative link frequencies similar to these expected values. The equilibrium braided reach may experience shifting bar forms and distributary abandonment, but on average, over a period of years, total sinuosity or wetted perimeter area is maintained. Locally, braided reaches may sustain this equilibrium to maintain transport continuity after chance bar formation or local widening at an area of erodible bank material. More generally, high-energy streams may create braided channel patterns as a result of the instability of bedload transport in wide, shallow channels (Parker, 1976). The hydraulic inefficiency of this pattern then permits the excess energy to be dissipated.

Braiding and the energy continuum

Several studies have discriminated between straight, meandering and braided streams on the basis of discharge and channel slope (s_b). Lane (1957) defined an intermediate 'transitional' pattern between meanders and braids, and showed that braids replaced 'transitional' streams if

$$s_b > 0.004Q_m^{-0.25} \tag{7.28}$$

where Q_m is the mean annual discharge. Using bankfull discharge data, Leopold and Wolman (1957) produced the discrimination defined by

$$s_b = 0.013Q_b^{-0.44} \tag{7.29}$$

which also predicts braids at higher slopes and discharges (Figure 7.12a). Antropovskiy (1972) developed a similar function for bankfull discharge conditions; the steeper gradient of these two functions compared with Equation (7.28) may reflect a tendency for the peak flood:mean flow ratio to be higher in smaller, flashier catchments. These results all imply a higher power expenditure rate in braided streams, a conclusion reinforced by Schumm and Khan's (1972) flume experiments. Here, equilibrium conditions were maintained by balancing sediment feed with sediment output, and at constant discharge but changing slope, thresholds occurred in the sediment transport–flume slope relationship at critical gradients when the pattern changed from straight to meandering and then to braided (Figure 7.12b). If sinuosity is plotted as a function of flume slope (Figure 7.12c) an artificial threshold occurs when the initial moulded channel begins to meander, and then another pattern threshold is reached when braiding begins. Because Schumm and Khan (1972) defined a braided channel as one of low sinuosity, this diagram indicates a reduction of sinuosity; had 'total sinuosity' been measured, a monotonic increase (cf. Figure 7.1d) might have resulted (slope variation measuring power variation at constant discharge).

One improvement in Schumm and Khan's (1972) study relative to the various channel slope–discharge discriminant functions is its use of flume (= valley) slope as the independent variable. Maximum power expenditure per unit length depends on valley gradient (Table 1.3), while channel gradient is not an independent variable, being altered by adjustments of channel pattern. However, none of these investigations recognizes the control of channel pattern by sedimentology. Since bed material transport and bar formation is necessary in both meander and braid development, the threshold between patterns should relate to bedload calibre. Henderson

Figure 7.12 (a) Discriminant function distinguishing meanders and braids on the basis of bankfull discharge (Q_b), channel slope (s_b) and median bed material size (after Leopold and Wolman, 1957). The broken lines are defined by Equation (7.30). (b) Thresholds of sediment transport and channel pattern in a flume channel at increasing slopes (after Schumm and Khan, 1972). (c) Variation of sinuosity with flume slope (after Schumm and Khan, 1972). (d) Single-thread and multi-thread channels discriminated by a stream power index (Ω') and median bed material size (D_{50}). (e) Straight, meandering and braided streams defined by the slope: Froude number (s_b/Fr) and depth:width (d/w) ratios (after Parker, 1976). (f) A cusp catastrophe model of the response of channel pattern to total stream power (Ωp) and bank resistance (B = silt–clay percentage) controls.

(1961) re-analysed Leopold and Wolman's (1957) data to derive an expression including median grain size (mm),

$$s_b = 0.0002 D_{50}^{1.14} Q_b^{-0.44} \qquad (7.30)$$

which suggests a higher threshold slope is necessary for braiding in coarse bed material (Figure 7.12a). An alternative illustration of this is the

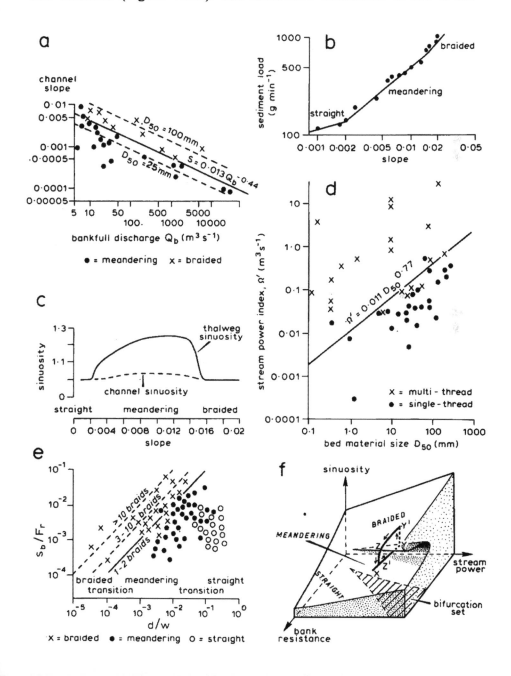

discrimination between single-thread and multi-thread channels (Figure 7.12d) based on the bankfull discharge–channel slope index of power per unit length, and median grain size. The threshold is

$$\Omega' = 0.011 D_{50}^{0.77} \tag{7.31}$$

which implies that the power needed for braiding to occur increases with bed material size.

Bank material resistance affects rates of channel migration and should also influence the threshold, although its effect may be difficult to quantify and may also be nonlinear since greater stream power is required to erode clays and cobbles than sands. Parker's stability analysis (1976) indirectly illustrates the effects of bank material resistance by defining the meander–braid threshold

$$s_b / Fr = d/w \tag{7.32}$$

where Fr is the Froude number. Deep, narrow channels maintain double cell secondary circulation which may lead to meander development and are indicative of resistant banks (Figure 7.12e). Wide, shallow streams (weak, sandy banks) tend to be braided (Engelund and Skovgaard, 1973). Parker's threshold is geomorphologically unhelpful as a criterion of channel pattern, since it requires channel depth and width to be known. However, depth, width and Froude number may all be expressed in terms of discharge and bank silt–clay percentage by using the equations based on Schumm's (1960b, 1968a) data (see p. 170). Ferguson (1984) has suggested substituting these and other alternative hydraulic geometry relationships into Equation (7.32), to give a threshold equation of the form

$$s_b = 0.0028 Q_b^{-0.34} B^{0.90} \tag{7.33}$$

which predicts steeper threshold slopes for braiding in channels with resistant silty banks. Thus the pattern continuum begins at low power with 'straight' streams in resistant sediments. Ackers and Charlton (1970b) and Nagabhushanaiah (1967) have both identified a straight-meandering threshold in flume channels with straight streams occurring at lower slopes and discharges. Intermediate power and width:depth ratios characterize meanders, and high-power streams in erodible sediments that develop wide, shallow cross-sections are likely to be braided. In fact, braid intensity increases above the braid threshold: Howard *et al* (1970) demonstrate this by using various topological indices (see Table 1.2 and Figure 1.4d) which reveal that braided reaches show greater complexity (more bars and distributaries) in the highest-energy environments. This confirms the suggestion that the braided pattern is optimal for the dissipation of excess energy in high-energy streams, since the enhanced total flow resistance of the multi-thread channel results in rapid energy loss.

The continual spatial variation of channel pattern in response to varying energy conditions suggests the need to quantify pattern by a continuous variable, such as total sinuosity, in contrast to the use of qualitative discrete pattern definitions. This continuous system state (dependent) variable may then be related to two control (independent) variables measuring total stream power and sediment resistance. For a particular combination of control variables, the equilibrium total sinuosity minimizes an energy criterion (Chang, 1979). For example, the power per unit length is at a maximum in a straight stream, but meandering or braiding initiated by the bed material transport resulting from the high power expenditure reduces the gradient and therefore the actual (as opposed to potential maximum) power expenditure. Since width and depth also vary as channel pattern changes (Figures 7.12e), unit stream power (Yang, 1971c; Table 1.3) is a more appropriate energy criterion, but the nonlinear variation of width and depth with total sinuosity may mean that, for some combinations of power and resistance, two alternative equilibrium sinuosities exist. Thus, the surface of equilibrium sinuosities may be double-valued at some points and resembles a cusp catastrophe (Poston and Stewart, 1978). This is the type of surface developed in catastrophe theory, which is a branch of mathematics concerned with discontinuous system behaviour, to describe such behaviour in systems characterized by a single dependent and two independent variables. Figure 7.12(f) shows a hypothetical cusp catastrophe model of channel pattern. Projection of the total sinuosity surface onto the $\Omega p - B$ plane defines a 'bifurcation set' outlined by the folds in the surface, where meandering and braided patterns overlap (see Figure 7.12d). This bimodal system behaviour arises from the delayed response to changes in control variables. At high stream power, a reduction in bank resistance results in increased sinuosity, slowly at first while single-thread channels persist, then abruptly when the fold is reached (path X–Y–Y'). Hysteresis in behaviour arises if the bank resistance increases again, with sinuosity variation following the path Y'–Z–Z'. This qualitative model could explain the abrupt change in the pattern of the Cimarron River (Schumm and Lichty, 1963) and the delayed response in reverting to the original state (pp. 253–4).

Braided stream sedimentology

Braided streams are active, high-energy sediment transport systems which frequently re-form and destroy sedimentary structures. Classic proglacial gravelly braided streams such as the White River (Fahnestock, 1963) and the Lewis River (Church, 1972) constantly switch from channel to channel, although main channels may be relatively fixed compared to secondary distributaries. In Alaska, the adjacent Knik and Matanuska Rivers display behavioural contrasts (Fahnestock and Bradley, 1973): the former has finer sediment and suffers catastrophic floods following proglacial lake drainage, but the greater instability of the latter braided stream arises because lower-magnitude floods generate deeper flows over the narrower outwash plain, and the outwash is five times steeper. Werritty and Ferguson (1980) demonstrate rapid migration in the gravelly River Feshie, Scotland (see Plate 7), partly through bank erosion but mainly because of channel switching; lack of fines prevents infill of old channels which remain topographic lows in the floodplain. Downstream bar migration in the main channel eventually encourages scour of the sediment blocking the entrance to the old channel, which then becomes re-occupied. Distributary closure arises initially because at different flow stages there are changes in the location of maximum competence to transport bedload between the distributary and main channels (Cheetham, 1979). At low flow, when competence is generally low, it tends to be higher in the distributary channels from which sand is therefore scoured. At high flow, when cobbles can be moved, competence is

Plate 7 A gravelly braided stream: the River Feshie, Scotland.

higher in the main channel, so that bedload entering a distributary is deposited to produce upstream bar accretion and the eventual choking of the channel entrance. In the sandy braided stream (e.g., the Platte River), instability arises because of wholesale bar dissection and destruction, since the bedload is here mobile over a wider range of flows.

Sediment sizes are extremely varied in proglacial braided streams. Rust (1972) quotes a variation from -7 to $+8\,\phi$ units arranged in two trends – the proximal–distal and the active zone–stable zone trends which both involve sediment fining. For the Donjek River, Williams and Rust (1969) identify seven facies ranging from silt–clay low-energy deposits, such as those in abandoned channels, to high-energy gravels on active bar surfaces and in migrating channels (Doeglas, 1962). Miall (1977) provides a detailed review of braided stream alluvial facies which are defined in Table 7.3 together with the sediment transport and deposition processes mainly responsible for each one. Cyclic sedimentation patterns in braided streams are more varied than

Table 7.3 Sedimentary facies in braided stream alluvium. Based on Miall (1977).

Facies types	Process types	Typical successions
Gm: massively bedded gravel	1	1. *Gravelly braided stream facies*
Gt: trough cross-bedded gravel	3	ancient environments[†]
Gp: planar cross-bedded gravel	2	/Gm ⇌ Sh, St ⇌ Sr, Fl/
St: trough cross-bedded sands	2	modern environment (Alaskan outwash)
Sp: planar cross-bedded sands	2	Gm → Sh → Sr → Fm
Sr: ripple cross-laminated sands	4	2. *Sandy braided stream facies*
Sh: horizontally bedded sands	2, 4	ancient environments[†]
Ss: scour–fill sands	3	/St ⇌ Sp → Sr, Fl/
		↑
		↘ Ss
Fl: laminated sand–silt–mud	4, 5	
Fm: mud or silt drapes	4	modern environment (ephemeral streams)
		/St → Sp → Sp, Sh, Sr
		↑
		↘ Fm

Processes responsible for each facies are: 1, longitudinal bar formation and sediment accretion; 2, bedform generation and migration; 3, channel scour and fill; 4, low-water accretion; 5, sedimentation in overbank areas.

Typical cyclic sedimentary successions are defined by † if based on Markov chain analysis. / = scour surface; broken arrows are less common successions.

in meandering streams, and include coarsening-upwards cycles caused either when bar head gravels migrate downstream and overrun the bar tail sands, or when re-occupation occurs of abandoned channels. Typical vertical sequences in Table 7.3 reflect gravel-dominated proglacial braided streams and sandy, semi-arid ephemeral (flash-flood-controlled) streams. The examples from ancient environments are based on Markov chain analysis of several cycles (Miall, 1973), whereas the modern cycles are based on the limited evidence of contemporary deposition. The relative import-ance of each facies changes downstream as the gravel:sand ratio is altered by sorting processes, and the nature of vertical sequences reflects this adjust-ment. Proximal–distal trends are often complex in sandur deposits because recent meltwater supply during glacier retreat has resulted in a slight incision to leave terrace surfaces; the occurrence of degrading braided streams such as the Donjek gives the lie to the association of braiding and aggradation. Ballantyne (1978) showed that 40% of clast size variance in a small Canadian Arctic sandur is explained by down-sandur fining, with further variance explained by local elevation-controlled differences. Channel deposits are coarser than the local average, and bar deposits finer, but the difference lessens downstream as overall sorting improves. This pattern is repeated on a high-level surface, but not on a seasonally inundated intermediate surface.

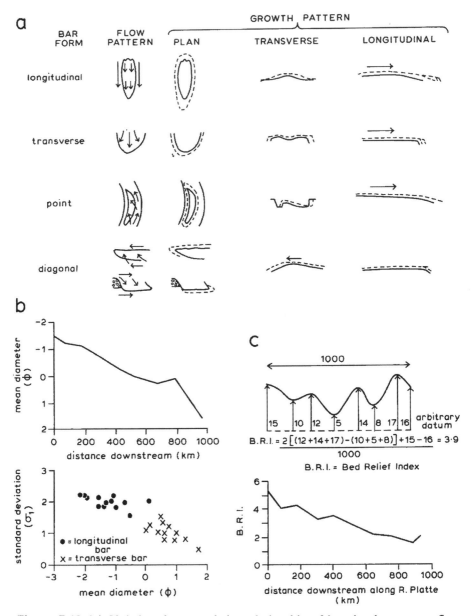

Figure 7.13 (a) Unit bar forms and the relationship of bar development to flow patterns (after N. D. Smith, 1974). (b) Downstream trends in sediment size and sorting and bar type in the River Platte (after Smith, 1970). (c) Derivation of the bed relief index (BRI) and its downstream variation in the River Platte (after Smith, 1970).

Braided channel bars influence local facies change, but themselves reflect proximal–distal sediment size trends. Bars in gravel-bed rivers are predominantly longitudinal (95% in the Donjek River), although Church (1972) identifies spool bars in the Lewis River which are similar to longitudinal bars after lateral accretion and dissection (cf. the final stage in Figure 7.11a). Transverse bars occur in the sandier Platte River (Smith, 1970) and develop from one bank by avalanche face extension on downstream and lateral margins. Active transport over the bar surface creates a within-bar size decline and a change from dune to ripple bedforms and associated stratification, reflecting a down-bar decline in stream power. Diverging channels are cut as flow stage drops and the bar is dissected. Diagonal bars are aligned perpendicular to the flow and are formed of horizontal gravel strata. These forms of simplified 'unit bars' (N. D. Smith, 1974) are summarized in Figure 7.13(a). Other types include linguoid bars (Collinson, 1970) which are large-scale ripple-like forms in sandy sediments which migrate downstream as the bedload avalanches down the lee face. Because of the reverse flow that may occur in the lee eddy downstream from these large features, small ripples may form which travel upstream (counter-current ripples) until they are buried by the advancing linguoid bar. Directional structures in bar sediments reflect the variations of current direction at different stages (Bluck, 1974). Bar long axes and gravel imbrication orientations parallel dominant high-stage flow directions, whereas sand cross-strata orientations vary in relation to falling-stage currents diverted by the main bed structures (e.g. deltaic wedges oriented into secondary channels). Ripple cross-laminae, however, may be less variable since they form at low flows confined to the main channel. Directional structures thus vary in dispersion with particle size and bedform type, overall bar structure and downstream sediment size trends. Smith (1970) illustrates the correlated downstream trends in sediment size and sorting, bar type, bed relief and stratification in the Platte River. Cross-section bed relief is measured by the index (Figure 7.13c).

$$\text{BRI} = \frac{2[(T_1 + T_2 + \ldots + T_n) - (t_1 + t_2 + \ldots + t_n)] \pm T_{e1}, T_{e2}}{L} \quad (7.34)$$

where the T_i are height maxima between hollows, the t_i are minima between peaks, L is transect length and the T_e are end heights (added if they adjoin a low, subtracted if they adjoin a high). Figures 7.13(b) and (c) show the downstream sediment size decline and improved sorting, and the replacement of longitudinal by transverse bars, which are associated with less variable bed topography. The proximal–distal trends in sediment properties are therefore reflected in the sediment structures and in the bar and channel morphology.

8

Channel gradient
and the long profile

Long-profile characteristics, which constitute the third adjustable dimension of alluvial channel morphology, are strongly scale-dependent. Bedforms such as the riffle–pool sequence and large-scale dunes cause local, within-reach gradient variation (Figure 1.4e). However, mean reach gradients are averaged over this local variability and are mutually adjusted to cross-section and plan-form properties in the medium-term steady-state time scale. Consecutive reach slopes combine to create the complete long profile which, at broader temporal and spatial scales, reflects long-term geological development influenced by tectonic and morphogenetic histories as well as the recent channel pattern adjustments. For example, the valley fill gradient inherited from palaeochannel deposition controls the present rate of potential energy expenditure, and therefore the channel pattern, sinuosity and gradient. Figure 8.1(a) illustrates the resulting profile and plan-form inter-relationship between valley and channel gradients and sinuosity (Schumm, 1968a). In normal valleys which widen downstream, equilibrium channel gradients are less than or equal to valley gradients (Figure 8.1b) and the $P = 1$ sinuosity curve in Figure 8.1(a) is therefore a limiting curve. Channel gradients steeper than the valley gradient only occur in entrenched streams which are flushing the valley fill deposits as they approach a new equilibrium (Salisbury, 1980). Meander development and increased sinuosity on a steep valley surface lengthens the long profile and alters the overall long-profile shape, which is therefore both an independent control in that it records the history of fluvial processes, and a dependent variable in that it incorporates contemporary steady-state adjustments.

The long profile is the least transient expression of fluvial processes, reflecting as it does geological influences such as the effect on available relief

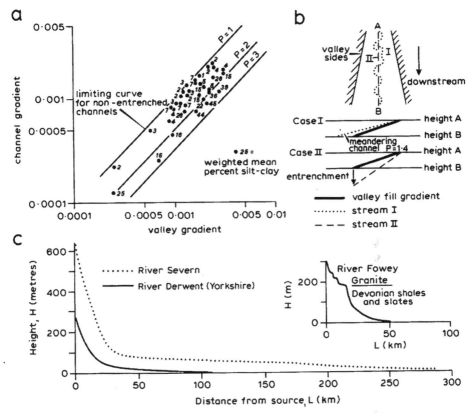

Figure 8.1 (a) The interrelationship between channel and valley gradients and sinuosity. Each point is labelled with its silt–clay percentage (after Schumm, 1968a). (b) Meandering and entrenched channel bed gradients in relation to valley gradient. (c) Height–distance diagrams of the long profiles of three British rivers. The profile of the River Fowey in Figure 8.3(b) is that part confined to the granite outcrop.

of tectonic history and base-level change, of climatic change on the processes of erosion and deposition and of the distribution of outcrops of different lithologies. However, perennial rivers tend to develop a broadly concave-upward form (Figure 8.1c), and a concavity index describing profile form illustrates some of these influences. For example, Wheeler (1979) used the index C_w (Table 1.2) to show that profile concavity increases with relief (R, m) in a sample of 115 British rivers according to

$$C_w = 0.28 + 0.00039R. \tag{8.1}$$

Excessively concave profiles (positive residuals from this regression) occur in rivers with extensive Quaternary or Flandrian estuarine deposition. 'Underconcave' profiles have mid-profile convexities associated with resistant outcrops, or are markedly linear on chalk outcrops where discharge increases slowly downstream. Geological control of profile concavity is also shown by the direct correlation between concavity and the percentage of granite and gneiss outcropping along tributaries to Sandy Creek in central Texas (Shepherd, 1979). Here, one 'under-concave' tributary, Crabapple Creek, has not yet adjusted its profile to the geologically recent capture of head-waters from an adjacent basin.

The general concavity of perennial river profiles plotted in the Cartesian co-ordinates of 'height, H' and 'distance from source, L' (Figure 8.1c) may be described by a range of exponential functions. At any point the local slope s is the first derivative of the profile curve ($s = -dH/dL$), and an equation for the profile is therefore obtained by integrating an appropriate slope–length relationship (Hack, 1957, Brush, 1961). If slope is a decreasing power function of length,

$$s = -dH/dL = k_1 L^{-n} \qquad (8.2)$$

which on integration yields

$$H = \frac{-k_1}{1-n} L^{1-n} + C \qquad (8.3)$$

where C is a constant of integration. The slope–length relation is a rect-angular hyperbola in the special case where $n = 1$ (Hack, 1973a, Knighton, 1975b), and the profile equation after integration is

$$H = -k_1 \ln L + C, \quad \text{or} \quad L = \exp[(C-H)/k_1] \qquad (8.4)$$

which plots as a straight line on semi-logarithmic graph paper with a loga-rithmic length axis. Alternatively, the local gradient may vary as a function of altitude (Tanner, 1971), so that

$$s = -dH/dL = k_2 H. \qquad (8.5)$$

On integration, this gives the profile

$$\ln (H/C) = -k_2 L, \quad \text{or} \quad H = C \exp(-k_2 L) \qquad (8.6)$$

which plots as a straight line on graph paper with a logarithmic height axis, implying that the profile is asymptotic to the divide and intersects the profile base level – the converse of the profile defined by Equation (8.4).

Smooth, concave long-profile curves such as are implied by these expo-nential functions were considered diagnostic of the graded state (Davis, 1902b). However, the concept of grade has little relevance to the analysis of steady state alluvial channels (pp. 18–23). Firstly, this is because the graded time scale is one in which dynamic equilibrium occurs, since con-tinual progressive adjustments of the long profile take place as relief, sediment calibre and runoff decrease during the cycle of erosion (cf. Figure 1.6a): '. . . in virtue of the continual, although slow, variations of stream volume and load through the normal cycle, the balanced condition of any stream can be maintained only by an equally continuous, although small, change of river slope' (Davis, 1902a, p. 96). Secondly, the Davisian concept of grade fails to acknowledge the multivariate nature of river channel behaviour. Long-profile (gradient) adjustments are emphasized as being necessary to maintain sediment transport with the available discharge and given channel characteristics (Mackin, 1948). However, mutual adjustment of gradient, plan-form and cross-section properties characterize the true response of alluvial streams to the multivariate environmental controls of runoff, flood magnitude and frequency, sediment yield and sediment calibre. The interaction of these controls determines the gradient at the reach scale, and their downstream variation determines the spatial adjustments of gradient which create the complete long profile. Only in special, limited circumstances is this likely to result in a smooth, concave long profile capable of generalization by a bivariate mathematical function (Wolman, 1955).

Channel gradient and discharge

The inverse relationship between channel gradient and discharge recognized by Gilbert (1877) explains the concavity of the long profile of perennial rivers, since their tributary inflows cause a downstream increase of discharge which enables the sediment load to be transported on progressively lower slopes even if it remains constant in calibre and quantity. The long profile adjusts by a combination of spatially varied incision and channel pattern changes, the concavity being more accentuated if the downstream increase of discharge is rapid. Notwithstanding the temporal variability of natural discharge, which results in frequent short-term local re-adjustments of gradient to maintain transport capacity in spite of scour and fill (Kesseli, 1941), this relationship represents the *general* adjustment of gradient to *average* discharge conditions.

Single laboratory channels in uniform sand develop linear profiles, but if tributary junctions are created that cause discharge to increase along the main stream, a concave profile evolves (Lewis, 1945). At a larger scale, Langbein (1964b) notes that profile concavity is much less in exogenous streams such as the Nile and Colorado which do not increase appreciably in discharge downstream, than in normal perennial rivers. Their concavity increases with 'specific length', measured by $L/\sqrt{Q}$, where L is length and Q mean discharge, probably because longer rivers experience a greater downstream fining of sediment and therefore can maintain transport on lower slopes with a constant discharge. Enough gauging stations exist along some British rivers to allow estimation of a discharge–basin area power function whose exponent is positively correlated with the profile concavity (Wheeler, 1979). Thus, when discharge increases rapidly downstream with increasing contributing area, profile concavity is greater. Since the long profiles of

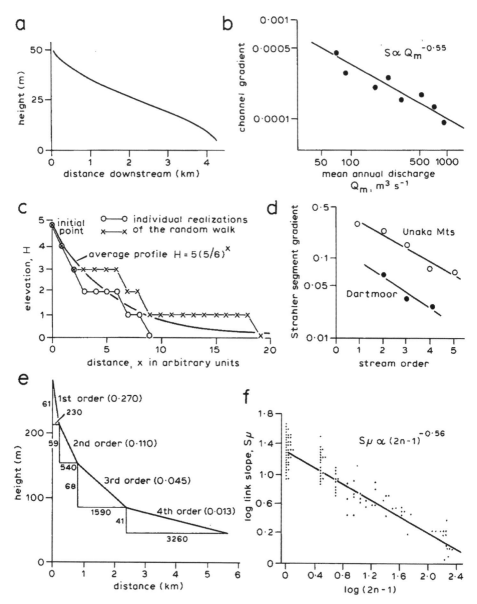

Figure 8.2 (a) The convex long profile of a small semi-arid stream (after Schumm, 1961c). (b) The slope–discharge relationship for an equilibrium alluvial channel, the Red River in Louisiana/Arkansas (after Carlston, 1968). (c) Individual random walks and the mean of all random walk profiles unconstrained in length (after Leopold and Langbein, 1962). (d) Slope–order relations for basins in the Unaka Mountains and Dartmoor (after Chorley and Morgan, 1962). (e) An average long profile of the Salt Run, Pennsylvania. Figures are mean stream segment lengths, falls (m) and gradients (after Broscoe, 1959). (f) Link slope–magnitude relationship for Cooks Run, Pennsylvania (after Flint, 1974).

these rivers have evolved over considerable periods, their apparent correlation to present hydrological conditions suggest that climatic gradients in Britain have been remarkably persistent. A final piece of evidence linking profile shape to discharge is the form of profile in some semi-arid streams (Figure 8.2a). Discharge often decreases downstream because of evaporation loss and infiltration into their sandy beds; transmission losses (Burkham, 1970) of up to 75% of flood discharges over 25–100 km occur in such perched streams (Thornes, 1977). In consequence, the profile form reflects the inverse discharge–slope relation by being convex-upward as a result of aggradation by the individual rare flood events (Schumm, 1961c).

If stream gradient varies with discharge, it may be treated as part of the downstream hydraulic geometry (pp. 155–61), being related to discharge by

$$s = gQ^z. \tag{8.7}$$

The long profile associated with a given slope–discharge relation will, of course, depend on the rate of increase of discharge with stream length, and so the exponent z is not in itself a measure of profile concavity. Woodford (1951) fitted curves with $z = -0.5$ to several sets of data, but exponents vary between rivers (Wolman, 1955). Carlston (1968) found that well-defined slope–discharge relationships with an average exponent of $z = -0.65$ are characteristic of alluvial channels, and distinguish these rivers from bedrock channels whose slopes are affected by geological structures. An example of such a relationship is shown in Figure 8.2(b). The modal exponent may be theoretically predicted by considering the alternative contradictory energy constraints which influence the development of equilibrium river morphology (Leopold and Langbein, 1962, Langbein and Leopold, 1964). Channel forms may adjust until the energy expenditure per unit bed area is equalized. For example, if cobble grain roughness is replaced downstream by sand bedform roughness, the width and depth may adjust so that the frictional energy loss is uniform, which requires that power expenditure per unit bed area ($\gamma_w Qs/w$) is constant. Since the average width–discharge relationship is $w \propto Q^{0.5}$, this implies that $\gamma_w Q^{0.5} s$ is constant, and therefore that $s \propto Q^{-0.5}$. However, minimization of total work in the fluvial system requires that the sum of power expenditure per unit channel length over a sequence of incremental reaches ($\Sigma \gamma_w Qs$) is minimized. This is easily achieved if power expenditure is the same in every reach, which requires that $s \propto Q^{-1.0}$. Both conditions cannot be satisfied simultaneously, and so streams develop on average a profile which provides a compromise, with $z \approx -0.75$, and the equilibrium is a shifting quasi-equilibrium as the relationship between slope and discharge fluctuates between the two opposing limits. A limitation of this theory is, however, its assumption that channel gradient is as adjustable a dependent variable as width, depth, sinuosity and meander wavelength, when in part it represents an inherited constraint on the development of equilibrium morphology – as suggested by Figure 8.1(a).

This predicted average long-profile form is one of maximum probability between two improbable extremes which may nevertheless represent individual profile forms developed under particular conditions of downstream discharge and sediment size variation. The average, most probable profile form may be simulated by a random walk model (Leopold and Langbein, 1962, Langbein, 1964b), in which the probability of a downward step from a given elevation H (Figure 8.2c) is a function of H_0, the initial profile height, that is $p = H_0/(1 + H_0)$. A single random walk long profile is one realization of a stochastic process, but on average the mean of all possible realizations is a smooth exponential concave profile with the equation

$$ H \propto H_0(p)^L \propto H_0 \left(\frac{H_0}{1 + H_0} \right)^L. \tag{8.8} $$

This form of random walk profile implies that height is an exponential function of distance, as in Equation (8.6) where slope is a function of elevation H. Individual realizations of the random walk (Figure 8.2c) vary in length, there being no constraint such as that provided in reality by the position of the river mouth. Random walks may, however, be constrained in length so that all realizations terminate at a fixed downstream point, in which case the probability of a downward step decreases with increasing total length L, and $p = L/(1 + L)$. This gives

$$ x = L(p)^H \tag{8.9} $$

in which elevation varies as the logarithm of distance downstream (x) as in Equation (8.4), the commonest form of long profile in which slope is an inverse function of length.

Slope–discharge relationships cannot themselves indicate the long-profile form without information on the variation of discharge downstream along the main channel, which necessitates the consideration of drainage network structure, tributary junctions, and catchment area increments (pp. 33–7). The implicit continuity of slope–discharge functions smooths the discrete changes of slope occurring at junctions. Miller (1958) showed that when comparably sized rivers join, the stream gradient below the junction averages approximately one-third of the sum of the two tributary gradients. If each stream segment obeys Equation (8.7), it follows that

$$ (gQ^z + gQ^z)/3 = g(2Q)^z \tag{8.10} $$

if two equal tributaries of discharge Q join; the solution of this equation is $z = -0.58$, which indicates rough agreement with the conclusions of Carlston (1968) and Langbein and Leopold (1964). More general channel slope relationships with the drainage network are summarized by slope–order plots (Figure 8.2d). Channel gradients in the Unaka Mountains, North Carolina, are consistently steeper than in Dartmoor, on an order-by-order comparison, because of their fine-textured networks which reflect higher runoff rates (Chorley and Morgan, 1962). However, the parallelism of the plots illustrates similar *relative* adjustments between orders in the two areas. These relationships apply to mean gradients for streams of each order, and by combining slope–order and length–order relations (Figure 8.2e) a smooth, generalized long profile may be defined for a given catchment (Broscoe, 1959). This reiterates the idea that smooth concavity is a generalized, average condition, whereas any single path from a headwater to a stream mouth may be irregular in gradient as sediment calibre, discharge and lithology (Morisawa, 1962) vary and as sediment supply from valley side slopes (Chapter 2, Figure 2.1b) alters and is accommodated by variations of channel cross-section form as well as gradient. Furthermore, a Strahler stream segment includes several minor tributaries which cause increased discharge along it, so that a more satisfactory channel slope–network relation is that between link gradient, s_μ, and link magnitude, n (Flint, 1974):

$$s_\mu = k(2n - 1)^{z_1} \tag{8.11}$$

where z_1 ranges from -0.37 to -0.83 in humid temperate basins. This implies that gradient is a link-associated variable (Figure 8.2f), like width (pp. 160–1), with slope adjustments along the main stream occurring predominantly where roughly equal tributaries join to cause a sharp increase in magnitude and therefore discharge.

Channel gradient and bed material

Since channel form, including the long-profile dimension, adjusts so that streamflow can maintain sediment transport, the calibre of sediment load is an additional variable related to gradient. However, whereas discharge is clearly an independent environmental control, causation is less obvious in the gradient–sediment size relationship. With uniform discharge conditions, a direct relationship exists between slopes and bed material particle sizes. This *could* arise because gradient adjusts to independently produced particle size variation; abrasion during transport reduces grain sizes, and the gradient required to maintain transport therefore decreases. Alternatively, however, sorting of particles would result in a similar direct correlation, since as gradient decreases downstream, coarser particles are deposited when they cannot be transported over the lower gradients. These alternative hypotheses suggest a feedback relationship between sediment size and gradient rather than simple causation, with neither being strictly an independent variable in relation to the other.

Sternberg's (1875) law of particle abrasion during transport, experimentally verified by Schoklitsch (1933), states that an exponential weight loss occurs with transport distance L from an initial weight g_0, according to

$$g = g_0 \exp(-aL) \qquad (8.12)$$

where a is a coefficient of abrasion. If slope is proportional to grain size (Shulits, 1941), it follows that

$$s = s_0 \exp(-aL) \qquad (8.13)$$

where s_0 is the headwater slope. On integration this yields an exponential profile form whose shape varies with the coefficient of abrasion. However, since this coefficient varies downstream with stream velocity, particle diameter and grain shape (Shulits, 1941), the profile is not necessarily a simple exponential. Krumbein (1941b), for example, showed that weight reduction is initially rapid as angular particles break along cleavage planes, then slows as abrasion gradually reduces the size of smaller, more spherical particles. The potentially complex relationship between grain size and gradient is illustrated by Yatsu's (1955) study of thresholds in the long profiles of some Japanese rivers. Cobbles collapse into their mineral grain constituents over a narrow transport distance, and particles of intermediate size are deficient. The result is a long profile characterized by a marked break between two exponential curves (Figure 8.3a), with a low-frequency riffle–pool oscillation about the upper curve and a high-frequency sand-dune variation about the lower curve (Bennett, 1976). The sudden reduction of slope correlated with the occurrence of unimodal sandy bed material creates a long profile of pronounced concavity.

In other cases, however, the downstream decrease in particle size is demonstrably the result of sorting, controlled by the varying gradient. The Knik river valley train decreases exponentially in height from the Alaskan glacier at its head (Bradley *et al*, 1972), and mobile bed material diameter decreases by 87% in a 25 km reach with no significant sediment or water inflow, the granite and greywacke pebbles becoming more platey downstream. This change in shape is because platey pebbles of a given intermediate axis are lighter and more transportable in such a turbulent environment. The decline in size is also interpreted as a sorting effect, and is supported by abrasion-tank experiments which only cause an 8% reduction in size in a comparable travel distance. Theoretical evidence of sorting as a mechanism causing grain size diminution is provided by Rana *et al* (1973). For given long profile, discharge and input sediment size distribution, point-to-point sediment transport is calculated for each size grade using the Einstein bedload function. The result is an exponential bed material size reduction in the downstream direction caused by the downstream decline in flow competence as the long-profile slope diminishes. This decline in size may be modelled by an equation similar to (8.12), but the coefficient a must obviously be interpreted as a sorting index.

In reality abrasion and sorting processes are interrelated, and the coeffi-

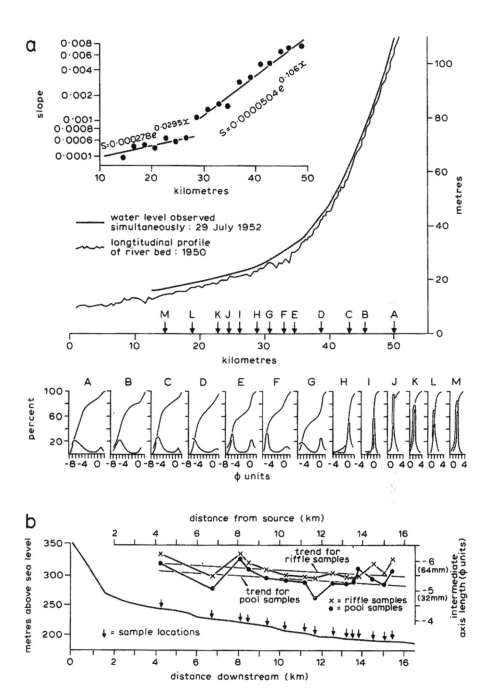

Figure 8.3 (a) The long profile of the Watarase River, Japan, and some examples of the bed material grain size distributions showing the bimodality of particle sizes in the upper reaches (after Yatsu, 1955). (b) The long profile of the River Fowey in Cornwall, and the mean intermediate axis length of bed material in riffle and pool sections.

cient a is the sum of two components, $a_1 + a_2$, which respectively define the size reduction caused by abrasion and sorting and which vary in relative importance (Tanner, 1971, Knighton, 1980). Since bed material transport is temporally discontinuous, local sorting and deposition allows particles to be stored in floodplain sediments and weathered between each transport episode. Rapid breakdown occurs on re-entrainment, so that the rate of downstream grain size decline reflects the speed of floodplain reworking (Bradley, 1970). Abrasion-tank experiments illustrate that weathered material disintegrates faster than fresh, unweathered rock particles.

Grain size trends along a main stream are complicated by inputs from tributaries and bank erosion. The rather gradual size reduction in Figure 8.3(b) reflects the effects of inputs from steep, short tributaries in a long, narrow catchment delivering bed material coarser than that in the main channel. This influence of tributary inputs affects the sorting characteristics of bed material, which becomes poorly sorted immediately downstream from a coarse input then improves as selective transport operates. Thus the bed material homogeneity fluctuates downstream in response to tributary sediment supply (Knighton, 1980). Particle size segregation particularly occurs at the local scale; the riffle–pool sorting process (Figure 8.3b) necessitates consistent selection of sampling sites in order to provide a satisfactory identification of downstream trends. Such local sorting reflects small-scale variations of stream power. For example, in semi-arid washes of uniform slope, Frostick and Reid (1979) identified an anomalous downstream *increase* in grain size as a result of the increasing flood discharge, which provides greater competence on the constant gradient at downstream locations. However, at a more local scale the bed material is coarser than average where the valley floor narrows to increase the flow depth and competence, and finer than average in wide valley sections. Also, in braided channel systems the high frequency of zones of efficient bedload transport allows coarser particles to travel further downstream, so that the size reduction is slow; in the basalt plateau gravels related to the drainage of Lake Turkana in Kenya, a linear reduction of only 3 mm km^{-1} occurs in median diameter (Frostick and Reid, 1980). However, associated lithologies behave differently. Feldspar, which is of poor durability, decreases in size rapidly and is virtually eliminated in 30 km, whereas agate increases in size downstream because few bed niches exist for the deposition of larger, lighter agate particles. Thus, the changing petrology of the deposit reflects the combined effects of abrasion and sorting.

Downstream trends in sediment size, shape and petrology are significant in two contexts. Firstly, rivers may be classified into gravel- and sand-bed types, separated by a threshold or sediment size discontinuity (Yatsu, 1955, Howard, 1980). The former have riffle–pool bedforms and experience bedload transport only at or near bankfull stages, while the latter have live beds with ripples and dunes at sub-bankfull discharges. Stream cross-sections are adjusted to the transport of finer sediments (Schumm, 1960a) and are modified even by single flood events, while slope is controlled by the gravel component and is therefore more stable. Secondly, the proximal–distal variations in sediment properties provide a key to palaeohydraulic interpretation. Along the direction of transport, grain sizes change consistently in a variety of depositional environments (Krumbein, 1937), and the varying textures, shapes and petrologies of alluvial sediments must be evaluated in order that fluvial deposits may be distinguished from aeolian, littoral, marine, lacustrine and glacial deposits.

Textural properties: the particle size distribution

The texture of a deposit is described by the mean, standard deviation (sorting), skewness and kurtosis of the grain size distribution in ϕ units. Frequently, however, fluvial deposits display a bimodal 'clast and matrix' distribution with a deficiency of 2–8 mm granules (Figure 8.3a), and so these statistics can be unreliable. Such bimodality, although shown by analysis of over 11,000 samples to be far from universal (Shea, 1974), arises because of the restricted sediment sizes supplied from the source areas, as well as abrasion during transport and local sorting. Skewness, kurtosis and bimodality result when two distinct log-normal distributions are mixed in varying proportions (Figure 8.4a; Spencer, 1963); the separate size grades may be generated by abrasion or sorting.

In basal tills, a clast (rock fragment) mode diminishes in importance with increasing transport distance relative to a matrix mode formed as clasts disintegrate into their constituent minerals whose 'terminal grades' vary according to the mineralogy (e.g. 2–4 ϕ units for feldspars) (Dreimanis and Vagners, 1971). The evolution of bed material textures evident in Figure 8.3(a) suggests that similar processes may occur in fluvial environments, although they may be most apparent when source rocks are granular rather than fissile. However, sediment size discontinuities may also be interpreted hydraulically in relation to sediment transport process (Visher, 1969, Middleton, 1976). Traction load is usually coarser than 1 ϕ unit, the saltation load and intermittently suspended bed material ranges from 1.75 to 2.5 ϕ units, and wash load is truncated at sizes of 2.75–3.5 ϕ units. The latter two components are deposited into the pores between the gravel–cobble component, and the entire grain size distribution is distinguished by segmentation of a log-probability cumulative percentage curve (Figure 8.4b). Local spatial sorting of particles also influences depositional textures. Folk and Ward (1957), for example, studied sediments ranging from bar platform open-work gravel to fine supraplatform sands in a single bar and showed that the size segregation reflecting source rock properties and local depositional conditions results in the systematic interrelationship of mean size, sorting, skewness and kurtosis (Figure 8.4c). Bulk samples of bed material may appear to be composed of a bimodal gravel and sand size-distribution. However, such bimodality may be difficult to interpret because the gravel and sand may be segregated spatially and with depth beneath the bed surface. This arises because of the varying effects of local depositional conditions, dispersive stresses (Bagnold, 1954) which bring coarse mobile sediment to the surface of a deposit, and armour development which occurs as fines are winnowed from the surface of a heterogeneous deposit containing an immobile coarse fraction.

Some textural properties allow discrimination between depositional environments, while others are more useful for tracing the evolution of texture during transport within one environment. River sands are usually moderately sorted and positively skewed, and so these parameters distinguish river from beach sands, which are well sorted and negatively skewed (Figure 8.4d; Friedman, 1961, 1962). The $C–M$ diagram which plots the coarsest one percentile, C, against the median grain size, M (Passega, 1964, Royse, 1968) uses arithmetic size data and indicates local sorting and transport processes, as well as discriminating turbidite, alluvial fan and fluvial deposits (Figure 8.4e). Within the band representing fluvial deposits, sediments from traction load lie in the zone NOP and form channel lag deposits. Graded suspension deposits (QR) represent suspended bed material load which accretes on point bars and channel bars. The silt–clay deposits from uniformly distributed wash load, whose maximum grain size is ~200 μm, form the zone SR and are found in floodplain overbank sediments and crevasse splay deposits. Finally, the plot of quartile deviation, $QDa = (D_{75\%} - D_{25\%})/2$, against the median (Figure 8.4f) is also environmentally sensitive and indicative of textural evolution. The general relationship implies improved sorting as deposits become finer, with different trends for dune, river, beach, marine and lake sediments – the gradient of the trend decreasing through this environmental sequence (Buller and McManus, 1972). However, within the fluvial envelope, sorting during transport is revealed by the migration of the data points as transport processes remove or add different grain sizes to the initial deposit. Migration from P to S implies

Figure 8.4 (a) Frequency distributions showing development of skewness and kurtosis by mixing populations, with cumulative frequency of distributions resulting from mixed sand and clay (after Spencer, 1963). (b) A segmented cumulative percentage frequency distribution for the Altamaha River (after Visher, 1969). (c) Sorting–mean size relationship for fluvial sediment; the sorting index is the 'inclusive graphic standard deviation', $\sigma_1 = [(\phi_{84} - \phi_{16})/4] + [(\phi_{95} - \phi_5)/6.6]$ (after Folk and Ward, 1957). Sorting categories; vws = very well sorted, ms = moderately, ps = poorly, vps = very poorly sorted. (d) Skewness–standard deviation plot for medium to very fine grained beach and river sands (after Friedman, 1961). (e) CM patterns for fluvial sediments; C = coarsest one percentile, M = median diameter (after Passega, 1964). (f) Quartile deviation (QDa) plotted against median diameter showing envelope for fluvial sediments (after McManus, 1979).

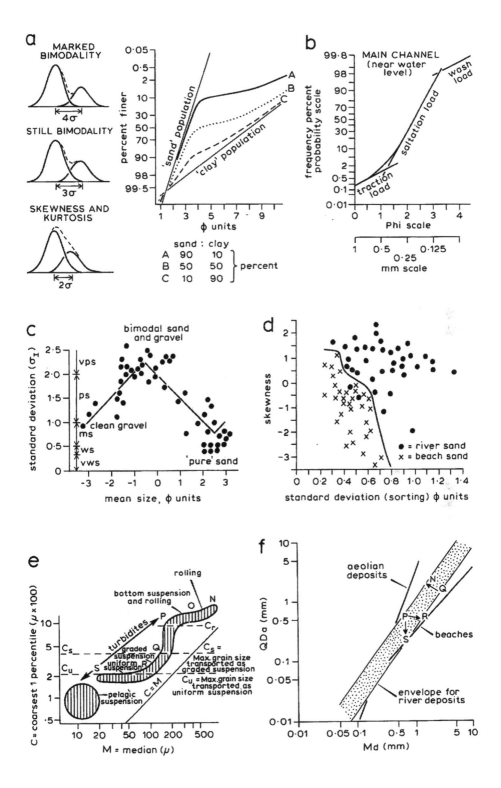

that both 'tails' of the size distribution are being removed, whereas coarsening (P to R) implies that fines are being carried away in suspension. If sandy saltation load is trapped in the pores of open textured gravels, the fining trend Q to N may occur (McManus, 1979).

Roundness and sphericity: the abrasion of fluvial sediments

The variation of depositional textures remains ambiguous, since both sorting and abrasion contribute to the observed downstream trends. Their relative importance has been interpreted from an analysis of inter-related trends in texture, particle shape (roundness and sphericity) and petrology. For example, Plumley (1948) calculated a chert ratio, $100X/(X + Y)$, where X is the number of resistant chert pebbles and Y the number of pebbles of a second lithology, for samples from Black Hills terrace gravels in South Dakota. Varying rates of down-current increase in this ratio illustrate the varying susceptibility of the other lithologies to destruction during transport, and changes in the texture and petrology of the deposits suggest that 80% of the downstream fining is accomplished by selective transport and 20% by abrasion. Figure 8.5(a) summarizes the normal downstream trends in particle size and shape (Krumbein, 1941b). Size decreases exponentially, while roundness, measured by visual comparison with charts (Krumbein, 1941a) or by a ratio of the radius of the sharpest corner to a particle length dimension (Table 1.3), increases from an initial low value to an ultimate limiting value (Plumley, 1948, Potter, 1955). However, after prolonged transport roundness may decrease again. This is suggested by Wentworth's (1919) tumbling-barrel experiments and by observations by Russell and Taylor (1937) of Mississippi River sands which, particularly in the coarser grade, become slightly less rounded because of chipping as they are transported towards the delta. Generally, Ouma (1967) concludes that a downstream trend consists of progressive rounding as gravels become finer, followed by decreased roundness as the sand grade becomes finer. However, this is partly a sorting effect; large particles round faster but are left behind by selective transport and therefore appear more angular than the smaller particles carried further downstream. The systematic trend implied in Figure 8.5(a) is somewhat misleading in that rapid changes of roundness often occur over short distances in high-energy fluvial environments such as rapids and waterfalls (Gregory and Cullingford, 1974), and distortion also arises when dissimilar material is introduced by tributaries.

Sphericity (Table 1.3) is an index of particle shape which is strongly dependent on lithology, and often only minor rapid increases occur as angular corners are rounded in the early stages of transport (Figure 8.5a). It is related to the qualitative particle shape classes (spheres, blades, discs and rods; Figure 1.5c) and influences sorting processes by controlling entrainment and the ease of transport. Disc-shaped particles of low sphericity are

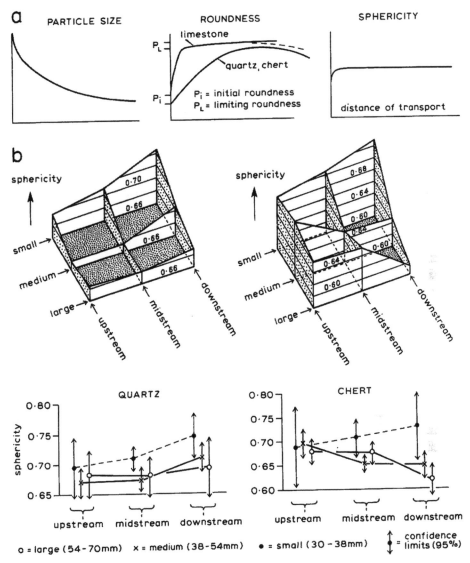

Figure 8.5 (a) The downstream trends of particle size, roundness and sphericity. (b) Downstream trends of particle shape for different pebble sizes and lithologies (after Sneed and Folk, 1958).

imbricated and difficult to entrain, but once in transport – especially if suspended – they travel further because of their low fall velocity (Lane and Carlson, 1954). The mode of transport strongly influences the development of sphericity. Pittman and Ovenshine (1968) suggest that only saltating particles have sufficient kinetic energy for particle breakage to occur, and hence intermediate grain sizes change their sphericity most markedly. This size and transport dependence is well illustrated by Sneed and Folk (1958), who analysed pebble shapes in the lower Colorado (Figure 8.5b). Large quartz pebbles tend to be rod-like, and become increasingly so as they roll on their intermediate axis, whereas small pebbles become increasingly spherical as they wear on their long axis during saltation; chert pebbles, however, behave differently. The processes of selective transport, sorting and abrasion are thus inextricably related and are controlled by lithology.

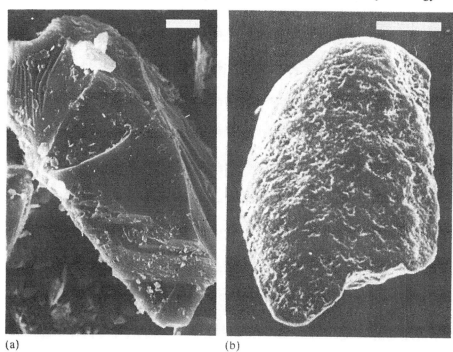

(a)   (b)

Plate 8 Scanning electron micrographs of quartz grains from: (a) a fluvioglacial stream, showing conchoidal fractures, sharp edges and crushing, typical of glacial erosion; and (b) a subaqueous environment, showing rounded edges and the characteristic mechanical impact pits (after Tovey, 1978). These impact pits occur in river and littoral environments, and vary in their density between and within environments as energy levels vary. Note that on (a) the scale bar is 4 μm, while on (b) it is 200 μm.

The precise mechanisms by which particles wear are difficult to establish. Chemical weathering may cause rounding, as some salt weathering experiments suggest (Williams and Robinson, 1981), but mechanical processes predominate. Experimental studies (Kuenen, 1956, 1959) in abrasion tanks illustrate that splitting, crushing, chipping, cracking and grinding all occur and vary in importance with the lithology and cleavage of individual particles. Experimentally determined abrasion rates are very slow (e.g. only a 1% weight loss of angular quartz sand over 20,000 km of transport) although they vary realistically with velocity, particle size, and the nature of the bed over which transport occurs. However, angular volcanic scree can be rounded within 3 km of its source (Pearce, 1971), which suggests that either sand-blasting occurs, as evidenced by the upstream rounding of large static boulders, or that *in situ* pothole action adds to the effective transport distance. Schumm and Stevens (1973) show that hydrodynamic lift forces cause pebble vibration, particularly of gravel sheltered by larger boulders under flow conditions that would otherwise cause transport, so that particles are being moved against one other without downstream displacement. The importance of weathering between discontinuous episodes of bedload transport must also be emphasized (Bradley, 1970, Tricart and Vogt, 1967). Analysis of the unconfined compressive strength of bed material grains in six size classes (Moss, 1972) suggests that when weak, large particles are being broken in transit, they result in stronger, smaller particles. A downstream strength increase indicates particle disintegration, whereas a decrease results when inputs occur of new material weakened by weathering. Scanning electron microscopy is another technique which may be brought to bear on the general problem of particle erosion in fluvial and other environments. Sand grain surface textures vary significantly according to the transport and weathering environment (Krinsley and Doornkamp, 1973) and record the provenance and history of the grain (Plate 8). In addition, the high magnifications attained may permit closer examination of the chemical and mechanical processes of grain destruction.

Discharge and sediment calibre: joint controls of gradient

Imposed discharges and sediment sizes represent the interacting controls of the reach gradient, the latter reflecting the combined influences of sediment supply from tributaries and bank erosion, selective sorting, and particle weathering and erosion in the upstream reaches. If sediment transport continuity cannot be maintained through the reach by the available streamflow (pp. 243–8), selective sediment size deposition or erosion will occur until the reach morphology, including its gradient, is adjusted and the establishment of continuity is reflected in a local equilibrium between grain size and gradient. Scale and feedback considerations complicate this relationship. At the long-profile scale, grain size sorting appears dependent on gradient, while at the reach scale, gradient is dependent on input sediment sizes. However, the reach gradients, once determined, combine to form the long profile.

The interacting effects of discharge and sediment properties are illustrated by the different slope–length relationships developed on various lithologies (Hack, 1957). In Pennsylvania, Brush (1961) identified the relationships

Sandstone: $\quad s = 0.063 L^{-0.67}$ $\hfill (8.14)$

Shale: $\quad s = 0.050 L^{-0.81}$ $\hfill (8.15)$

Limestone: $\quad s = 0.027 L^{-0.71}.$ $\hfill (8.16)$

The three lithologies have different downstream rates of change of slope (exponent) and different profile gradients at a given distance downstream (measured by the constant). The rate of change of slope reflects the combined effects of the downstream increase of discharge and the decrease in grain size. For example, on shales, discharge may increase rapidly and particles abrade easily, so that slope responds by decreasing sharply. The discharge per unit area and the average particle sizes generated from the bedrock initially by weathering then determine the actual gradients, which are lowest on limestones. If these slope–length relationships are integrated, each lithology will be seen to have a distinctive average long-profile form. One obvious implication of this is that a stream traversing several lithological outcrops will experience a range of discharge and sediment conditions which will result in an irregular equilibrium profile as successive gradient adjustments are made to accommodate these.

A summary of the bivariate control of channel gradient is provided by Hack's (1957) empirical relationship

$$s = 0.006 D_{50}^{0.6} A^{-0.6}. \tag{8.17}$$

This is illustrated in Figure 8.6(a) with examples of the slope–grain size–basin area (A) relationship plotted. If grain size is approximately constant downstream, as in the River Fowey (Figure 8.3b), slope varies inversely with basin area. This is because the larger discharges of bigger catchments are competent to transport the bed material on lower gradients. If particle sizes decrease downstream, the gradient reduction is more rapid, while an increasing grain size may result in profile convexity with slope steepening downstream (Brush, 1961). Hack (1957) found that

$$L \propto A^{0.6} \tag{8.18}$$

and by substituting this expression in Equation (8.17) it can be seen that slope varies as the reciprocal of length for a constant grain size. If $s \propto L^{-1}$, it follows that

$$sL = k \tag{8.19}$$

where k is a constant. If this slope–length product is constant along a river, it is because grain sizes are uniform, since slope will be decreasing to compensate for the increasing discharge. However, this 'gradient index' will increase or decrease downstream as bed material sizes increase or decrease (Hack, 1973a). Furthermore, values of sL along streams longer than 30 km may be mapped to identify areas where the geology encourages steeper slopes. For example, an anomalous zone of high values of sL occurs where resistant sedimentary and metamorphic rocks outcrop to the north-west of the main Appalachian watershed (Hack, 1973b).

The slope–grain size–discharge relationship is complicated, however, by

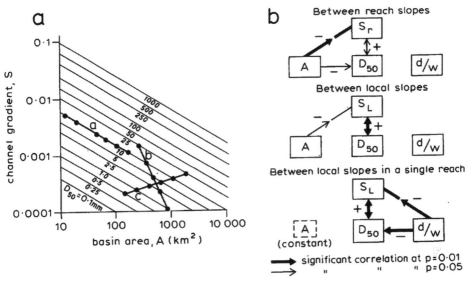

Figure 8.6 (a) The interrelationship between gradient, basin area and bed material grain size defined by Equation (8.17). Lines a,b,c are hypothetical slope–basin area relations for streams with constant, decreasing and increasing grain sizes in the downstream direction. (b) The correlation structures defining the interrelationship between gradient s, basin area A, bed material size D_{50}, and channel cross-section form (depth:width ratio) at different scales (after Penning-Rowsell and Townsend, 1978).

scale dependence and by variation in its strength between locations. Field-surveyed local bed gradient is more spatially and temporally variable than the generalized reach slope measured from cartographic sources, although the two are broadly correlated (Park, 1976). Penning-Rowsell and Townshend (1978) have shown that *local* slope is poorly correlated with basin area (used as a discharge index), but is more strongly related to bed material size. *Reach* slopes, on the other hand, show stronger correlation in general, but appear to be more closely influenced by basin area than bed material size (Figure 8.6b) when partial correlations are examined. This is because, between reaches, discharge and therefore competence vary and control differences in gradient, while within a single reach discharge is uniform and local variability in slope and particle size are more closely related. Wilcock (1967) emphasized that cartographic reach slope should be related to a grain size parameter less susceptible to short-term fluctuation than the median diameter, which varies with the time period since the last flood (Figure 7.2b; Hack, 1957). In the River Hodder, gradients are strongly correlated not with the median diameter of all bed material, but with the median diameter of the 'residual' bed material – that part estimated to be immobile at bankfull discharge. This suggests that the coarse component of bed material controls channel gradient, a conclusion supported by Knighton (1975b). In the River Bollin–Dean in Cheshire, gradient decreases markedly as bed material sizes decrease from 64 to 4 mm. However, when gravel is sparse and the bed is sandy, the relationship between slope and grain size breaks down and the gradient remains uniform as further size reduction continues. It is suggested that the competence of the stream is maintained by gradient adjustments where bed material is coarse, and by other hydraulic adjustments (of the cross-section, for example) where bed material is sandy. Thus, a distinction emerges in the behaviour of gravel-bed and sand-bed streams (Howard, 1980).

Cross-section adjustments therefore represent an additional mechanism by which streams may maintain sediment transport continuity, so that mutual interdependence with gradient occurs in response to particular environmental controls. This is exemplified by Rubey's (1933b) qualitative model derived from energy considerations in alluvial rivers

$$sF \propto G^a D_{50}^b Q^{-c} \qquad (8.20)$$

where F is the cross-section width:depth ratio and G the sediment load. This 'equation of grade' implies that slope increases with imposed calibre and quantity of load, and decreases with discharge. A wide, shallow channel provides the optimal cross-section for transport of a heavy, coarse sediment load, but the product sF means that a mutual adjustment of gradient and cross-section occurs to maintain transport continuity; if slope is too gentle, the form ratio is increased to compensate, and vice versa. Morisawa (1968) suggests that the roughness coefficient n may be added to the left-hand side of this equation. The significance of this is that roughness and slope control stream velocity, which in turn affects bank stability (pp. 163–73), which represents an additional constraint on the equilibrium form ratio and helps to determine the combination of slope and cross-section that maintains sediment transport. Wilcock (1967) notes that high bed roughness in a section with a dominant coarse residual sediment component reduces velocity and necessitates a steeper slope, a higher width:depth ratio, or a mutual adjustment of the two before transport continuity is possible. Under uniform discharge within a reach, Penning-Rowsell and Townshend (1978) identify an inverse correlation between local gradient and width:depth ratio (Figure 8.6b) which echoes this conclusion. At a larger temporal scale, Cherkauer (1972) has shown that ephemeral stream gradients in various lithologies are controlled by interacting effects of relief, basin area, bed material size and the channel form ratio. For concave long profiles on sedimentary rocks

$$s = 2.5 \times 10^{-4} R^{0.8} D_{50}^{0.2} F^{0.14} A^{-0.4}. \qquad (8.21)$$

The evident direct relationship between the form ratio and gradient at the between-reach scale reflects their mutual adjustment, with steep, wide, shallow channels being associated with coarser bed material for a given discharge. It is this mutual adjustment of long-profile and cross-section dimensions – assisted by the plan-form changes that relate to gradient – that Wolman (1955) noted would mitigate against the necessity for development of a smooth, concave long profile. As cross-section variations help to accommodate the need for sediment transport continuity, irregular adjustments of gradient to the environmental controls of discharge and sediment may occur even in equilibrium river channels.

Mechanisms of channel slope adjustment

In a general sense, long-profile adjustments include changes of channel bed elevation, channel gradient, and overall profile shape. Bed elevation changes result from aggradation and degradation, which reflect alterations

Table 8.1 Factors affecting changes in stream bed elevation.

I Aggradation; increasing bed elevation
  (a) Upstream control; e.g. glacio-fluvial sediments entering at headwaters
  (b) Downstream control; e.g. rising base level
  (c) Basin-wide control; e.g. increasing sediment yield: streamflow ratio, caused by climatic change, vegetation clearance, etc.

II Degradation; decreasing bed elevation
  (a) Upstream control; e.g. bedload entrapment
  (b) Downstream control; e.g. falling base level and nickpoint recession
  (c) Basin-wide control; e.g. decreasing sediment yield:streamflow ratio, caused by climatic change, conservation measures, etc.

---

in the river's transport capacity relative to sediment supply, and are associated with upstream, downstream and basin-wide influences (Table 8.1). Historically, net aggradation is required to provide the accumulated alluvial sediments in which self-formed channels can develop (pp. 52–5). Typical examples occur in Piedmont valleys in the USA where observed slope erosion (basin-wide control) compared to catchment sediment yield implies that 5–6% delivery ratios have obtained since European settlement and have resulted in up to 6 m of valley floor alluviation (Trimble, 1975, 1977). However, spatial relationships of aggradation and degradation are often more complex than Table 8.1 acknowledges, since upstream river incision liberates stored sediments which cause downstream alluviation. Walker Creek, California, is an intensively studied case (Leopold, 1978, Haible, 1980) where 1.5 m of incision in the headwaters in the last sixty years has caused downstream sedimentation which has necessitated the raising of buried fence posts and has buried a terrace formed by an earlier sequence of alluviation and incision. Significantly, the channel geometry, appearing similar to that of equilibrium streams, has continually adjusted to accommodate the changing inputs of water and sediment, while the overall long-profile gradient has varied by only 4%.

Thus, at a more detailed level, the mutual and often rapid adjustments of the three dimensions of river morphology render channel behaviour indeterminate and unpredictable. To what extent transport continuity is maintained by changes in slope (through aggradation, degradation or channel pattern adjustment) and cross-section properties remains unclear in particular cases. A check-dam built across a small stream to simulate the downstream control of a base-level rise (Leopold and Bull, 1979) resulted in the deposition of a wedge of sediment extending only a short distance upstream to intersect the original bed profile. Transport was maintained across the *lower* gradient of this alluvial fill by creation of a *wide*, shallow channel with reduced form resistance, there being few bars and bends. This adjustment contrasts markedly with that occurring as discontinuous gullies fuse (Leopold, 1978, Leopold and Miller, 1956). As an initially slot-like valley-floor gulley *widens*, its bed gradient *steepens* to maintain transport of the increasing sediment supply from bank erosion. This evident complex interaction of the aspects of channel geometry leads to a re-statement of Mackin's (1948) definition of an equilibrium channel as one '. . . in which over a period of years, slope, velocity, depth, width, roughness, pattern and channel morphology delicately and mutually adjust to provide the power and efficiency necessary to transport the load supplied from the drainage basin without aggradation or degradation of the channel' (Leopold and Bull, 1979).

Theoretical analysis of channel slope adjustment

A general model of the adjustment of reach gradient can be based on the sediment transport continuity equation defined in Chapter 1 (Equation 1.6) and expressed in the form

$$\frac{\delta z}{\delta t} = \frac{1}{\gamma_s} \frac{\delta q_s}{\delta x} + i. \tag{8.22}$$

The changing transport rate with distance along the reach, δq_s, expressed as a weight per unit time per unit width, is converted to a volume by dividing by the unit weight γ_s, and thence to a change in bed elevation. The success of application of this equation depends on its ability to model simultaneously the change of slope resulting from the altered bed elevation and the widening cross-section which causes the input from bank erosion, i. This is often assumed to be zero, which requires constancy of cross-section form during slope adjustment. The model is applied by defining the water surface slope for an initial profile using the gradually varied flow equation (3.52) beginning at a downstream control depth. The flow geometry is then used to predict the spatial variation of the bedload transport rate over a series of incremental reaches by using one of the transport equations discussed in Chapter 4. By integrating these variations over a suitable time period by applying Equation (8.22) bed elevation and cross-section changes (Pickup, 1977) are calculated for each incremental reach. This cycle is then repeated; the new bed slope (and cross-sections) are used to calculate a new flow geometry, the altered transport rates, and further changes in the bed elevation. Iteration continues until the flow conditions on a new bed slope are threshold conditions for the bed material, so that transport and bed adjustment cease.

This model is applied in various problems in which transport continuity is disturbed by upstream or downstream effects. Reviewing the case of upstream control, Gessler (1971) notes that degradation results from bedload starvation and aggradation from excess bedload inputs. During aggradation the downstream slope steepens as a wedge of accumulation extends downstream from the point of influx of the excess bedload. This process is relatively rapid, taking about one-third of the time required for a similar degree of degradation. Figure 8.7(a) illustrates the general process of adjustment. As the bed elevation increases, the original bed gradient is maintained upstream from the sediment input location. However, the downstream slope steepens to enable transport of the extra load supplied at this point (Soni et al, 1980). Degradation which follows reduction of bedload, such as downstream from a reservoir, is an asymptotic process which slows through time as the bed slope diminishes and the transport rate is reduced. Predicted and observed degradation below a dam are shown in Figure 8.7(b) (Hales et al, 1970), which diagram may be compared with Figure 1.2(a). A complication involved in this process is the occurrence of armouring on the degrading bed surface. If the bed material is hetero-geneous, fines are removed and a coarse surface veneer eventually protects the bed from further incision and causes a marked reduction of the transport rate. Komura and Simons (1967) calculated degradation below dam sites using a model based on the continuity equation but incorporating an estimation of the changing grain size distribution; the median bed material diameter after degradation ceases is approximately the D_{90}–D_{95} grain size before degradation begins. Livesey (1965) and Hammad (1972) have both shown that armouring minimizes the depth of incision by increasing bed roughness and bed stability, even when only a partial veneer exists. In the Nile below the Aswan dam, 2 mm particles require a threshold velocity just in excess of the actual velocity predicted for their relative roughness; they therefore constitute the armouring grain size. Since particles coarser than 2 mm constitute 4% of the bed material, only 1 m of scour is required to provide a 4 cm layer of grains capable of inhibiting excessive incision.

Downstream changes in base-level control also cause adjustments of channel gradient. The lowering of base-level creates a nickpoint and initiates incision. Pickup (1975, 1977) has modelled flow and sediment

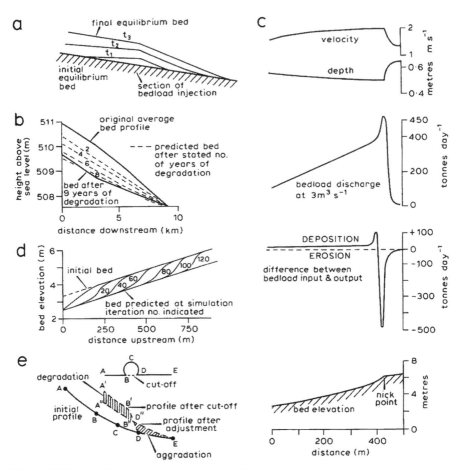

Figure 8.7 (a) Adjustment of river long profile by aggradation after an increased point input of bedload (after Soni *et al*, 1980). (b) Degradation of river profile below the Garrison dam site; observed and predicted (after Hales *et al*, 1970). (c) Flow conditions, bedload transport, and erosion/deposition through a reach containing a nickpoint (after Pickup, 1975). (d) Nickpoint retreat resulting in maintenance of bed gradient (after Pickup, 1977). (e) Adjustment of gradient as a result of a meander cut-off (after Shulits, 1936).

transport conditions at a nickpoint, where accelerating flow and steepening slope cause peak local rates of bedload transport. As a result, erosion occurs at the nickpoint, while deposition takes place downstream where the sediment yield from the nick accumulates (Figure 8.7c). The erosion causes the nickpoint to retreat, but as it migrates upstream past a cross-section, bed elevation decreases sharply then increases slightly as the initial incision is succeeded by the deposition of sediment from the now upstream nickpoint; the adjustment thus involves a damped cyclic oscillation. In the model developed by Pickup (1977), a narrow slot-like section is initially formed by incision in the bed of the existing channel. Then the channel width increases and the bank angle decreases as incision progresses until the initial cross-section geometry is re-established, mirroring the changes observed in reality, and the channel gradient after nickpoint retreat is the same as the initial gradient (Figure 8.7d). The nickpoint migrates parallel to itself leaving the bed slope required to transmit the load imposed from upstream. However, the final bed gradient may be steeper if bed material is hetero-geneous and an armouring process accompanies the incision (Begin *et al*, 1981). The development of a steeper final gradient is also observed in laboratory flume experiments in which artificial nickpoints are created at mid-flume locations (Brush and Wolman, 1960). The nickpoint retreats, reducing in height, and causes incision upstream from its initial position and deposition downstream so that the bed 'rotates' to a new steeper gradient. The rate of steepening increases with the ratio of the over-steepened gradient of the original nick to the average flume gradient, in the same way as in alluvial rivers with no resistant bedrock influences. This steepening may reflect armouring, or the relatively large magnitude of additional sediment supply resulting from bank erosion after the passage of the nickpoint, or the development of a cross-section morphology markedly different from that existing prior to the incision.

Generally it appears that the effect of a base-level *rise* on upstream channel behaviour is rather limited. If sea level rises, or a reservoir impoundment occurs, sedimentation in estuarine or lacustrine environ-ments continues by deltaic extension. Upstream growth of this depositional wedge (backfilling) might be expected, eventually producing a new valley surface at a higher elevation to provide the same channel gradient as was adjusted to the sediment supply prior to the base-level increase. However, Leopold *et al* (1964, pp. 261–6) quote unpublished data which illustrate that the depositional gradients developed above reservoirs are 30–60% of the initial gradients, and their upstream extent does not continue beyond the backwater curve generated by the new reservoir base level. Instead, as in the check-dam example quoted above, other channel adjustments occur which compensate for the reduced gradient in permitting the continuity of sediment supplied from upstream. This suggests that extensive valley-wide alluviation is more likely to reflect valley-wide control by slope erosion than downstream base-level control.

The complexity of gradient adjustment is exemplified by the case of the meander cut-off (see Figure 1.2c) analysed graphically by Shulits (1936, 1955). The existence of meander bends suggests that the gradient has been reduced by a lengthening of the channel path to provide the equilibrium-slope required for the transport of the given calibre of bedload. An artificial cut-off steepens the slope and results in adjustment of the bed gradient if bank protection prevents the re-establishment of a meander pattern. The higher cut-off slope delivers coarser sediment to the downstream end of the reach, which steepens by aggrading until this material can be transported.

However, at the upstream end of the reach the same gradient is required as before the cut-off since the input bedload calibre is unchanged. This is achieved by degradation (Figure 8.7e), and the final long profile, adjusted to the bedload particle sizes, is graphically predicted by 'sliding' the initial profile downstream through a distance equivalent to the shortening of the reach caused by the cut-off. In this instance, a local adjustment of gradient is achieved by a combination of degradation and aggradation which is both necessitated and caused by a spatial variation in the competence to transport the bedload imposed from upstream at each successive cross-section.

Alluvial fans

Alluvial fans are constructional landforms whose slopes are self-formed. They are characterized by concave radial profiles associated with a down-slope grain size reduction (Krumbein, 1937), and so their morphological and sedimentological properties display adjustments comparable to those of river long profiles. However, since they are continually reworked by un-confined flows and wide, shallow braided streams, those fans constructed by fluvial deposition are predominantly associated with gradients unaffected by the complex mutual adjustment of cross-section and long profile.

Fans are sedimentary accumulations of varying size, with a thickness of at least one-hundredth their length (Bull, 1977). They result from deposition when sediment-laden flows issue from an erosional or tectonically con-trolled mountain front onto a low gradient basin. They tend to be common in basin-and-range (horst-and-graben) areas, especially in semi-arid environ-ments where sediment removal from the depositional basin is inhibited (Plate 9). The variation of fan areas (A_f) reflects their drainage basin areas (A_d) according to

$$A_f = cA_d^n \qquad (8.23)$$

(Denny, 1967, Hooke 1968). Here, the constant c tends to be larger where potential basin depositional areas are large relative to the size of the mountain source area, since under such circumstances fan deposition may spread laterally. It also tends to be larger for fans deriving sediment from areas of easily eroded rock, and may even be used as a quantitative index of relative erodibility (Hooke and Rohrer, 1977). The exponent tends to be just less than unity, which reflects the greater storage capacity on slopes and floodplains in bigger source basins and their resulting lower sediment yield per unit area.

The three-dimensional form of alluvial fans was first modelled by Troeh (1965) using the equation

$$Z = P + sR + LR^2 \qquad (8.24)$$

where Z is the point elevation, P is the peak elevation, s is the slope at the

Plate 9 A large alluvial fan in Death Valley, California, built by braided stream deposition. A recent episode of fan-head trenching has occurred, and the stream is depositing on the right-central flank (lower left in the photograph). (Spence Air Photos)

centre and L is one-half the rate of change of gradient with radial distance R. The first two terms define a right circular cone, while the term in R^2 gives profile curvature; with s negative and L positive the form is radially concave with convex contour curvature (Figure 8.8a). This figure, however, is uniform with azimuth from the axis. Hooke and Rohrer (1979) have shown that fans of coarse sediment are steeper on their flanks because dominant (fan-forming) discharges are more extreme and sedimentation is mainly axial. In finer sediment, weaker discharges are effective, but are more likely to divert laterally and build up the fan margins to give similar gradients on all radials. The *average* axial gradient of fans decreases with fan size (and basin area), usually according to the relationship

$$s_f \propto A_f^{-0.3} \tag{8.25}$$

(Bull, 1964). This reflects the higher discharges, and therefore greater competence, generated over bigger fans (Hooke, 1968)). Coarser sediments are associated with steeper slopes, although this particularly reflects the mode of deposition (see below). In tectonically active areas, fan profiles are commonly segmented, with three or four straight or concave segments separated by breaks of slope (Bull, 1964). The varying locus of deposition, and the relationships between stream and fan slopes, reflect the relative speeds of mountain uplift and stream downcutting. If downcutting rates exceed uplift rates, fans develop in association with progressively gentler stream gradients, and the depositional locus shifts down-fan as the stream cuts into the old fan head and emerges onto the surface at a mid-fan location (Figure 8.8b). Where uplift rates exceed downcutting, the stream gradient steepens and the youngest area of deposition on the fan migrates towards the mountain front (Figure 8.8c).

Morphological relationships of an allometric nature (Church and Mark, 1980), such as (8.23) and (8.25), are best developed on steady state fans where input to the fan surface is balanced by basal removal (Denny, 1967). In arid areas, fans are rarely in a steady state, since they grow progressively

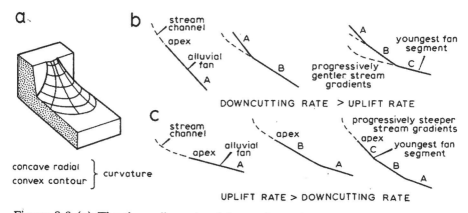

Figure 8.8 (a) The three-dimensional form of a typical alluvial fan (after Troeh, 1965). (b) Alluvial fan profile formed when the stream downcutting rate exceeds the mountain uplift rate (after Bull, 1964). (c) Segmented fan profile resulting when uplift rate exceeds downcutting rate (after Bull, 1964).

as spasmodic, extreme debris flows deposit heterogeneous sediments in the absence of basal removal (Beaty, 1970, 1974). The morphology of fans is therefore closely related to the depositional processes constructing them (Bull, 1977)). Braided stream fluvial deposits give the gentlest slopes for a given grain size, and the random switching of flow to local areas starved of sedimentation for a period tends to 'smooth' the fan surface morphology in the long term (Hooke and Rohrer, 1979). Mudflows result in coarser, more poorly sorted sediments and steeper slopes, because deposition occurs nearer the fan head (Bull, 1962, Bluck, 1964). Debris flows result in coarse deposits on steep slopes and produce an irregular unsorted surface morphology. 'Sieve deposition' (Hooke, 1967) is a mode of deposition characteristic of fans composed of coarse, permeable sediment with a paucity of fines and high infiltration rates. During a fluvial event, water is lost into the bed until transport ceases, when coarse sediment accumulates at a frontal lobe through the riser of which the water drains (hence 'sieve' deposition) and behind which a 'tread' of sediment builds, to produce a 'stepped' depositional surface. As alternation between these processes occurs during the development of a fan, the interpretation of its morphology and particularly its gradient is complicated. For example, mudflows and debris flows build up the fan head, which is then trenched by fluvial erosion (Hooke, 1967). Such fan-head trenching may reflect environmental changes, with increased runoff caused by climatic trends or overgrazing (Bull, 1964), but may also represent an inherent cyclic behaviour as fanhead slopes steepen until a threshold is exceeded and incision is initiated (Schumm, 1979).

9

River channel changes: adjustments of equilibrium

Channel morphology adjusts in the long, medium and short term to changing water and sediment discharges. Long-term influences – climatic, hydrologic and gradual tectonic effects – cause gradual, progressive and mutual adjustment of the three-dimensional form of alluvial channels to maintain dynamic equilibrium (Figure 1.6a), with morphological changes keeping pace with the environmental controls. Medium-term adjustments are often 'forced' by human activities creating a temporary disequilibrium in the channel, which then passes through 'transient states' in approaching the new equilibrium. The change in sediment transport is in this case a step-function or threshold (Figure 1.6a) rather than a continuous trend. Such man-induced changes include *direct* effects caused by deliberate management of the river to control streamflow, regulate water supply or improve navigation, and *indirect* changes caused when catchment land use is altered and sediment yield and runoff are inadvertently affected to create environmental conditions to which the river adjusts (Park, 1981). Finally, the channel may be affected in the short term by individual random extreme events. These may cause catastrophic changes of cross-section, plan and gradient over periods of days or even hours, particularly in sensitive semi-arid streams, where the lack of stabilizing bank vegetation may inhibit recovery of the channel morphology (Wolman and Gerson, 1978).

Absolute ranges for these arbitrary time scales vary with the channel's sensitivity to change. Stable humid-zone streams absorb the effects of catastrophic floods, and measurable *natural* adjustment may require hundreds of years. In semi-arid areas, however, the time scale of systematic adjustment is closer to that of random response to extreme events. Inevitably, the equations between natural adjustment and long-term processes, and man-induced change and the medium term, are over-generalized. For instance, ancient map and documentary evidence suggests that direct channel management occurred in mediaeval times in Britain; reaches of the River Don in South Yorkshire lack alluvial deposits and existed prior to the early seventeenth-century Fenland drainage schemes of Vermuyden. They probably result from the early land drainage of the mid-fourteenth century (Gaunt, 1975). Climatic control is often inseparable from the indirect influence of human effects on catchment processes during the Flandrian. There is general evidence of wetter conditions in the Atlantic (Zone VIIa of the Flandrian) and associated valley alluviation in north-west Europe (Starkel, 1966). However, aggradation in south-east Wales was associated with deforestation by Iron-Age man (Crampton, 1969), and a phase of

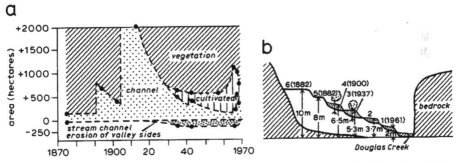

Figure 9.1 (a) Historical changes in the valley bottom of one reach of the River Gila, Arizona, showing the areas occupied by the channel, naturally vegetated and cultivated floodplain, and laterally eroding channel at various times (after Burkham, 1972). (b) A flight of terraces, dated dendrochronologically and historically, on Douglas Creek, western Colorado, formed by discontinuous downcutting punctuated by aggradation by tributary valley sediment yield (after Womack and Schumm, 1977).

alluvial filling in a small valley in the North York Moors is dated by carbon-14 as post-6270 years BP (Richards, 1981) in an area where other evidence of Mesolithic and Neolithic deforestation is considerable. Sequential rather than coincident natural and man-induced influences are more apparent in America, where meander cut-banks in Wisconsin display a stratigraphic hiatus dated at about 6000 years BP as basal gravels give way to finer sediments of aggrading rivers in more humid climatic conditions (Knox, 1972). Laminated silts at the top of the stratigraphy reflect the continuing alluviation associated with post-1830 deforestation and thus parallel the effects of forest clearance in the Piedmont valleys (Trimble, 1975).

The problems of interpreting channel adjustment are exemplified by the catastrophic widening of the Cimarron River in Kansas from an average of 15 m before 1874 to 365 m between 1914 and 1942, and its subsequent narrowing to 170 m following floodplain reconstruction by bank-attached bar growth and vertical accretion (Schumm and Lichty, 1963). The widening may have been caused by a secular reduction of mean precipitation and an associated decline of in-channel vegetation density which destabilized the bar sediments. Secondly, overgrazing may have increased the basin runoff, although livestock densities actually peaked after 1942 when the channel began to narrow, perhaps aided by increased sediment yield. Finally, the key controls of the initial channel change were two random extreme events which swept the bed material downstream, straightened the main channel alignment, reduced its sinuosity and steepened its slope. This conversion of the river from a narrow, deep, sinuous channel to a wide, shallow, braided channel suggests that it was initially close to the meandering–braiding threshold (see pp. 213–17). Similar changes on the Gila River, Arizona (Burkham, 1972) were triggered by floods in 1891 and between 1905 and 1917; the width increased from 45 to 610 m, sinuosity decreased from 1.2 to 1.0, and slope steepened by 20%. Figure 9.1(a) shows that, notwithstanding slight widening in 1941, bar growth and floodplain accretion in one reach had restored the channel through an exponential decrease in width in about 45 years after the devastation wrought by the 1905 floods.

Channel metamorphosis: approaches and methods

No comprehensive quantitative process–response model of channel adjustment exists because of the multivariate and indeterminate nature of river equilibrium. The interdependence of the nine or more adjustable morphological variables (Hey, 1978; pp. 24–8) which respond to changes in sediment quantity and calibre and water discharge precludes deterministic explanation and prediction. Furthermore, environmental changes commonly alter discharge and sediment yield simultaneously but to different and variable degrees, often with secondary responses (Howard, 1965). Increased rainfall in a dry region increases runoff and erosion, with commensurate channel changes, but after a time lag an increased vegetation density may cause a secondary reduction in runoff which damps the initial channel adjustment. These complex changes in the independent variables cause complex channel metamorphoses, the analysis of which requires one of two basic approaches. The first seeks to establish the direction and magnitude of the change of equilibrium form, while the second is concerned with the succession of transient states experienced by the channel morphology in approaching a new equilibrium.

The former approach is exemplified by the qualitative process–response equations (Table 9.1) proposed by Schumm (1969b). These show the direction in which the dependent morphological variables should change (+ being an increase, − a decrease) for given combinations of changing discharge (Q) and bedload yield (G_b). According to case (a) in Table 9.1, following a rise in discharge, width, depth and therefore channel capacity all

Table 9.1 Qualitative models of channel metamorphosis, illustrating the direction of morphological response to particular combinations of changing discharge and sediment yield (after Schumm, 1969b).

(a) Increase in discharge alone   Decrease in discharge alone
  $Q^+ \sim w^+ d^+ F^+ L^+ s^-$ $Q^- \sim w^- d^- F^- L^- s^+$

(b) Increase in bed material discharge Decrease in bed material discharge
  $G_b^+ \sim w^+ d^- F^+ L^+ s^+ P^-$ $G_b^- \sim w^- d^+ F^- L^- s^- P^+$

(c) Discharge and bed material load increase together;
  e.g. during urban construction, or early stages of afforestation
      $Q^+ G_b^+ \sim w^+ d^\pm F^+ L^+ s^\pm P^-$

(d) Discharge and bed material load decrease together;
  e.g. downstream from a reservoir
      $Q^- G_b^- \sim w^- d^\pm F^- L^- s^\pm P^+$

(e) Discharge increases as bed material load decreases;
  e.g. increasing humidity in an initially sub-humid zone
      $Q^+ G_b^- \sim w^\pm d^+ F^- L^\pm s^- P^+$

(f) Discharge decreases as bed material load increases;
  e.g. increased water use combined with land-use pressure
      $Q^- G_b^+ \sim w^\pm d^- F^+ L^\pm s^+ P^-$

Q = a suitable streamflow index; G_b = bedload transport (expressed as a percentage of total load); w = width; d = depth; F = width:depth ratio; L = meander wavelength; s = channel gradient; P = sinuosity.

increase, as does the form ratio because width increases faster than depth (p. 157). Meander wavelength increases and slope decreases. The predicted morphological adjustment to a change in bedload discharge reflects the inverse relationship between bedload and the silt and clay content of channel perimeter sediment. Hence, the form ratio increases when bedload transport becomes more significant because of the relatively sandier nature of the sediments. Bedload streams also tend to have longer meander wavelengths, lower sinuosities as they tend to become braided, and steeper gradients (see Table 6.2). If both independent variables increase together, width, meander wavelength and form ratio clearly increase since they are directly related to each variable separately. Slope and depth changes are indeterminate, and probably only occur to a minor degree. Rango (1970) attempted to quantify this model in order to predict the channel changes expected in a semi-arid region where cloud seeding was proposed to augment rainfall. However, inconsistent results were obtained when higher discharges were inserted in the empirical multiple regressions for three catchments, and no attempt was made to model the effects of simultaneous changes of discharge and sediment yield.

A general, again qualitative, summary of *transient* behaviour during channel adjustment is that of Hey (1979), in which positive and negative feedbacks are considered to be triggered by drawdown and backwater effects after the initial incision caused by a local excess shear stress. Incision steepens the slope and drawdown triggers upstream erosion, which supplies an input of sediment to the site of initial degradation, thus reducing the rate of slope steepening. As the point of maximum erosion migrates upstream, aggradation at the first site occurs because of bedload deposition and bank collapse, which reduces the local slope until backwater effects trigger upstream deposition. This theory generalizes Pickup's (1975) model of nickpoint retreat (pp. 245–7). The implied oscillation of erosion and deposition continues, gradually damping down, and the temporal variation of sediment yield after an initial incision is a mixture of exponential decline and damped sinusoids (Figure 1.6b). This complex response (Schumm, 1973) to an initial stimulus means that a flight of spatially discontinuous terraces can be generated by the successively less extreme phases of deposition and erosion (Figure 9.1b; Hey, 1979, Womack and Schumm, 1977). The Truckee River entering Pyramid Lake, Nevada, is an example (Born and Ritter, 1970). Rejuvenated by a lowering of the water level in the lake, the river created six terrace levels as a complex response occurred in the form of a pseudo-cyclic variation of sediment yield.

A variety of methods and sources of data are employed in the identification of both the transient states and ultimate adjusted channel equilibrium. These are considered below.

(a) *Calibrated river sections* The most successful approach monitors channel change at monumented sections through the transient states to the eventual restoration of equilibrium in the new environmental regime (Leopold, 1973a).

(b) *Secondary data sources* To study gradual adjustment, old large-scale (1:2500) maps (Hooke, 1977) and air photographs are used. However, the irregular sequence of observations thus obtained may give a misleading impression of long-term trends, depending on their timing relative to random events (Figure 1.7a).

(c) *Space–time substitution: the control catchment* Data from an experimental catchment subjected to change (e.g. of land use) are compared with an adjacent, similar control catchment to enable a direct evaluation to be made of natural and adjusted runoff, sediment yield and channel morphology (Figure 9.2a).

(d) *Spatial interpolation in the experimental catchment* Variation of channel form downstream is used to predict morphology in disturbed areas for comparison with observed channel properties (Figure 9.2b; Park, 1977b).

Methods (c) and (d) are often the only possible methods if adjustment has already begun, but suffer important limitations. First, calibration via

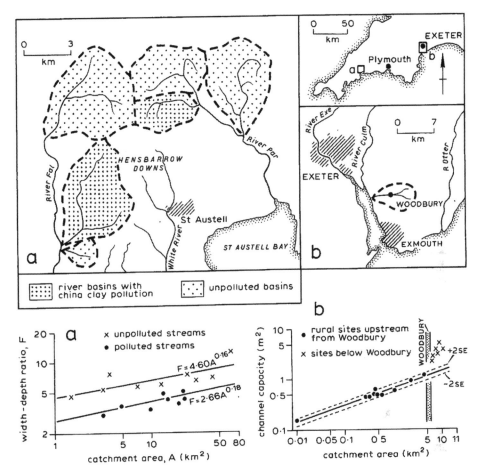

Figure 9.2 (a) Assessment of channel changes by comparison with similar control catchments; the reduction of channel form ratio by suspended sediment pollution by the china clay extraction industry in some Cornish catchments (see pp. 265–6). (b) Assessment of channel changes by comparison of observed characteristics downstream from a disturbance (here a small town) with trends established for natural upstream sections (after Park, 1977b). (See pp. 262–3.)

spatially separated undisturbed natural river reaches assumes constancy of potential environmental controls between reaches. Problems arise if, for example, geology or land use change downstream from the town in Figure 9.2(b). Second, channel form variables in the control reaches are measured relative to basin area, a discharge surrogate. This makes it difficult to assess the contributions of runoff and sediment changes to river adjustment. Third, in the absence of evidence of the mechanisms of channel change, measured morphological adjustment cannot contribute to a *general* model, being specific to the basin investigated. Finally, when adjustment is gradual (when caused by deposition, for example), it may be difficult to establish whether the channel is fully adjusted or still in a transient state.

(e) *Theoretical modelling*   Changes in the multivariate channel morphology may be modelled by methods of channel design such as tractive force theory or regime theory (pp. 281–94). Transient states in the adjustment of channel gradient to disturbances in bedload transport have been successfully modelled by combining the continuity equation and a transport equation (pp. 244–7), as long as channel cross-sections remain invariant.

(f) *Sedimentology and stratigraphy*   Longer-term Late Devensian and Flandrian channel changes are interpreted by using sedimentological evidence to establish depositional environments, stratigraphy to provide relative dating, radiometric methods (^{14}C, ^{40}K–^{40}Ar) for absolute dating, and floral and faunal remains to determine general environmental conditions.

Adjustments in the medium term; the effects of man

Inadvertent human effects on basin hydrology and sediment yield, and therefore on river sediment transport and morphology, have more widespread spatial extent and temporal significance than deliberate management and channel design (Chapter 10), with side effects which are potentially of economic significance but are rarely evaluated. Connectivity in the fluvial system often transmits the deleterious channel changes to the innocent occupants of the downstream floodplain, who are rarely protected against loss. For example, accelerated bank erosion and channel migration in the River Bollin in Cheshire probably reflect upstream urbanization, and are associated with permanent loss of a significant acreage of productive floodplain farmland. In spite of the costs incurred to the local farmers, the provision of bank stabilization works is generally not economically feasible (Mosley, 1975b).

Although weather modification (Rango, 1970) and interbasin water transfer (Lane, 1955a) are two examples of environmental disturbances which may have an impact on river channels, the following review concentrates on two catchment-wide influences which indirectly affect river processes – land-use changes and urban development – and two direct influences on channel sediment transport – hydraulic mining and reservoir impoundment – which result in unintentional channel adjustment.

Land-use changes

Runoff and sediment yield vary markedly with land-use differences between catchments of similar lithology and climate. Major differences occur between forest, pasture and cropland, and the contrasts in sediment yield are greater than those in runoff. For example, cropland runoff rates are two–three times those from catchments under pasture (Sartz, 1973), while an order of magnitude difference is apparent in sediment yield per unit area. Thus, in a sample of Mississippi catchments, if sediment yield from woodland basins is defined as the unit rate, that from pasture is approximately 30 units and that from corn cropland about 350 units (Ursic and Dendy, 1965). Management practices are also important. Lusby's (1970) paired watershed experiment showed that the increased bare area in an overgrazed basin causes a 30% increase of runoff and a 45% increase in sediment yield. Al-Ansari *et al* (1977) have even shown how turnip harvesting on the floodplain of the River Almond in Scotland causes a clearly identifiable increase in sediment yield under conditions of constant discharge. Clearly, therefore, changes of land use or management will affect runoff and sediment yield, as is confirmed by analysis of data from experimental catchments (Leopold, 1973b, Rodda, 1976), particularly those used to test forest management techniques.

Partial controlled logging in a forested catchment increases water yield and peak discharge, especially in the growing season because transpiration loss diminishes with the forest cover (Sopper and Lynch, 1973). The effect may be relatively short-lived, however, and young regrowth may transpire more than the mature forest it replaces, to cause lower runoff after logging than before (Langford, 1976). Water quality is also affected. Suspended sediment yield increases discontinuously by 40% over four years after a watershed is 25% patch-cut, mainly because of local mass movements near forest roads, whereas after an 80% clear-cut forest operation, it increases three-fold then gradually decays (Beschta, 1978). Pierce *et al* (1973) report a nine-fold increase of sediment yield for a five-fold increase in runoff after clear-cutting in a watershed in the Hubbard Brook experimental forest, New Hampshire. The inorganic particulate yield increased relative to the organic suspension, and this increased erosion was paralleled by an increased loss of dissolved nutrients (Ca^{2+}, Mg^{2+}, K^+, Na^+, Al^{3+}, NO_3^-). Although changes reflect the forestry procedures, the general increase in water and sediment output after logging suggests that, conversely, afforestation should reduce flooding and erosion. However, the land drainage designed to increase timber productivity often has the opposite effect. For instance, erosion in drainage ditches floored with colluvium on steep slopes results in a bedload yield from an afforested subcatchment of the River Severn of three to four times that from a similar nearby catchment under upland pasture (Newson, 1980).

Because land-use changes are areally widespread, their effects on river channels cannot often be judged by spatial comparison. Stratigraphic investigations reveal, however, the aggradation in Piedmont valleys after removal of forest and intensified cropping (Trimble, 1975). Typical morphological adjustments caused by this land-use change may be judged from

Table 9.1 assuming that discharge (Q) and bedload yield (G_b) both increase. Vegetation conversion may have several interrelated effects however: it alters runoff and sediment yield; the stability of the stream banks is reduced when the anchoring effect of root networks is removed; and the stream velocity may increase if the roughness of bank vegetation is diminished (Parsons, 1965). A reduced vegetation cover may therefore increase discharge, increase the velocity (and erosional capacity) at a given discharge, and reduce the bank resistance – all of which favour bed and bank erosion.

Urbanization

Although usually a relatively localized effect, hydrologically the most extreme land-use change is urban development. Impervious surfaces reduce infiltration, a higher percentage of rainfall becomes runoff, depression storage is reduced, overland flow velocities are faster on smoother surfaces and runoff enters the channel system more quickly, whereupon high flow velocities are maintained in a dense and efficient network of sewerage channels. Thus the runoff regime is flashier with shorter lag times and time bases, and with higher peaks; the unit hydrographs for stages in the expansion of Harlow New Town illustrate this (Figure 9.3a; Hollis, 1974). The increase in the peak discharge varies with the percentage of the basin urbanized; Gregory (1974) quotes two–two and one-half times for 12% urbanization. Figure 9.3(b), however, illustrates that the increase depends on the nature of urban development, being dependent on the percentages urbanized and sewered (Leopold, 1968). Natural catchment properties are also important, and Anderson (1970) demonstrates that the lag times of fully developed basins vary with length and slope, and that mean annual floods then depend on lag time, basin area, and the extent of impervious cover. Finally, flood peaks increase to varying degrees according to their return periods. Hollis (1975) shows that smaller frequent floods increase by up to ten times after 20% urbanization in a basin, while extreme events are barely increased at all (Figure 9.3c). This is because extreme floods in natural basins are generated by high-magnitude storms over fully saturated catchments with maximally

Figure 9.3 (a) Mean unit hydrographs for three stages in the urbanization of the Canon's Brook catchment by Harlow New Town (after Hollis, 1974). (b) Effect of urbanization on mean annual flood for a 2.5 km² catchment, showing proportional increase as a function of percentages impervious and sewered (after Leopold, 1968). (c) Effect of urbanization on floods of different recurrence interval (after Hollis, 1975). (d) Average suspended sediment rating curves for basins in Maryland under different land uses: rural, undergoing urbanization and fully developed (after Wolman, 1967). (e) Stream bed sedimentology and sediment yield downstream from an urban area. Each plotted point represents an average for a specified distance class (after Fox, 1976).

extended networks, and therefore with a hydrological response similar to that of an urban basin.

Urbanization causes a more complex cyclic variation of sediment yield, which is extremely high in the construction phase (Figure 2.6a) but often markedly reduced by complete urban development (Wolman, 1967) when suspended sediment concentrations may decline below levels in natural

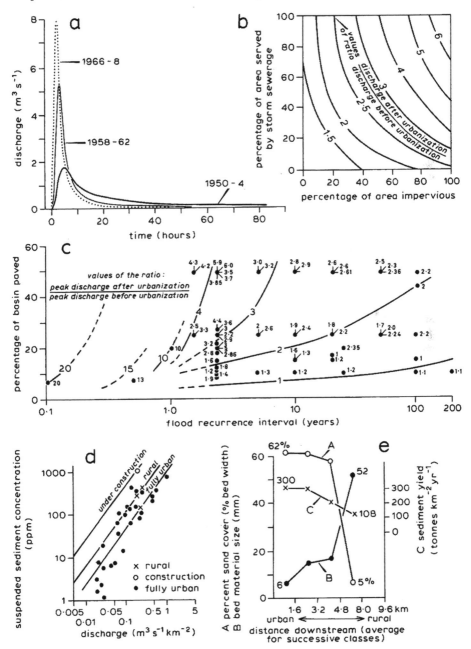

catchments (Figure 9.3d). Sediment yields in urbanizing basins are from 2–200 times the natural yield, depending on the degree of dilution of supply from construction sites by rural conditions in the complete catchment. Studies in Maryland show a three-fold increase by sediment budgeting between two gauging stations on one river separated by an urban development in a basin of 80 km^2 (Keller, 1962), a fourteen-fold increase by comparing urban and rural basins of about 3 km^2 (Yorke and Davis, 1971), and yields of up to 50,000 t km^{-2} yr^{-1} from construction sites themselves (Wolman and Schick, 1967). The continual availability of sediment in an urbanizing basin tends to suppress seasonal variation of sediment yield, which is increased most markedly in smaller, more frequent flood events (Walling, 1979). The cycle of urban sediment yield is an oversimplification, however, since concentrations of polluted suspended sediment of up to 4500 mg l^{-1} have been observed in the 'first flush' of storm runoff from fully urbanized catchments (e.g. Ellis, 1979). This primary peak represents sediments trapped by fungus and algal mats in the culvert system on the falling limb of the last flood, while a secondary peak occurs as fresh dust from roads and rooves is washed into the system.

The cyclic variation of the sediment load:discharge ratio affects urban rivers by causing a cycle of alluviation and incision downstream. For example, in a suburban catchment near Denver, Colorado, a two and one-half- to three-fold increase in floodplain area, followed by exposure of 1 m of accretion deposits by incision, has been observed (Graf, 1975). Channels are choked with sandy sediments, and gravel riffle beds become sand-dune beds during the construction phase (Wolman and Schick, 1967), although the sand cover diminishes downstream as rural conditions dilute the local effect of urbanization (Figure 9.3e). In time, the construction phase sand wave migrates downstream, and eventually gravel-floored channels are re-established in the post-construction period of low sediment load:discharge ratio. Wolman (1967) estimates a 7 year lag before the fine sediments are shifted, and Robinson (1976) considers 15 years is required to allow the channels to adjust their equilibrium to the post-urbanization gravel-bed state. The channel bed may, however, remain unstable if heavily polluted organic sediments escape into the natural channels from inefficient, elderly, combined sewerage systems.

Channel cross-sections eventually enlarge to accommodate the increased urban flood peaks. If channel capacities (C, m^2) are predicted using a basin area–channel capacity relationship for adjacent natural basins, then the channel enlargement ratio (R = observed capacity:predicted capacity) is seen to increase with the proportions of the basin covered with paved surfaces and older, sewered streets (Hammer, 1972). Ratios are smaller in younger urban areas (<4 years old) where the sedimentation cycle is incomplete, and are partly controlled by natural watershed characteristics such as the slope. Width and depth increase to different degrees as the cross-section enlarges. Park (1977b) quotes average enlargement ratios of $R = 1.43$ and 2.61 on two small urbanized Devon streams, mainly because depths increased by 36% and 82% while widths increased by 3% and 41%. Wolman (1967), however, emphasizes the increase in width, as does Mosley (1975a), whose study of the River Bollin indicates that the 8.27 m wide channel of 1872 was 12.87 m wide in 1969. Variations of response reflect the perimeter sediment properties: bed armouring inhibits bed degradation, while resistant clay banks minimize widening. The effects of increased flood magnitudes on channel adjustment are also complicated by other urban influences on flood hydrology. Cross-sections of natural rivers tend to be relatively narrow and deep where the flood regime is flashy, all else being equal. The increased *peakedness* of urban floods may therefore tend to encourage bank stability by reducing the period of wetting, and this effect may partially counteract that of the increased flood *magnitude*. Further-more, bank erosion is normally associated with more extreme events in natural rivers than is bed material transport. Since these extreme events are less markedly increased in magnitude by urbanization, it may be that the channel enlargement will be dominated by incision into the bed rather than lateral erosion of the banks.

Channel enlargement ratios vary spatially, declining downstream away from the urban area. They also vary temporally at a section as the urban sedimentation cycle progresses. During the phase of excessive sediment yield cross-sections may not enlarge significantly (Hollis and Luckett, 1976) and may even decrease in area (Leopold, 1973a) as bed elevations rise faster than the floodplain accretion fostered by increased overbank flooding. There is thus a delayed response in that a critical stage in urban expansion is required before significant enlargement of the channel occurs, and Morisawa and Laflure (1979) suggest that this is when about 25% of the basin is more than 5% impervious. The increased discharge increases competence which eventually erodes perimeter sediments, enlarges the channel and by negative feedback eventually causes a reduction of velocity. The delay may also arise because the bed and bank erosion need to be triggered by an extreme, high-magnitude flood to which urban channels respond more rapidly than their rural counterparts (Fox, 1976). The effects of floods of different frequency are suggested by Knight's (1979) analysis of adjust-ments downstream from two English new towns. Channel capacity enlarge-ment ratios of 1.5–2.3 occur in the Stevenage Brook, while in the River Tawd below Skelmersdale (Figure 9.4a) they are 3–4. This larger increase in cross-section area is thought to reflect adjustment of the channel in this flashy catchment to more frequent events of return period averaging 1.5 years, which are increased by a greater percentage on urbanization. The Stevenage Brook, however, is adjusted to slightly less frequent events which are increased to a lesser degree. Consequently the channel is less enlarged.

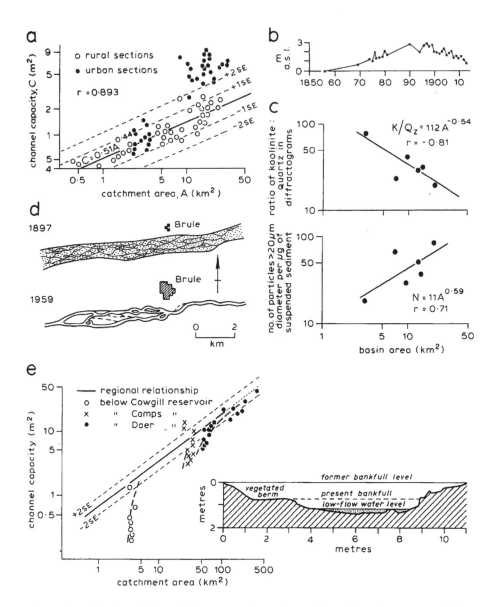

Figure 9.4 (a) The relationship between channel capacity and catchment area in the River Tawd basin above, below and within the new town of Skelmersdale, Lancashire (after Knight, 1979). (b) Fluctuations of low water level during the passage of hydraulic-mining debris deposits along the Sacramento River at Sacramento (after Gilbert, 1917). (c) Variation downstream of the kaolinite: quartz ratio of peak heights in X-ray diffractograms and the particle sizes of suspended sediment in a small Cornish stream polluted by china clay waste. (d) The South Platte River at Brule, Nebraska, in 1897 and 1959 (after Schumm, 1971). (e) Channel change downstream from three Scottish reservoirs (after Petts and Lewin, 1979). The inset shows a channel-in-channel section below Clatworthy reservoir on the River Tone, Devon (after Gregory and Park, 1974).

Extractive industries

Extractive industries generate enormous quantities of waste, and runoff and gully erosion on inert, unvegetated and unstable tailings dumps carries sediment to the river system when the waste tips are sited close enough. In the Susquehanna River, 10% of the spring flood suspended load is coal-mining debris washed from dumps (Meade, 1969). The impact of mine debris varies with the pattern of mining activity and the nature of the affected river, however. For example, lead mines at the heads of some narrow Welsh valleys provide point inputs of heavy metals whose concentrations in active bed sediments decline exponentially downstream as dilution by 'clean' tributary sediment input occurs (Wolfenden and Lewin, 1978). This pattern is less evident where mining is widely distributed and in valleys with active downstream floodplain reworking. In the River Rheidol, for example, metals stored in the floodplain during the intensive mining phase of the late nineteenth century are now being released to the river, and concentrations of lead and zinc are locally higher downstream than upstream nearer the mines. Zonation of alluvial sediments by age demonstrates a marked decrease of lead concentration in post-1900 deposits (Wolfenden and Lewin, 1977).

Hydraulic mining not only generates tailings dumps, but also uses the diverted streamflow to effect heavy mineral separation of detrital ore deposits. Its influence on river channels depends on the dominant grain size imposed. Hydraulic gold-mining waste in the Sierra Nevada, generally sandy and gravelly in calibre, choked the river systems which became steepened, braided and aggraded before the material could be transported (Gilbert, 1917). However, when some measure of control was instituted and the gold rush subsided, the input of debris declined, and because the external inputs of sediment decreased the rivers began to excavate the earlier deposits. The effect on a given reach was that a wave of mining debris passed through towards San Francisco Bay, with bed elevations rising then falling as aggradation preceded incision (Figure 9.4b). Hydraulic tin mining in Devon, England, produced a direct impact locally when boulder banks were built to confine and accelerate the flow. However, streamed and unstreamed reaches of the River Teign appear to show little evidence of the effects of fine sediment inputs, probably because of the continuity of transport from reach to reach (Park, 1979).

Surface wash on tips and overflows from settlement lagoons have supplied kaolinite and micaceous waste to those Cornish streams in whose catchments hydraulic china clay extraction is undertaken (Richards, 1979). Such fine-sediment pollution, which has continued since about 1750, has reduced channel width:depth ratios (Figure 9.2a) mainly because widths have declined. Examination of particle sizes by Coulter Counter and mineralogy by X-ray diffraction of suspended sediment in a still polluted stream shows that it becomes coarser and less dominated by the kaolinite clay component downstream (Figure 9.4c). The kaolinite clay flocculates, becomes incorporated in the channel perimeter sediments, and gives rise to channel adjustment. This change of equilibrium caused by the man-induced increase in suspended sediment transport illustrates that channels do adjust to the prevailing fine-sediment transport conditions, in this case as a result of accretion of fine sediments on the banks and narrowing of the section. This was predicted in Chapter 6 (see pp. 170–3, Table 6.2 and Figure 6.6c).

The downstream effects of reservoir impoundment

Reservoirs, like natural lakes, provide storage for sediment and water which disrupts the normal downstream trend in sediment transport and results in channel adjustment after they are impounded. The water storage capacity of the reservoir damps down flood peaks to a degree dependent on the size of the reservoir relative to its catchment area, and on the operating and management rules. A flood control reservoir with planned releases may diminish the downstream frequency of overbank events but increase the frequency of exceedance of the bed material transport threshold. Peak events are reduced in magnitude more in summer when storage capacity is available than in winter, so that the most marked influence is on high-frequency low-magnitude events (as in the urban hydrology case). On the River Hodder below Stocks reservoir in Lancashire, the mean annual winter flood is decreased by 10%, while the mean annual summer flood is decreased by 62% (Petts and Lewin, 1979). Sediment storage is measured by the reservoir 'trap efficiency', that is the percentage of incoming sediment stored. This is higher in reservoirs on bedload streams than on suspended load streams, is reduced by sluice operation in flood control reservoirs, and increases with the water capacity:inflow ratio (Brune, 1953). If the trap efficiency is high, clear-water erosion may occur downstream.

Downstream channel changes depend on the relative sizes of the discharge and sediment load reductions, and on the initial sedimentology of the river. If bed material is a sand–gravel mixture and discharge peaks are not too depressed, bed degradation continues until the reduced discharges become incompetent to move the armour layer (p. 245). On suspended load streams with pebble beds, a reduction of channel capacity is often identified. On the Rivers Nidd and Burn in Yorkshire, Gregory and Park (1976) noted observed:expected channel capacity ratios of 0.55 and 0.40 immediately downstream, associated with depositional berms within the old channel section which create a channel-in-channel section (Figure 9.4e) whose inner element is eventually adjusted to a new, smaller bankfull discharge (Gregory and Park, 1974). When streams with heavy sandy bedloads are regulated, more extreme morphological adjustment occurs as flood peaks become smaller and less variable and bedload is trapped. The South Platte River was braided and 800 m wide in 1897, but became confined to a single channel and changed to a meandering stream 65 m wide by 1959 (Figure 9.4d), because of upstream regulation and water diversion (Schumm, 1971). The importance of reservoir operating policy is also more apparent in rivers with heavy bedload transport rates. For example, the consequences of impoundment on Chinese rivers such as the Yellow and Han differ between storage and flood detention reservoirs. The former are associated with bed degradation, but because sediment flushing occurs in the latter during drawdown, they cause downstream aggradation (Chien, 1985).

Although reduced channel capacity is the commonest adjustment in upland pebble-bed British rivers subjected to regulation, this response is complex (Petts, 1979). If there is no available sediment supply to allow a depositional adjustment to a channel-in-channel section (Figure 9.4e), there is simply a reduction in the frequency of overbank flooding, the old flood-plain effectively becoming a terrace. When tributaries bring bedload into the main channel, however, there may be an adjustment, as on the River Derwent below the Ladybower reservoir in Derbyshire downstream from the River Noe, a major right-bank tributary. Trees on the old floodplain have a maximum age of 101 years, while on in-channel benches created by

Table 9.2 Adjustments of the River Derwent, Yorkshire, in the Forge valley, because of flow regulation by the Sea Cut.

Observed channel capacity in the Forge valley = 7.89 m^2

(a) Channel capacity–basin area relationship for sections above the Sea Cut

$$C = 0.59A^{0.79}$$

Predicted capacity in the Forge valley = 26.25 m^2
Ratio of observed:predicted capacities = 0.30

(b) Channel capacity–discharge relationship for sections above the Sea Cut

$$C = 4.98Q_{5\%}^{1.02}$$

Predicted capacity in the Forge valley = 4.02 m^2
Ratio of observed:predicted capacities = 1.96
($Q_{5\%}$ is the discharge equalled or exceeded 5% of the time)

---

the deposition of sediment from the River Noe, the maximum age of 51 years is less than the age of the dam (97 years; Petts, 1977). When capacity is reduced it may involve changes of depth on larger rivers, of width on intermediate rivers and of both on small rivers (Figure 9.4e; Petts and Lewin, 1979). The adjustment may be localized below tributary inputs in stable rivers, but widely distributed downstream on actively migrating meandering streams. It always diminishes with distance downstream, however, as the effect of the reservoir is 'diluted' by the increasing size of the unaffected part of the catchment. In semi-arid areas, the conversion of ephemeral to perennial flow downstream may lead to vegetation growth and stabilization of a previously shifting channel (Bergman and Sullivan, 1963).

Such a complex range of effects emphasizes both the need for observation of the mechanisms of adjustment and the limitations of indirect analysis of morphological change by comparison with control data. Consider the case of the River Derwent in Yorkshire, which leaves the North York Moors via a glacial meltwater channel (Forge valley) to enter the low-lying Vale of Pickering. To minimize flooding in the vale, a diversion channel (the Sea Cut) was established upstream from the Forge valley, and this regulation of flow has encouraged channel adjustment within the Forge valley. Table 9.2(a) shows that, according to a channel capacity–basin area relationship established upstream from the Sea Cut, the actual capacity in the Forge valley is reduced. However, discharge data are available for the Forge valley and a range of sections on nearby rivers. A channel capacity–discharge relationship suggests that Forge valley sections are too large for their discharge and may actually still be in a transient state, a conclusion which cannot be suggested without reference to the discharge data.

Long-term adjustments to environmental change

An interpretation of longer-term progressive river metamorphosis during general Devensian and Flandrian environmental change rests on a wide range of information. Morphological, sedimentological, stratigraphic, palaeoecological and chronometric data combine to enable the progressively more detailed reconstructions of environmental, hydrological and hydraulic conditions, as exemplified by Schumm's (1968a) analysis of the Murrumbidgee River palaeochannels (pp. 6–7; Figure 1.3b). Here, prior bankfull discharges were estimated from the cross-section dimensions and gradient of channel fills by applying Manning's equation with an appropriate roughness coefficient (Equation 5.1) and from a meander wavelength–discharge relation (Equation 7.9). The sandy, cross-bedded fill of the older palaeochannel correlates stratigraphically with saline palaeosols and deposits lacking organic material, and the wide, shallow, steep, bedload-transporting river is interpreted as the product of a semi-arid environment dated palynologically to approximately 36,000 years BP.

The three discrete channels indentifiable on the Murrumbidgee riverine plain suggest that threshold conditions are required before new channels are

Figure 9.5 (a) Traces of meandering and braided channels on terrace surfaces in the Vistula river valley, Poland. 1, valley side; 2, dunes; 3, alluvial fans. A, Late Devensian braided channel; Ba to Bc, three generations of meanders formed from the Late Devensian through the Flandrian; C, present braided valley bottom. After Mycielska-Dowgiałło (1977). (b) Manifestly underfit and Osage-type underfit stream patterns. (c) Aggradational terraces showing inset and overlapping deposits. (d) The warping of the Oakview Terrace of the Ventura River, California, where it crosses an anticlinal axis (A) and the active Red Mountain thrust fault (B). After Putnam (1942). (e) Long profile of the River Ystwyth, Wales, illustrating extrapolation of curves fitted to upstream segments through valley bench remnants to prior base levels (after Brown, 1952). The inset illustrates how terrace and bench elevations change upstream if the profile is actively degraded above the nickpoint while it is retreating.

formed, rather than a continuous adjustment occurring to environmental change. However, post-Devensian river adjustments in Eastern Europe display greater continuity. In the Vistula valley (Mycielska-Dowgiałło, 1977) and the Carpathians (Froehlich *et al*, 1977), in Poland, thick gravelly fills accumulated after about 40,000 years BP. Extreme, variable meltwater discharges transported heavy sediment yields from unvegetated uplands experiencing periglacial conditions, and aggradation associated with

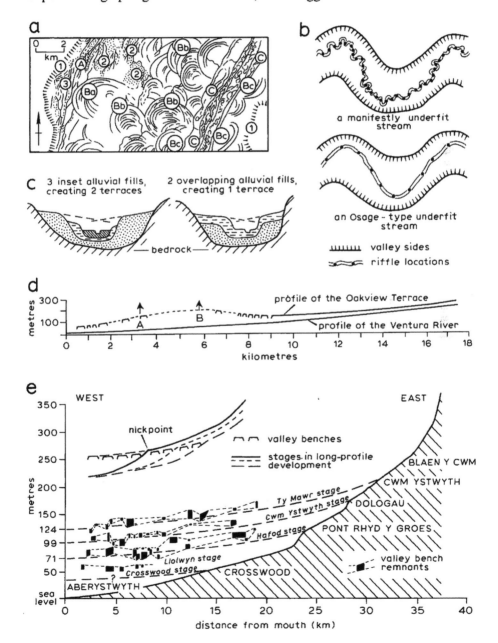

braided channels occurred in the valleys. Cold conditions are indicated by interdigitation of fluvial sediments with solifluction deposits at the valley sides. Post-Devensian climatic amelioration allowed an improvement of upland vegetation cover, and the discharge:sediment yield ratio increased so that progressive incision into the previously aggraded sediments occurred and terrace flights evolved. Braided channels were converted to meander patterns with large wavelengths and radii of curvature, associated with wide, shallow channels still transporting bedload. However, the calibre of load declined with the sediment yield, and channel traces on successively lower terrace surfaces display smaller bend sizes which were formed as the rivers became dominated by suspended sediment transport (Figure 9.5a). The most recent channel development is a reversion to braided conditions associated with increased sediment yield relative to discharge which has been caused by deforestation and intensified agriculture since the eighteenth century.

Reconstruction of past fluvial conditions from the sedimentology and sedimentary structures associated with palaeochannels is inhibited by several problems. First is that of scale; palaeohydraulic interpretations of current strength and direction are based on grain parameters and bedform structures (Figure 1.3b), while palaeohydrology involves retrodiction of discharge regimes. These distinct scales of analysis are often inadequately distinguished. Palaeohydraulic reconstruction is an essential adjunct to the interpretation of *local* depositional conditions (velocity, shear stress or power), for example in placer investigations. The techniques applicable at this scale of analysis, while often theoretically sound in themselves, are not necessarily the most appropriate for palaeo*discharge* estimation. A second problem lies in the interpretation of vertical sedimentary logs, and the assessment of the temporal context of the sequence of sedimentary structures. Vertical variations due to general changes in the fluvial environment must be distinguished from those due to changing site location relative to a laterally-migrating river of uniform regime. A third, related, issue concerns the distinction between secular trends and short periods of accelerated sedimentation. It may be tempting to relate the sedimentary history of valley fill to long-term variations of climate or land-use, but individual sediment units are often the result of single, random, extreme events. For example, in the upper Derwent valley in North Yorkshire, two 0.5m sediment units are separated by a thin (about 5 cm) peat layer outcropping in the river bank; dates for peat at the top and bottom of this sequence are statistically indistinguishable, indicating rapid floodplain accretion about 200 years ago. Perhaps the most dramatic examples of extreme events having sedimentological consequences are the Pleistocene glacial lake outbursts such as that of Lake Missoula in Washington State; this was responsible for massive, gravel, current ripples formed during an event whose peak discharge is estimated at $21 \times 10^6 \, \mathrm{m^3 s^{-1}}$ (Baker and Nummedal, 1978). This illustration points to a final problem, namely that many relationships between flow properties and sedimentary structures and channel forms, are based on uniform flow conditions and equilibrium states. Most of the evidence used in palaeoflow studies was created by transient, non-uniform flows or by rivers which, because they were necessarily aggrading to produce the preserved sedimentary structures, were *not* in equilibrium.

Palaeohydrological studies encompass a range of approaches (Starkel and Thornes, 1981). Quantitative retrodiction usually requires preliminary classification of the type of river responsible for creating alluvial depositional structures, and even this qualitative analysis requires care. For example, Dawson (1985) has investigated one fragment of the River Severn Main terrace in detail, showing from sedimentary logs and palaeocurrent directions that the terrace is a composite feature in which main valley sedimentation by a low sinuosity river has been capped by fan gravels from a tributary source, and therefore for which a single palaeodischarge estimate would be invalid. Simplistic assumptions relating sands to single-thread channels and gravels to braided rivers, or associating particular channel facies to one broad environmental regime – such as gravelly braided streams to Arctic environments – must be avoided when classifying river types (Bryant, 1983).

Quantitative reconstruction of *local* flow depths and velocities involves identification and measurement of the sizes of the largest cobbles or boulders which show evidence of fluvial transport (imbrication, or upstream dips of their *ab* planes). A range of methods (Costa, 1983) then allows maximum flow depth prediction using the Shields entrainment function (pp. 80–2) or an empirical particle size–shear stress relation (Baker and Ritter, 1975; p. 82), or velocity estimation using uniform flow formulae incorporating an appropriate friction factor. There are clearly several potential traps for unwary palaeohydrologists here; the largest boulders found may simply have entered the channel from adjacent eroding banks, or conversely may be the largest sizes supplied from the erosional sources although significantly smaller than the river's competence. Also, in assessing the flow resistance during palaeofloods by measuring boulder sizes, an overestimate of roughness may be made because the boulders were 'drowned' in a carpet of sandy bedload in the depositing event, although subsequent winnowing leaves no evidence of this. In the right circumstances – in straight, single-thread channels with rock walls – these palaeohydraulic estimates can form the basis for discharge calculation. For braided alluvial channels partially preserved on terrace surfaces, they are highly equivocal (Maizels, 1983). Thus the alternative approach – retrodiction from channel dimensions and morphology – may be more useful in the palaeohydrological context of discharge estimation if the plan, section and slope characteristics of the channel can be identified by air photograph analysis and augering to the former bed sediments (Rotnicki and Borowka, 1985). Different methods are required for meandering (single–thread) and multi–thread channels. Williams (1984) has successfully estimated prior discharges from bend shape properties, developing an approach introduced by Dury (1965) to estimate valley meander discharges (see below). For braided streams, the relationship between total sinuosity and stream power illustrated in Figure 7.1d offers some potential for discharge estimation from the traces of braided palaeochannels on terrace surfaces. In such reconstructions the key problems lie in converting fragmentary morphological evidence into appropriate morphological statistics for the entire palaeochannel (as, for example, Ferguson (1977a) has reconstructed reach sinuosity from a random spatial sample of channel directions), and in interpreting the magnitude frequency properties of the reconstructed discharge.

Some of these problems of palaeohydrological and palaeohydraulic reconstruction are illustrated by the cases of river channel change discussed below.

Underfit streams

The total meander pattern of a river includes hydraulic alluvial meanders and large-scale topographic bends (pp. 198–9). Occasionally the latter become sufficiently systematic and regular to be defined as 'valley meanders', whose wavelength between inflections is from five to thirteen times that of the normal stream meanders found on the valley floor (Dury, 1955, 1965). Subsurface exploration reveals that beneath the valley fill the valley bed topography mimics that of alluvial channels (Figure 2.7a), with symmetrical cross-sections in the inflections and asymmetric cross-sections at valley bend apices (Dury, 1962, 1964a). These large valley channels contrast with the smaller alluvial channels on the valley fill surface which appear to be 'misfit' or 'underfit' streams. Underfit streams may wander across a wide but relatively straight valley floor (Figure 7.2a), but the 'manifestly underfit' and 'Osage' type underfit streams are of particular interest (Figure 9.5b). In the former, valley and alluvial meandering are clearly combined, but in the latter the alluvial channel simply follows the valley bends. Australian examples of Osage-type underfitness include the lower Colo river (Dury, 1966b) and the Port Hacking river (Shoobert, 1968). On these rivers, riffles are spaced 'correctly' at five to seven times the channel widths, but the apparent meander bends have wavelength:width ratios of the order of forty to fifty. However, this simply reflects a stream power level which is insufficient to enable significant bank erosion, channel migration and alluvial meander development to occur. Osage-type streams are underfit streams which are effectively straight (low alluvial sinuosity) but are forced to follow valley bends. For example, Osage reaches of the River Severn have stream powers of one-half to one-third those of nearby actively meandering but manifestly underfit reaches (Richards, 1972).

An explanation of underfitness must account for the two-phase valley and channel system, which appears to imply some reduction of flow. Some meandering valleys – the Yare, Wensum and Nar in Norfolk, for instance – have been interpreted as 'tidal palaeomorphs' (Geyl, 1976), or infilled fossil estuaries. These are distinguished, however, by a rapid and systematic bend-to-bend increase in meander wavelength and amplitude. Normal valley meandering is attributed to higher past runoff caused by periglacial meltwater or higher precipitation. Dury (1965) attempted to reconstruct the valley-formative discharge using the meander wavelength–bankfull discharge relationship of Equation (7.4). Since valley wavelengths are five to thirteen times alluvial meander wavelengths, and Equation (7.4) shows discharge to be proportional to the square of the wavelength, the valley-formative discharges predicted are 25–169 times the normal, present alluvial bankfull discharge. The differences in cross-section and gradient of valley and alluvial channels led Dury (1965) to favour the lower end of this range, but even though some valley fill deposits in central England began to accumulate at approximately 9000 years BP, after valley bend formation, the late Devensian and Flandrian decline in runoff has probably not been of this magnitude.

Three basic limitations to this reconstruction may be identified. First, meander wavelength is dependent on sedimentology as well as discharge (Equation 7.9), and longer bends occur in wide, shallow streams with a heavy sandy bedload. Hack (1965) showed that alternation between entrenched and alluvial meanders occurs in successive reaches of streams cut in glacial deposits in Michigan in the last 4000 years, reflecting the varying effects on meander bends of different sediments. Bedrock valley bends might therefore naturally be expected to have larger wavelengths than alluvial bends. Second, observations of actively eroding valley meanders in south-central Texas (Tinkler, 1971) suggest that extreme events of 20–50 year return periods are responsible for valley bend development. Valley bends are not related to bankfull discharges of 1–2 year return periods, but to extreme events of the order of twenty-five times their magnitude. Much of the apparently increased discharges needed to form valley bends arises because they are related to different events on the magnitude–frequency scale from alluvial features. Finally, Palmquist (1975) has demonstrated that maximum bed scour during floods approximately equals the normal bank height, so that modification of the bedrock topography can occur, unless the valley fill is partially relict and substantially thicker than twice the stream depth. Streams migrate across the valley floor at valley bend inflections, uniformly scouring the bedrock to create a symmetrical section. Their preferred location at valley bend apices is at the outside of the curve, so that deeper scour occurs here and the bedrock topography becomes asymmetric; the bedrock and alluvial channel sections therefore evolve together, not sequentially.

River terraces

River terraces are lateral benches between a river channel and its valley sides (see Plate 10). They are formed when river incision elevates the prior floodplain above the channel bed to a height where its frequency of inundation is reduced significantly below the normal distribution of bankfull frequencies (Wolman and Leopold, 1957). Partial destruction of this elevated floodplain by actively migrating streams renders most terrace features longitudinally discontinuous, and in the case of patchy remnants the internal sedimentology and sedimentary structures must often be investigated to distinguish river terraces (characterized by typical floodplain fining-upwards sequences) from kame terraces, solifluction benches and raised beaches. River terraces may be rock-defended features covered only by a veneer of alluvial sediments, or aggradational features entirely composed of sediments (Figure 9.5c). Successive phases of alluvial filling and incision in a valley create flights of terraces if each younger depositional set is *inset* in the trench formed by the previous incision. *Overlapping* alluvial fills may bury previous terrace surfaces, and if successive fills are deposited on different gradients, they may change from inset to overlapping terrace types along the river valley, thus making it difficult to trace particular terrace surfaces. Terraces formed by rapid incision may be *paired*, with benches at matching elevations on each valley side, but the classic form of terrace is the *unpaired* variety formed by meanders sweeping from side to side as slow incision continues (Davis, 1902b). Unpaired discontinuous terraces can also be produced by braided streams, however. Small (1973) describes examples in the Val d'Herens, Switzerland, caused by a braided stream which locally aggraded when partially blocked by alluvial fan growth from tributary valleys, then 'broke through' the fans and incised the alluvium so accumulated.

Plate 10 River terraces bordering the River Rakaia at the inland edge of the Canterbury Plain, near Christchurch, New Zealand.

Terrace formation fundamentally reflects one or both of two controls: relative base-level changes; and variations in the discharge:sediment yield ratio imposed by the catchment. River incision may be triggered by a base-level fall caused by tectonic or isostatic uplift, by a eustatic sea-level fall or by a local base-level fall. For example, Sissons (1979) describes a sequence of more than twenty fluvial terraces formed in Glen Spean and Glen Roy in Scotland by rivers entering an ice-dammed lake, whose level fell intermittently as glacier wastage progressed. The fluvial deposition in still-stand periods extended further downvalley as the lake receded, and the terrace downvalley limits therefore record approximately the changing levels of the lake.

Pseudo-cyclic variations in catchment runoff and erosion cause alluviation when the discharge:sediment yield ratio is low and incision when it is high. Such variations occur on different time scales. For example, the terraces of the River Thames (Briggs and Gilbertson, 1980) and the River Avon (Shotton, 1953) generally contain evidence of cold-period deposition by braided streams: they merge laterally with solifluction deposits from the hillslopes, contain ice-wedge casts, display braided stream sedimentary structures, and include cold-climate non-marine Mollusca. Cold-stage aggradation reflected the high sediment yield from a periglacial landscape, and although discharges declined in the succeeding temperate stages, sediment yield decreased further and the discharge:sediment yield ratio rose, leading to incision and terrace formation. In the absence of absolute dates, it is unwise to correlate the successive fills of terrace flights to glacial stages, because stadial–interstadial climatic fluctuations may have been of sufficient magnitude and duration to promote fill and cut 'cycles'. In the shorter term, small-scale, discontinuous and possibly short-lived terraces arise in small catchments as a result of the complex response of main-stream and tributary sediment yield to initial stimuli (Figure 9.1b).

Terrace morphology provides a key to the history of a river's development, but one fraught with practical problems of interpretation. The terrace elevation is not the prior river elevation; terraces often have lateral slopes, and the prior stream elevation can only be estimated if the mid-valley intersection can be found of the continuation of the lateral slopes of paired terraces (Figure 9.5c). An additional complication arises because of the often unknown amount of post-formational erosion of a terrace surface. The height and thickness of the accumulated sediment beneath an aggradational terrace reflects the discharge and sediment *volume* generated during its period of formation, but its longitudinal slope reflects the discharge and sediment *calibre*, which control the prior stream gradient (p. 240; Hack, 1957), and the prior stream sinuosity, which determines the ratio between the stream gradient and the valley gradient preserved by the terrace. Warping, of course, may complicate considerably the longitudinal correlation of terrace remnants and interpretation of gradients (Figure 9.5d). In some cases, however, terraces tilted by isostatic adjustment may grade into tilted raised beaches (Sissons and Smith, 1965), in which case the terrace gradient may be approximately re-established by assuming the raised beach to have been horizontal at the time of its formation. Longitudinal continuity of a terrace must be established by a range of appropriate stratigraphic and sedimentological criteria, including evidence of palaeosols (Leopold and Miller, 1954), floral and faunal remains, and archaeological artefacts. A particularly valuable basis for correlating and separating discontinuous terrace fragments is provided by pedogenic and weathering indices, such as the visual evidence of reddening of the B-horizons of terrace surface soils, or the quantitative chemical and physical evidence of relative iron enrichment or clay-film development in illuvial horizons, and the horizon and solum thicknesses. Indeed, recent studies of soil chronosequences in the Ventura basin, California (Rockwell *et al*, 1985) have shown that the terraces interpreted as being warped by Putnam in 1942 (Figure 9.5d) are in fact displaced by neotectonic faulting. The displaced remnants of former valley fills cannot be correlated morphologically, but relative ages are identifiable from soil properties and occasional radiometric dates provide an absolute time scale for both soil development and faulting rates.

Relative terrace heights above the present river may vary considerably downstream as terrace gradients change in response to discharge, sediment size and stream sinuosity variations during their formation. Terraces may cross, with older surfaces buried by younger alluvium; relative heights alone thus provide a questionable basis for correlation. Mathematical curves fitted to upstream segments of river long profiles have been extrapolated to suggest correlation with downstream valley benches or terraces (Jones, 1924, Bull *et al*, 1934, Brown, 1952) and to estimate prior base levels (Figure 9.5e). However, upstream long-profile shapes may have adjusted to altered discharge and sediment calibre since the formation of the downstream terraces, and different mathematical curves (which fail to model successfully the potential complexity of equilibrium profiles anyway) may fit equally well statistically, but give different extrapolated heights (Miller, 1939). Furthermore, upstream valley lowering continues gradually as nickpoints recede. The inset in Figure 9.5(e) suggests, therefore, that terrace remnants successively further upstream are increasingly lower relative to the original valley floor (Culling, 1957), so that, again, the assumption of morphological continuity is possibly misleading.

The aggradational terraces produced by cut-and-fill epicycles in semi-arid valleys present classic examples of the problem of interpreting the environmental significance of inherited fluvial morphology and sediments. The history of multiple phases of alluviation and degradation may sometimes be dated accurately by archaeological evidence – for example, as in the Greek Kokkinopilos badlands (Harris and Vita-Finzi, 1968). However, the fundamental causal influences are less clear. Valley floor gullies (Brice, 1966) might be expected to be initiated at times of higher discharge and wetter climate. However, incision begins locally when the shear stress exerted by the flow over the valley floor exceeds a threshold which is largely dependent on the shear resistance imparted to the sediments by the vegetation cover, which may be quantified by the biomass per unit area (Graf, 1979). Thus, trenching of valley fill is associated with depletion of its vegetation cover in dry periods (Antevs, 1952). Quite subtle climatic changes may be involved. For example, the late-nineteenth century trenching in south-west and Midwest American valleys may reflect a period of deficient light summer rain from 1865–1900, which resulted in poor grass cover (Cooke, 1974). During this period annual rainfall totals show no systematic variation, because the deficiency of the light rain which is essential for the growth of grass in this area is countered by a slight increase in the frequency of heavy rains. Gully initiation and terrace formation may have resulted either when the frequency of light rains increased but the surface was still unprotected *after* the period of deficiency, or when individual extreme random events occurred *during* this period. A further complexity arises because during the 1860–1900 period, when many gully cycles were initiated, the rangelands were being overgrazed. This may have triggered cutting, but its influence on valley alluviation and degradation depends on the relative effects on runoff and sediment yield from the catchment hillslopes.

Finally, the epicyclic nature of alluviation followed by incision and terrace formation may be inherent in the semi-arid erosional system. Schick (1974) suggests a random mechanism in which the magnitude–frequency properties of successive 'spotty' floods are crucial. Small floods build up the valley floor, while large floods cut into the accumulated sediments (Figure 5.1b). A more systematic cycle is envisaged by Schumm and Hadley (1957), who argue that aggradation in the main valley gradually steepens its slope until shear stress exceeds the threshold and cutting begins. This is transmitted to the tributaries, whose sediment yield begins a new phase of main valley alluviation until the threshold is again reached. The actual initiation of gullying as the threshold is approached depends on the random occurrence of a suitably extreme event. If these inherent epicycles are common, the environmental significance of river terraces cannot rest on morphological evidence alone; palaeoecological and chronometric data are crucial.

10

Channel management and design

Water resource development requires multi-purpose planning at the catchment scale. Various complementary and conflicting objectives include water supply, power generation, irrigation, flood control and pollution abatement, the optimum mix of which depends on specific local needs, benefits and costs, and the necessity to conserve wildlife, landscape aesthetics and recreational facilities (O'Riordan and More, 1969). Deliberate, direct channel management often represents an important component of such broad schemes for water resource development, but is generally achieved at the more local scale of individual river reaches. Channel management includes the design and maintenance of stable artificial channels for drainage and irrigation, channelization for flood control, river training and bank stabilization in rivers used as boundaries or for navigation, channel reconstruction after extreme flood damage, and the design of bridge crossings. These objectives are increasingly constrained by the demands for environmental conservation of aquatic ecology, fisheries, and river channel aesthetics.

The interaction between catchment and channel management is illustrated by the American flood control controversy (Leopold and Maddock, 1954). This debated the relative merits of flood *protection* by downstream channel-phase management in areas subjected to flooding (by dyking, channelizing and building large dams), and flood *abatement* by upstream land-phase management (reduction of runoff by small headwater reservoirs and land-use control). This controversy is paralleled by that between structural or 'interventionist' engineering solutions and the view that natural river morphology should be maintained wherever possible, with minimal widening, deepening, straightening, clearing, dyking and artificial bank stabilization (Keller, 1975, 1976). Four general problems illustrate the advantage of this latter viewpoint.

(a) *Downstream and upstream effects*   Local, piecemeal channel 'improvement' frequently has undesirable effects outside the reaches directly involved. For example, the accelerated transmission of water and sediment through reaches channelized for flood control (straightened, steepened and deepened) causes increased flooding and aggradation downstream (Emerson, 1971) and may trigger incision and gullying upstream (Figure 1.2b). Floodplain storage naturally attenuates flood peaks, and so artificially raised levees only accentuate flooding downstream. In such a way increased flooding in the Sacramento river valley between 1850 and 1900 was partly caused by the aggradation of mining debris on the channel beds and partly by the reclamation of floodplain basins (Gilbert, 1917). More locally, artificial structures themselves often create conditions in which expensive maintenance becomes necessary, as when bank erosion undermines concrete urban culverts at their downstream end.

(b) *Instability and readjustment*   Channelized reaches often undergo natural readjustment, particularly when their design has been based on inadequate data. As an example, the River Tame downstream from Birmingham was locally widened from 12.2 to 21.4 m to accommodate urban flood peaks. However, natural channel enlargement had already occurred, and shoaling and in-channel vegetation growth promptly narrowed the river to 15.3 m (Nixon, 1959). Similarly, the Pequest river in New Jersey was dredged in 1950 to control flooding (Dunne and Leopold, 1978, pp. 707–9). The natural channel was 13.4 m wide, 1 m deep and 13 m^2 in cross-section area at the bankfull discharge of 15.6 m^3 s^{-1}. The dredged channel was 27.4 m wide at the top, 11 m wide at the bed and 3.2 m deep. By 1969, deposits had accumulated on one bank to a thickness of 1.2 m, and a discharge of 15.6 m^3 s^{-1} was carried at the bankfull level of this section-within-section, whose water surface width was 10.7 m with a cross-section area restored to 13 m^2. These examples may appear less damaging than underdesigned sections experiencing rapid bank erosion, but the sedimentation occurring in over-designed urban streams may be toxic or organic, and therefore of concern in relation to the third issue, the river ecology.

(c) *Aquatic ecology*   Ecological status in river channels is commonly measured by species diversity, with a healthy state being associated with moderate populations of several different species. Insects such as mayflies (*ephemeroptera*), stoneflies (*plecoptera*) and caddis flies (*trichoptera*) characterize healthy streams which have low nutrient status and which also contain fish, crustaceans and snails but lack algae and rooted aquatic plants (Dunne and Leopold, 1978, p. 776). Nutrient enrichment by runoff from agricultural areas or by sewage results in higher populations of certain species, particularly of rooted aquatics, snails (e.g. *physa*), waterweed (*elodea*) and pondweed (*potamogeton*). Bacteria, algae and flatworms become dominant in the less diverse ecosystems affected by organic and toxic pollution.

Natural channels provide a diversity of habitat capable of supporting ecological diversity. Trees and bank vegetation provide shade which controls the diurnal temperature variation. Riffle gravels allow aeration which encourages the incubation of salmon and trout ova (Stuart, 1953), and the diversified flow conditions provide ample opportunity for fish breeding, feeding and shelter. Pools provide resting areas during peak flows and sufficient depth to support fish in the dry season. Man-made channels,

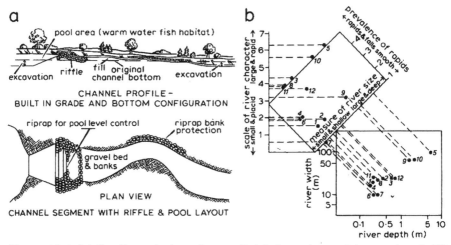

Figure 10.1 (a) Profile and plan of an artificial channel containing pool and riffle elements designed to conserve fish populations (after Keller, 1976). (b) Derivation of a 'scale of river character'. The axes of the criteria are oriented at 45°, implying an equal weighting to each; the plotted points identify values for twelve reaches in Idaho, of which number 5 is Hell's Canyon, on the Snake River (after Leopold, 1969).

however, commonly lack shade and experience high daytime temperatures and marked temperature variation. The unsorted bed material provides no habitat diversity, and velocities are uniformly high during floods while depths are uniformly shallow during low flows (Keller, 1976). Channelization thus reduces the ecological health and may reduce fish populations by more than half. Nevertheless, natural gravel sorting can gradually recreate the requisite diversity of bed and flow conditions, and gravel accumulations deliberately dumped at intervals of five to seven times the channel width generally remain stable (Stuart, 1953). Channel design can therefore encourage the maintenance of ecological status by including pool and riffle features, in which self-cleansing of pollutants is achieved by the bed-divergent flow in pools (Figure 10.1a; Keller, 1976).

(d) *River aesthetics* The important recreational role played by river channels reflects both their ecology and appearance. Urban channels involve costly construction and maintenance, and concrete channels are unsightly and dangerous during floods. Linear waterside parks, however, with natural vegetation and river morphology, provide flood wash zones which attenuate flood peaks and protect downstream areas. This combined flood control and recreational facility, coupled with sophisticated flood-warning techniques, is exemplified in the new town of Milton Keynes (Davies, 1974).

Methods exist which permit comparison of the aesthetic value of channelized and natural reaches and which allow evaluation of the quality of natural reaches scheduled for alteration. These methods are based either on measurable physical attributes, or on subjective personal responses. Table 10.1 lists an abbreviated set of physical, biotic and human use criteria which have been used to quantify the 'uniqueness' of samples of river reaches (Leopold and Marchand, 1968, Melhorn *et al*, 1975). Each reach is classified into one of the five categories of a given criterion, and its 'uniqueness index' is the reciprocal of the number of streams in that category. The value of the resultant 'total uniqueness score' for the river depends on whether the initial sample of streams is representative of the variety of types in the area. According to this inventory, a polluted man-made channel may be as unique as a mountain torrent. Melhorn *et al* (1975), however, have simply discounted the 'unpleasant' categories to generate an 'aesthetic river index'. Leopold (1969) used three criteria (width, depth and prevalence of rapids) to derive a 'scale of river character', which was applied to twelve river reaches in Idaho to determine objectively the aesthetic quality of Hell's Canyon on the Snake River where a dam site had been proposed (Figure 10.1b). This reach had the highest score for river character, as well as for valley character, which reflects valley width, the height of nearby hills, the extent of scenic outlook and the degree of urbanization. These techniques, in which the supposedly significant criteria are identified by experts, may be supplemented by an appeal to the general public – the potential users of the river as a recreational resource. A random sample of individuals may be asked to assign scores on a predetermined scale to photographs of different reaches (Morisawa, 1971). Various psycho-social factors condition the individual scores, but averages may correlate with particular physical attributes, which can then be identified as the significant determinants of aesthetic quality.

The design of stable artificial channels

A stable artificial channel excavated in natural sediment must transmit the design discharge without experiencing bed or bank erosion (scour) or sedimentation of any load introduced at the upstream end (Lane, 1937, Shen,

Table 10.1 An inventory of physical, biotic and human use criteria which may be used to define 'river uniqueness' (after Melhorn *et al*, 1975).

Criteria	Categories 1	2	3	4	5	Typical uniqueness ratios for one river
Physical features						
1. Channel width (m)	<3	3–9	9–30	30–90	>90	0.17
5. Channel pattern	sinuous (pools & riffles)	meandering (pools & riffles)	sinuous (no riffles)	meandering (no pools or riffles)	braided	0.50
9. Bedslope	<0.0005	0.0005–0.001	0.001–0.005	0.005–0.01	>0.01	0.17
10. Width of valley flat (m)	<30	30–150	150–300	300–1500	>1500	0.13
11. Erosion of banks	stable——————————————————————————————slumping——————————————eroding					0.11
13. Sinuosity	<1.25	1.25–1.50	1.50–1.75	1.75–2.0	>2.0	0.08
Biologic and water quality						
16. Floating material	none	vegetation bed & bank partly covered	foam	oil	varied	0.07
17. Algae	none		mostly covered	mostly covered	entirely covered	0.07
18. Land plants (floodplain)	open	wood & brush	wood	cultivated	mixed	0.06
20. Water plants	absent——abundant					0.08
Human use and interest						
21. Trash per 30 m	<2	2–5	6–10	11–50	>50	0.50
23. Artificial control	free & natural	partial control	partial channelization	complete channelization	dammed	0.50
24. Utilities, bridges, roads	none	<4	5–10	11–20	>20	0.50
28. View confinement	open——————————————————————————————————————closed by hills, cliffs					0.08
29. Rapids and falls	none——abundant					0.06
30. Land use	agriculture	recreation	urbanization	recreation & urban	agriculture & urban	
					Total uniqueness = 3.14	

Uniqueness ratios here assume a sample of 20 streams: thus on criterion 1, the hypothetical sample stream shares category 2 with 5 others, giving a uniqueness ratio (UR) of 1/6 = 0.17.

1971a). Natural equilibrium river channels provide an analogue, since they also undergo changes of width, depth, velocity, slope and pattern when imbalance occurs between the available stream power and the thresholds required to initiate and maintain sediment transport. However, artificial channels usually have fewer 'degrees of freedom', in that they lack riffle–pool variations and commonly are straight. Project requirements (irrigation or flood control) determine the economic or hydrological considerations which define the design discharge, which is the 'dominant' or 'formative' event (Inglis, 1949a; pp. 135–45) having the same morphological consequences as a sequence of flows in a natural channel. The design discharge, to which the channel design is adjusted, is conceptually equivalent to a natural bankfull discharge.

In a straight regular channel the design variables are limited. Frequently, the discharge, slope and perimeter sediment sizes are fixed by project requirements and valley conditions, and a non-scouring combination of width (w) or wetted perimeter (p), depth (d) or hydraulic radius (R) and mean velocity ($\bar{v}$) is determined for clear-water flow. For channels carrying a sediment load, the discharge, sediment transport rate and sediment sizes are imposed and slope (s), $\bar{v}$, d or R, and w or p are selected to avoid silting or scouring. In each case an economic constraint exists as well as the physical limitations; minimal excavation costs, compatible with physical stability, are desirable. Methods of channel design appropriate to these two problems are the physically based 'tractive force theory' and the essentially empirical 'regime theory'.

Tractive force theory

A channel may be designed to carry a given maximum discharge of clear water without scour by ensuring that the tractive force exerted by the flow on the perimeter is everywhere equal to or less than the threshold for initiation of particle motion. Such a channel may transport suspended wash load as long as the flow velocity exceeds the fall velocity of the particle sizes involved, to inhibit silting. In this respect, physico-chemical properties of both sediment and water are important, as they control flocculation into aggregates of higher fall velocity. A channel which is at incipient motion over its entire wetted perimeter is a 'threshold channel', and natural gravel- and cobble-bed streams are probably in this state at or near bankfull discharge. In fact, Li et al (1976) suggest that such channels must be at the threshold condition at bankfull to be in geomorphic equilibrium.

Tractive force design for non-cohesive sediments involves four basic assumptions concerning the force balances on the grains (Lane, 1955b). First, at and above the water surface the channel side slope is equal to the angle of internal friction of the sediment; Figure 10.2(a) may be used to estimate this angle (ϕ) for grains of a given size and shape. Second, at

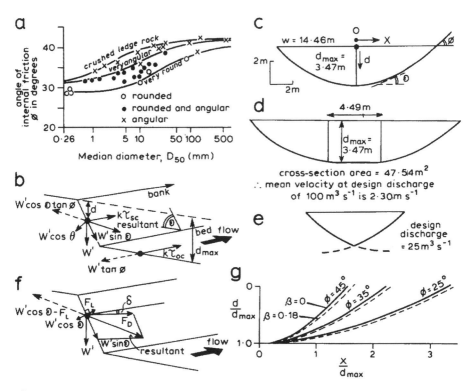

Figure 10.2 (a) Relationship between the angle of internal friction and size and angularity of sediments (after Stevens and Simons, 1971). (b) Forces acting on grains on a horizontal stream bed and a sloping bank. (c) Threshold channel designed to carry 50 m^3 s^{-1} without scour. (d) Threshold channel for a design discharge of 100 m^3 s^{-1} for the same sediment and slope as that in (c). (e) Threshold channel for 25 m^3 s^{-1}, again in the same material, and at the same slope as (c). Both (d) and (e) are drawn at the same scale as (c). (f) Forces acting on a grain on a sloping stream bank. F_L is the lift force, which acts normal to the surface in the opposite direction to the body force W' cos θ. F_D is the drag force acting in a direction deviating by $\delta°$ from the general channel direction. (g) Dimensionless threshold cross-sections for zero lift (β = 0) and lift = 0.18 drag (β = 0.18). After Lane *et al* (1959).

incipient motion traction is opposed by the component of the submerged weight (W') of the grain acting normal to the bed, multiplied by the tangent of the friction angle (see Figure 10.2b). Third, at the channel centre where the side slope $\theta°$ is zero, motion is initiated only by the shear stress exerted by the flow, whereas fourthly, on the sides where $\theta > 0°$, this is augmented by the downslope component of the submerged weight of the grains. At limiting equilibrium on a horizontal river bed

$$k\tau_{0c} = W' \tan \phi \qquad (10.1)$$

where τ_{0c} is the critical mean bed-shear stress and k is a factor (the reciprocal of a grain packing index) converting the shear stress to a drag force per grain. On the river bank, grain movement is determined by the resultant of the shear stress on the bank (τ_{sc}) and the downslope grain weight component:

$$\text{resultant} = \sqrt{(k^2\tau_{sc}^2 + W'^2 \sin^2 \theta)} \tag{10.2}$$

and since the normal force on the bank is $W' \cos \theta$, the limiting equilibrium equation is here

$$\sqrt{(k^2\tau_{sc}^2 + W'^2 \sin^2 \theta)} = W' \cos \theta \tan \phi. \tag{10.3}$$

By dividing this equation by Equation (10.1) and simplifying, we obtain an expression which defines the ratio between the critical shear stresses required for movement of grains with a given ϕ angle, on a bank of slope $\theta°$ and on a horizontal surface:

$$\frac{\tau_{sc}}{\tau_{0c}} = \cos \theta \sqrt{\left(1 - \frac{\tan^2 \theta}{\tan^2 \phi}\right)} = K. \tag{10.4}$$

Assuming further that the local shear stress is solely due to the component of the weight of overlying water acting in the direction of flow, which ignores the lateral transfer of momentum caused by secondary currents, at any point $\tau_s = \gamma_w ds$, where d is the local depth. At the centre, $\tau_0 = \gamma_w d_{max} s$ (Figure 10.2b). It thus follows that

$$\frac{d}{d_{max}} = \cos \theta \sqrt{\left(1 - \frac{\tan^2 \theta}{\tan^2 \phi}\right)} \tag{10.5}$$

and since $\tan\theta = -dd/dx$, this represents a differential equation for the channel section which can be solved to give

$$d = d_{max} \cos\left(\frac{x \tan \phi}{d_{max}}\right) \tag{10.6}$$

(Glover and Florey, 1951). The term in parentheses is in radians, and x measures the distance from the channel centre, where $d = d_{max}$ (Figure 10.2c). Table 10.2 contains an example of the application of this equation in the derivation of the threshold channel section required to carry a discharge of 50 m^3 s^{-1} in a given sediment and on a specified slope. This cross-section involves the minimum wetted perimeter, minimum top width and minimum excavation for the given discharge.

The design method outlined in Table 10.2 identifies a *maximum* stable depth for the sediments in the channel perimeter. If the design discharge exceeds the capacity of the theoretical section, the extra flow cannot be accommodated by increasing depth. Stability is maintained by inserting a segment of constant depth in the channel centre (Figure 10.2d). Similarly,

Table 10.2 An example of the derivation of a stable threshold channel cross-section for a design discharge of 50 m^3 s^{-1}.

Problem: to design a stable threshold channel section to carry 50 m^3 s^{-1} of clear (sediment-free) water through slightly rounded quartz gravel of median diameter $D_{50} = 35$ mm on a slope of 0.001

   (i) From Figure 10.2(a), the angle of internal friction for slightly rounded grains of $D_{50} = 35$ mm is $\phi \approx 37°$.
  (ii) From Strickler's equation (3.21), Manning's roughness coefficient for grains of this size is $n = 0.0151 D_{50}^{1/6}$; here, $n = 0.027$.
 (iii) From the Shields diagram (Figure 3.8c) estimate the critical shear stress. In the fully rough phase, $\tau_{0c} = 0.06 \; (\rho_s - \rho_w)gD$ (Equation 3.67). Thus, for this sediment, $\tau_{0c} = 0.06 \, (2650 - 1000) \, (9.81) \, (0.035) = 33.99 \, \text{N m}^{-2}$.
 (iv) The maximum depth acceptable in this sediment at a slope of 0.001 is found from $d_{max} = \tau_{0c}/\rho_w gs = 33.99/(1000) \, (9.81) \, (0.001) = 3.47$ m.
  (v) The shape of the section is given by Equation (10.6) and is

$$d = 3.47 \cos (0.217x).$$

This section is shown in Figure 10.2(c).
 (vi) By manipulating Equation (10.6), the following properties of the cross-section can be obtained (Graf, 1971, p. 120):

$$\text{Width, } w = \frac{d_{max}\pi}{\tan \phi} = 14.46 \text{ m}$$

$$\text{Cross-section area, } C = \frac{2d_{max}^2}{\tan \phi} = 31.96 \text{ m}^2$$

$$\text{Mean velocity, } \bar{v} = \frac{1}{n}\left(\frac{2d_{max} \cos \phi}{\pi(1 - \frac{1}{4}\sin^2 \phi)} \right)^{2/3} s^{1/2} = 1.82 \text{ m s}^{-1}.$$

(vii) The discharge carried by this section is $Q = C\bar{v} = 58.2$ m^3 s^{-1}. This may provide an acceptable factor of safety and freeboard to be chosen directly as an appropriate section for the design discharge.

---

when the design discharge is less than the theoretical maximum a segment of constant depth is removed (Figure 10.2e; Graf, 1971, pp. 120–1); this saves considerably on unnecessary excavation costs. The capacity of the theoretical section in Table 10.2 is about 16% larger than the design discharge and would thus provide a satisfactory freeboard. It would be undesirable to run an irrigation canal completely full, and the freeboard provides a safety factor. The side slope of the theoretical section equals the sediment friction angle at the water surface. Thus, the bank above the water is potentially unstable if disturbed and may avalanche into the channel. It would be highly susceptible to erosion at a flow slightly above the theoretical maximum. Thus, design would normally incorporate a safety factor such that the side slope is less than the friction angle at the water surface (Lane, 1952).

Trapezoidal sections are both more akin to natural channels and easier to excavate than the cosine-function section of Figure 10.2(c). The abrupt 'corners' in trapezoidal sections cause secondary currents and a non-uniform distribution of shear stress, which has been theoretically predicted (Olsen and Florey, 1952) and experimentally verified using Preston (pitot) tubes to measure the pressure distribution close to the channel wall (Ippen and Drinker, 1962). These results show that when width:depth ratios exceed 3, the maximum bank shear stress is approximately three-quarters of that on the bed (Lane, 1952). Equation (10.4) predicts that the maximum shear stress withstood on banks of slope 30° is 56% of that withstood on the bed by the material involved in the problem considered in Table 10.2, that is, 19.0 N m^{-2}. An appropriate trapezoidal channel depth would therefore be one giving a bank shear stress

$$\tau_{sc} = 0.75\rho_\omega gds \qquad (10.7)$$

no greater than this, that is, $d = \tau_{sc}/0.75 \rho_\omega gs$. For the data in the problem, this depth is 2.59 m. In an artificial, stable trapezoidal channel so designed, threshold conditions are only experienced locally on the banks.

The particle stability analysis from which the fundamental equation (10.4) is derived is only partially complete. To ascertain the stable grain size on a stream bed or side slope it is necessary to consider lift as well as drag forces (Stevens and Simons, 1971). Furthermore, secondary currents, even in straight channels, may deflect the direction of the drag force from the mean flow direction. By combining these two effects, the force balance is (Figure 10.2f)

$$(W' \cos \theta - F_L) = \sqrt{(W'^2 \sin^2 \theta + 2F_D W' \sin \theta \sin \delta + F_D^2)} \qquad (10.8)$$

where F_D and F_L are the drag and lift forces and δ is the deflection angle. Both Lane et al (1959) and Li et al (1976) have developed versions of Equation (10.4) which incorporate a parameter β which measures the ratio of lift to drag; the effect of secondary currents is usually negligible in straight channels. Figure 10.2(g) shows that, by reducing particle stability, lift forces result in wider sections with shallower side slopes than those predicted by the simple method. Christensen (1972) further extends the analysis by showing that β is dependent on the grain size distribution, and by incorporating the effects of instantaneous shear stresses and lift forces due to flow turbulence; the method based on Equation (10.8) only considers the time-averaged shear stress and lift force. Failure to consider the effects of lift and the random instantaneous peaks of both drag and lift which exceed the time-averaged means results in unsafe design if the design method is a purely theoretical analysis of particle stability. The problem is less serious if the empirical Shields criterion for the threshold shear stress is used, as in Table 10.2. This threshold is based on projecting a bedload transport–mean shear stress relation to a condition of zero transport, and thus reflects the instantaneous lift and drag occurring at, and measured by, the critical mean bed shear stress, which therefore incorporates information on these factors.

Artificially designed channels based on these methods are similar to natural channels in coarse non-cohesive gravels and cobbles whose perimeter sediment only becomes mobile at bankfull stage. Li *et al* (1976) show that the friction angle ϕ is a dominant influence on channel shape, with the width:depth ratio increasing as ϕ decreases (i.e. for smaller and less angular grains). Material with a low friction angle requires a less steep bank angle for it to remain stable, and this therefore increases the width. The channel shape is therefore a nonlinear function of perimeter sediment size; width:depth ratios are low in clay-rich (very fine) sediments, are high in sands (Schumm, 1960a) and are lower again in coarser cobbles. However, there is no analytical method which can predict channel shapes in cohesive clay-rich sediments whose resistance to erosion is dependent not on body forces reflecting grain size and gradation, but on the Atterberg limits, clay mineralogy, cation exchange capacity, pore water chemistry and bulk mechanical strength. Empirical design criteria are therefore necessary for the construction of stable channels in cohesive sediments. For example, Flaxman (1963) defines channel stability in terms of the unconfined compressive strength of bank materials (Figure 6.5e), and Strand (1971) has successfully applied this criterion to channel design in cohesive loessial soils. In sands, of course, channels are unlikely to exist at the threshold state, and design must allow for continuing sediment transport.

Finally, it may be necessary to design bends in artificial channels on steep valley slopes which cause high shear stresses and therefore demand very wide, shallow threshold channels. The bends reduce the slope and allow design of deeper, narrower stable channels. Ippen and Drinker (1962) measured shear stress distributions in 60° bends of various curvature:width ratios and showed that peak local stresses twice the average occur downstream from the bend at the base of the outer bank. To compensate, bank angles must be very gentle, or be protected by coarser sediment than exists naturally in the banks (rip-rap). However, the sharp distinction between the hydraulically rough banks thus created and the smooth bed may actually accentuate secondary currents and enhance meandering tendencies (Shen and Komura, 1968). The design of stable meandering channels is therefore exceptionally difficult.

Regime theory

Channel design must frequently allow bedload transport; tractive force theory is less appropriate for large channels in sandy sediments supplied from heavily loaded rivers. For example, the Punjab irrigation canals are supplied by Himalayan rivers carrying up to 10,000 mg l^{-1} of sediment during summer snowmelt and autumn monsoon floods. The link canals from the

Indus, Jhelum and Chenab rivers carry up to 610 m^3 s^{-1} (nearly twice the bankfull discharge of the River Thames at Teddington weir). These canals have been recently designed under the Indus Waters Treaty (1960) to transfer water to eastern West Pakistan where irrigation was previously supplied by rivers assigned to India (Mao and Flook, 1971). Such canals are usually of lower gradient than their source rivers, since they cross valley axes and 'lift' irrigation water to the valley surface. Their transport capacity per unit discharge is therefore lower, and their headworks must minimize bed-load inputs. This is achieved by combining ideal location, orientation and engineering design.

Canal headworks are ideally situated on the outside of meander bends in supply rivers, since the secondary flow is here transferring bedload to the inner convex bank and therefore away from the canal entrance (Garde and Ranga Raju, 1977, pp. 384–7). Thus although the Rice and Rohri canals, fed by the Indus above the Sukkur barrage, were designed to carry similar discharges on a gradient of 0.00008, excessive sediment input to the inner bank canal (the Rice) caused aggradation to a slope of 0.0001, and sediment deficiencies in the outer bank Rohri caused degradation to 0.00005 (Inglis, 1949a). Figure 10.3(a) shows a similar case on the North Platte River in Wyoming. The orientation of the canal offtake relative to the supply river also controls the sediment input. Suspended sediment tends to bypass an offtake channel, but secondary currents at the bed may carry a high percentage of bedload into a channel having a small diversion angle (Figure 10.3b). Finally, ejectors such as curved vanes may be built at the river bed to divert the flow near the bed away from the canal entrance, where settlement ponds may be used to extract excess load. As a result of these headwork controls, bedload concentrations rarely exceed 500 mg l^{-1} in stable canals. Headwork management can never exclude the effects of extreme natural conditions, however. For instance, 2 m of deposition occurred over 12 km of the Ganga canal in July 1970 after several Himalayan landslides raised sediment concentrations to 30–40,000 mg l^{-1} in the River Ganga. The result was a two-month canal closure during the crop season.

Control of sediment inputs to canals may destabilize the supply river downstream, since it is required to carry a relatively heavier sediment concentration. A convenient index of a 'regime state' is the 'specific' gauge, that is the stage reached by a specified discharge (Blench, 1952, 1969). Figure 10.3(c) shows that degradation followed the closure of the Sukkur barrage, but as a result of sedimentation behind the barrier, the bed elevation rose to a final height above the initial level. This aggradation reflects the greater loss of water than of sediment from the river to the canals above the barrage as a result of sediment control at the canal headworks.

Regime theory provides the basic design criteria for stable sediment-bearing artificial channels. A canal 'in regime' has a sediment transport capacity which matches the input sediment transport rate, so that, although

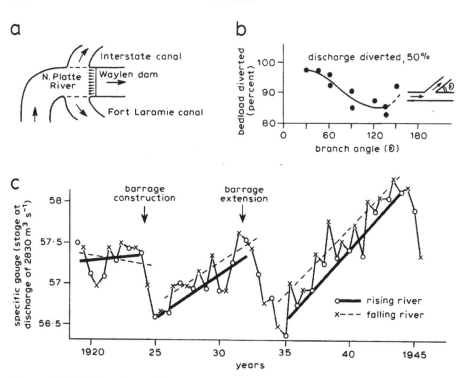

Figure 10.3 (a) Canals supplied by the North Platte River in Wyoming. The Interstate canal has functioned satisfactorily, but severe sedimentation occurred in the Fort Laramie canal. (b) Diversion of bedload controlled by branch angle of a distributary channel (after Garde and Ranga Raju, 1977). (c) Specific gauge record for the Indus River downstream from the Sukkur barrage (after Inglis, 1949a).

local, temporary scour and fill occur, over a period of years the morphology and the 'specific gauge' are on average constant (pp. 18–20). Such channels may mature after construction, undergoing self-adjustment of slope, width and depth to eliminate initial shortcomings in design and to accommodate differences between local sediments and those distributed eventually downstream from the headworks. Criteria for stability in regime canals (Blench, 1969) include steep, straight, hydraulically smooth cohesive banks, a steady operating discharge, steady bedload transport in the dune phase under subcritical flow and a large relative 'smoothness' (depth:grain size ratio). The design problem requires, for given discharge, sediment size and concentration, a stable combination of width, depth (which together determine velocity through $Q = wdv$) and slope (which may be fixed by valley and project conditions). A unique solution is provided by three equations, usually a resistance law, a sediment transport equation and a channel shape equation (see pp. 24–7). However, a unique solution may not be essential. Instead, there may be limiting conditions for channel stability, such as a maximum Froude number, sediment concentration, slope or stream power

(Mao and Flook, 1971). Above a critical slope, a stable *straight* channel becomes impossible (Schumm and Khan, 1972, Charlton, 1975; pp. 213–17). Furthermore, for a given material a range of velocities exists between scouring and silting velocities, as indicated by the Hjulström curve (Figure 3.8b); this range increases as sediment sizes decrease. The evolution of regime theory has involved the recognition of these limiting conditions, of the number of degrees of freedom for self-adjustment, and of the important control of the stable channel morphology exercised by the perimeter sedimentology and the sediment transport rate, which according to Schumm (1968a) are two related properties. The approach is empirical and based on observations of the morphology of canals known to be stable. The equations summarizing the data therefore have coefficients which vary according to the geotechnical conditions of the observed channels.

Punjab canals constructed after 1850 had arbitrarily selected widths and slopes, and often suffered severe sedimentation problems during self-adjustment. However, Kennedy (1895) attempted to rationalize design by gathering data from the stable Upper Bari Doab canal (in sandy silt) to produce a relationship for a velocity just capable of avoiding silting (Table 10.3). Although he also used a Manning-type resistance equation, the lack of a third equation meant that wide, shallow and narrow, deep channels were wrongly deemed equally stable under given conditions of discharge and sediment load. It was Lindley (1919) who first recognized the need for three independent relationships, with the non-silting velocity being a function of both depth and width so that a width–depth relation could be derived. Lindley also introduced the term 'regime' to describe stable canals.

Lacey (1930) was the first to introduce explicitly an external sedimentological constraint – the 'silt factor' (Table 10.3) – to a set of regime equations based on data from canals mainly in non-cohesive sediments. These regime relations have been variously modified since their introduction (Mahmood and Shen, 1971) and are expressed in Table 10.3 in a form derived by Inglis (1949a) which is convenient for design purposes, with each dependent variable expressed as a function of the discharge and the silt factor. The wetted perimeter (width) is evidently independent of sediment size, but depth and velocity vary inversely with grain size while slope varies directly with grain size but inversely with discharge (note the discharge exponents). The width:depth ratio (p/R) increases with discharge, but is inversely related to the grain size for the observed range of grain diameters. Inglis (1949b) further generalized these equations by introducing a parameter for the bedload concentration, C_b. Since $pRv = Q$, the Lacey equations are overdetermined; Equation (C) is unnecessary. Thus, in many channels, if a silt factor f_{vR} is calculated from Equations (B) and (C) it differs from f_{vs} derived from Equations (C) and (D), using measured velocity and slope. Chien (1957) showed that f_{vR}, obtained essentially from a sediment transport relation, is strongly related to the bedload concentration (Figure

Table 10.3 Selected regime equation systems for straight sand- and gravel-bed streams or canals. All coefficients have been converted to SI units; in the Lacey system the silt factor has been assigned the dimension $L^{1/2}$ (see Appendix).

Equations for straight sand-bed channels

Kennedy (1895)	Lindley (1919)	Lacey (1930)	Blench (1952)
A. Manning-type resistance equation	A. Manning-type resistance equation	A. $p = 4.84Q^{0.5}$	A. $w = (F_b/F_s)^{0.5}Q^{0.5}$
B. $v_0 = 0.55d^{0.64}$	B. $v_0 = 0.57d^{0.57}$	B. $R = 0.47Q^{1/3}f_L^{-1/3}$	B. $d = (F_s/F_b)^{1/3}Q^{1/3}$
	C. $v_0 = 0.28w^{0.36}$	C. $v = 0.44Q^{1/6}f_L^{-1/3}$	C. $s = \dfrac{F_b^{5/6}F_s^{1/12}\,v^{1/4}}{3.63gQ^{1/6}(1+C_b/2330)}$
	D. By combining B and C, $w = 7.2d^{1.61}$	D. $s = 0.0003f_L^{5/3}Q^{-1/6}$	D. $F_b = 0.58D_{50}^{0.5}(1+0.012C_b)$
		E. $f_L = 1.76D_{50}^{0.5}$	E. $F_s = 0.009$ (sandy loam) to 0.028 (clay loam)
			$F_s = F_b^2/8$ (gravel-bed rivers)

v_0 is Kennedy's non-silting velocity; f_L is Lacey's silt factor; F_b and F_s are Blench's bed and side factors

Equations for straight gravel-bed channels

Kellerhals (1967)

A. $w = 3.26Q^{0.5}$

B. $\dfrac{v}{\sqrt{(gRs)}} = 6.5\left(\dfrac{d}{D_{90}}\right)^{1/4}$

C. $\tau_{0c} = \rho_w gRs = 15.79D_{90}^{0.8}$

Derived from these:

D. $d = 0.183Q^{0.4}D_{90}^{-0.12}$

E. $v = 1.68Q^{0.1}D_{90}^{0.12}$

F. $s = 0.086Q^{-0.4}D_{90}^{0.92}$

Charlton *et al* (1978)

A. $w = 3.74Q^{0.45}$

$w_0 = KW$, where K ranges from 0.7 to 1.3 depending on bank vegetation

B. $d_e = 0.066D_{65}s^{-1}$

$d_0 = 0.9K^{-1.82}d_e$

C. $\dfrac{v}{\sqrt{(gds)}} = 5.49 + \log 2.77\left(\dfrac{d_e}{D_{90}}\right)$

D_{65} and D_{90} are here measured in metres, making d/D dimensionless; w_0 and d_0 are width and depth adjusted for bank vegetation effect; d_e is effective depth (surveyed channel depth + median bed material small-axis length).

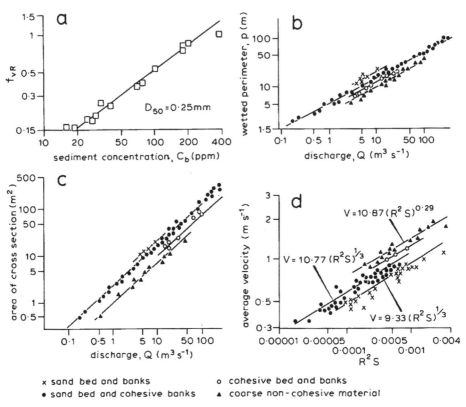

x sand bed and banks   o cohesive bed and banks
• sand bed and cohesive banks   ▲ coarse non-cohesive material

Figure 10.4 (a) Relationship between the velocity–hydraulic radius silt factor, f_{vR} and the bedload sediment concentration (after Chien, 1957). (b) Variation of canal cross-section wetted perimeter with discharge and perimeter sediments. (c) Variation of canal cross-section areas with discharge and sediment type. (d) Relationship of velocity to the product of hydraulic radius and slope for canals in different materials. (Figures b–d, after Simons and Albertson, 1960.)

10.4a), while f_{vs}, which is derived from a resistance relation, is therefore, as might be expected, correlated to the bed material grain size.

The recognition that regime channel forms reflect the intensity of bedload transport is also apparent in Blench's (1952, 1969) regime equations (Table 10.3), which further expressly define two sedimentological constraints, the bed (F_b) and side (F_s) factors, which attempt to measure the distinct influences of non-cohesive bed and cohesive bank materials. Initially estimated by

$$F_b = v^2/d \quad \text{and} \quad F_s = v^3/w \tag{10.9}$$

it appears that the bed factor is essentially the Froude number and the side factor is a measure of the shear stress on a hydraulically smooth bank (Blench, 1969). However, the bed factor may also be defined in terms of grain size and bedload concentration by weight (C_b, ppm), and limiting values of the side factor have been specified for different bank materials. Thus the dependent design variables are estimated from the design discharge and external material constraints (although these are not well defined) and, when calculated from Equation (10.9), they fail to correlate satisfactorily with specific sedimentological properties (Charlton *et al*, 1978). Consequently, Simons and Albertson (1960) have adopted a simpler classificatory approach. Data from American and Indian canals in varying geotechnical environments were classified into four sedimentological types: sand bed and banks, sand bed and cohesive banks, cohesive bed and banks, and coarse non-cohesive materials. For each type relationships exist between wetted perimeter and discharge (Figure 10.4b), cross-section area and discharge (Figure 10.4c) and velocity and the product $R^2 s$ (Figure 10.4d). For design purposes, sedimentology and discharge are known, and wetted perimeter and cross-section are defined from the relevant curves in Figures 10.4(b) and (c). Then hydraulic radius R and velocity v are estimated from $R = A/p$ and $v = Q/A$, so that slope can be obtained from Figure 10.4(d).

The curves defined for different sediments in Figures 10.4(b)–(d) suggest that exponents are constant but intercept coefficients vary with geotechnical conditions. However, it is apparent in Chapter 3 that a general resistance law (e.g. Equation 3.23) is a logarithmic function, not a power function (i.e. it is a curve on a log–log plot). Thus a power function only approximates the law over a narrow range. Lacey's (1930) resistance equation (Figure 10.4d)

$$v = 10.77 R^{2/3} s^{1/2} \qquad (10.10)$$

in his original regime system, although confirmed as valid for rippled sand beds by laboratory experiments (Ackers, 1964), is of restricted applicability. Lacey and Pemberton (1972) demonstrate, therefore, that the functional form of the relationship varies with bed material size as the logarithmic resistance equation is approximated by different power functions:

$$\left.\begin{array}{ll}
\text{fine sand,} & v \propto Rs \\[4pt]
\text{medium sand,} & v \propto R^{3/4} s^{1/2} \\[4pt]
\text{coarse sand,} & v \propto R^{2/3} s^{1/2} \\[4pt]
\text{gravel,} & v \propto R^{5/8} s^{1/4}
\end{array}\right\} . \qquad (10.11)$$

Thus considerable care is necessary in ensuring that regime equations are applied to the same geotechnical conditions as those for which they were derived, since both intercepts *and* exponents vary.

The classic regime equations developed by Kennedy, Lindley, Lacey, Inglis and Blench, based on data from the Indian canal systems, are mainly applicable to sand-bed channels. In straight, gravel-bed streams the bed material is only mobile at bankfull, or dominant discharge, and Kellerhals

(1967) in Canada and Charlton *et al* (1978) in Britain have developed systems of gravel-bed regime-type equations that incorporate a threshold criterion (Table 10.3). The depth relation (Equation B) of the latter authors is, for example, based on the Shields criterion for incipient motion. These approaches are more physically based and include 'relative smoothness' relations which are more general than the Lacey (1930) resistance equation (10.10). That adopted by Charlton *et al* (1978) – their Equation (C) in Table 10.3 – is a logarithmic function valid over a wide range of relative 'smoothness' values. Notably, this system of equations also allows approximately for the effect of bank vegetation on width and depth. If the bank vegetation factor (K) in Table 10.3 implies a wider than expected channel, it is also shallower than expected to compensate. However, as Henderson (1961) notes, there is a basic incompatibility between the regime concept, which implies stability with sediment transport, and the tractive force threshold concept. Charlton *et al* (1978) overcame this by replacing the Shields criterion by the Meyer-Peter and Muller bedload transport equation for cases where the gravel bed experiences significant transport at sub-bankfull discharges.

Regime theory defines canal geometries which are similar to those of equilibrium natural rivers. Thus, for constant values of the silt factor, the Lacey (1930) equations indicate discharge exponents of $\frac{1}{2}$, $\frac{1}{3}$, $\frac{1}{6}$ and $-\frac{1}{6}$ for width, depth, velocity and slope. Only the last exponent differs significantly from the downstream hydraulic geometry exponent characteristic of rivers ≈ -0.6 according to Hack, 1957; p. 227). A major difference, however, is the smaller number of degrees of freedom experienced by canals. Rivers are free to vary their roughness over a wider range, whereas canals are designed for a limited range of stream power or Froude number conditions, so that they do not experience transitional and supercritical flow with plane bed and antidune bedforms in sand. Rivers are also free to vary their cross-sections spatially, developing riffles, pools and meanders. As Charlton (1975) notes, limiting slope criteria may be used to ensure that shoaling and meandering do not occur in regime canals. Although it might be feasible to design a stable meandering channel, by defining sinuosity and wavelength from relationships such as those in Figure 7.1(d) and Equation (7.9), the bend shapes must also be correctly designed to maintain stability. This would necessitate selection of a radius of curvature:width ratio and identification of the optimal sine-generated curve or circular bend morphology in order to minimize rapid energy dissipation.

Some aspects of river management

Regulation and control are essential elements of many river uses and management schemes. Various examples are discussed below.

(a) *Rivers as boundaries* Although water resource management is complicated by the administrative division of a basin, rivers have formed boundaries between land holdings, counties, states and countries. At the land-holding scale, ownership disputes arise if a migratory channel centre line forms a boundary. Normally, under Common Law, land ownership remains fixed after a sudden channel change, whereas ownership of accreted land passes to the owner of the land added to if the change is gradual (Hodges, 1976). However, after river training along the Missouri River created new, stable areas of productive floodplain, prolonged litigation occurred over land titles (Ruhe, 1971). Where unstable international river boundaries occur (e.g. the Rhine and Rio Grande) disputes may be complicated by different legal systems, and river training by international co-operation is desirable. For example, the stabilization of the Rio Grande between the USA and Mexico (Figure 10.5a) involved exchanges of land and apportionment of costs in relation to benefits received (Beckinsale, 1969).

(b) *River navigation* The viability of commercial river traffic is controlled by the hydraulic geometry, with inadequate depth and excessive velocity being particularly critical. Dredging may be necessary to maintain navigability through the shallow 'crossings' at bend inflections. For example, of 95 crossings in 180 km of the Mississippi River near Vicksburg, Mississippi, 62 were dredged at least once between 1950 and 1968 and 3 needed more than 20 dredgings (Winkley, 1971). However, the expense of dredging is often more than offset by the commercial benefits generated by the extra traffic. For instance, maintenance of the Columbia River, Washington, at a depth of 11 m, compared with an undredged depth of 3.5 m in 1885, produces a benefit:cost ratio of 1.78 (Hyde and Beeman, 1965). Langbein (1962) derived an index, the 'specific tractive force', which relates the horsepower of a boat to the power required to maintain its design speed; barges have values of 0.0026. For any combination of river depth and velocity, the minimum specific tractive force required to sustain upstream transport is defined, and if it exceeds approximately 0.002 river transport is not feasible (Figure 10.5b). Thus it is possible to classify rivers in terms of their suitability for navigation and to identify the combination of depth and velocity required after channelization and flow confinement by training works.

(c) *Flood control* A highly sinuous meandering river such as the Mississippi is prone to flooding because of the backwater effect of bends. Flood protection and prevention is necessary for the fertile, populous agricultural floodplain areas. Levees exclude floods up to their design discharge, but Figure 1.2(d) shows that deterioration of the channel capacity since construction has largely eliminated the freeboard 'safety factor'. The levee width, cross-section and internal material composition must be designed to minimize seepage, which under a high hydraulic gradient may create wash-

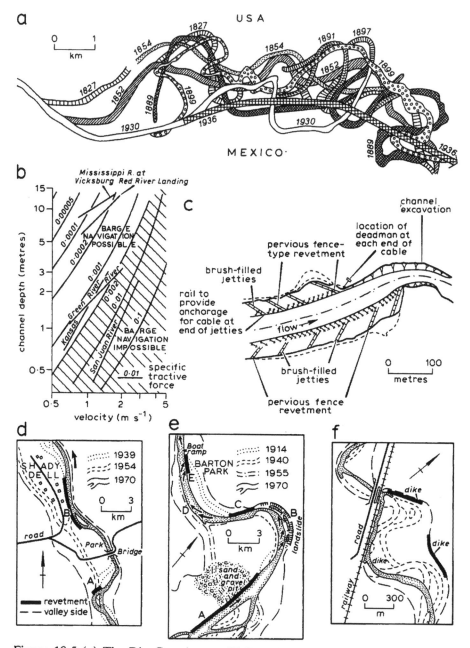

Figure 10.5 (a) The Rio Grande near El Paso–Ciudad Juarez, showing the main recorded shifts of the channel prior to regulation (after Beckinsale, 1969, and Boggs, 1940). (b) River depth and velocity surface contoured by values of the minimum 'specific tractive force' required for upstream navigation; the Green and Kansas Rivers are at the margins of navigability (after Langbein, 1962). (c) A typical example of training works on the Fivemile Creek (after Miller and Borland, 1963). (d) River management problems at Shady Dell on the River Molalla, Oregon. (e) Revetment and bank erosion on the River Clackamas, Oregon. (f) Bridge approach problems on the River Molalla, Oregon. (Figures d–f after Palmer, 1976.)

outs and sandboils which progressively undermine the structure. Between the levees, the river continues to erode its banks, a process contributing about 900,000 m³ of material per annum to the Mississippi sediment load before bank protection (Winkley, 1971). On large rivers with steep cohesive banks, mass failures occur especially during groundwater drawdown as flood levels drop. Chapter 6 (p. 167) suggests that banks more than 10 m high above the low-water level may be potentially unstable. Thus, to maintain levee flood protection it is necessary to preserve channel capacity and protect banks against erosion and migration so that the levees are not attacked. Meander bend cut-offs aid flood control by fostering bed degradation, which increases the channel capacity and accelerates the flood travel velocity. On the lower Mississippi, 16 cutoffs between 1929 and 1942 reduced the river length by 240 km (35%) and lowered flood stages by 2–2.5 m. Bank protection is a necessary concomitant of river shortening in that it prevents translation of the extra energy into accelerated channel migration. It also follows that great care is required in controlling flooding by deliberately-fostered bed lowering through dredging or removal of old control structures such as weirs, because the increased bank height can lead to instability and enhanced rates of bank failure and therefore of channel migration.

The extent of deliberate manipulation of river processes is increasing and the planning approach is becoming more integrated. For example, after the closure of an upstream dam on the Arkansas River for flood control and hydro-electric power generation, downstream clear-water degradation was actively encouraged by dredging to inhibit armour development, because the degraded profile needed fewer low-lift navigation locks which are expensive both to construct and operate (Madden, 1965). This level of management demands sophisticated modelling techniques which are capable of predicting both the short-term and long-term consequences of manipulation of river processes in the face of the uncertainties of random meteorological and hydrological influences. Nevertheless, these techniques must always be applicable to the practical circumstances of management (Simons and Li, 1979), being developed with the limitations of computational facilities, appropriately trained personnel, and available data for calibration purposes all in mind. Physical simulation models are preferred to empirical statistical models whose coefficients have limited generality. However, the river training and bank protection works which are clearly essential aspects of the river uses discussed above involve pragmatic combinations of empirical rules of thumb and theoretical models with structural and non-structural solutions, depending on the specific problems presented by each river.

River training

Tortuous meanders, rapidly migrating unstable banks, and wide shallow sections with shifting sand bars all inhibit navigation, flood control and riparian agriculture. River training attempts to control channel alignment, by defining and stabilizing the bank line, and to contract the width to provide a deeper navigable channel. Permeable and impermeable dykes, wooden and steel jacks, and quick-growing well rooted plants (e.g. willows) have all been used on the Rio Grande, Missouri, Mississippi and many smaller rivers. The general aim is to slow the bank velocity to encourage natural bed and bank sedimentation along the predetermined bank line, behind the dykes or jacks. This requires a heavy sediment load, which is likely to be available in a river requiring stabilization. However, rates of bank stabilization on the Mississippi between the Missouri and Ohio tributaries were slowed by land conservation and reservoir construction in the Missouri basin, which reduced the main source of sediment supply to the Mississippi (Tiefenbrun, 1965). Wooden pile dykes projecting into the flow are susceptible to rotting and underscour, and stone dykes are normally used on the Mississippi today (Winkley, 1971). Jacks linked by wire (Figure 10.5c) follow the desired bank line completely, but are aesthetically unsatisfactory especially if they fail after an extreme event.

The key to successful river training is to mimic the natural river geometry. Miller and Borland (1963), faced with a problem of rapid bank erosion on two streams draining into a reservoir and causing severe sedimentation, developed an empirical model for the design width and then checked that tractive forces in such a channel would lie between the limiting values which would cause scour or fill. After bank stabilization, sediment yields decreased by 90%, but Miller and Borland note that this could reflect the achievement of a natural equilibrium rather than being the result of river control. A major requirement is that slope is not steepened by shortening the river course drastically, and that channel bends are therefore maintained. The minimum bend radius for successful stabilization of Frenchman River, Nebraska (Frogge, 1967) was found to be 16 m as long as velocity was not excessive, and on the Mississippi severe bank attack occurs if the bend radius is less than 1830 m (2.5 times the width). Smooth meander curves should be designed with normal bar spacing, and structures should take advantage of naturally occurring secondary currents. As long as these factors are considered, training works will bring considerable benefits by stabilizing the banks and creating new, usable land (Ruhe, 1971), although possible reductions may occur in the channel's flood capacity and in aquatic biological productivity (Keller, 1976) which to some extent offset these benefits.

Bank protection

Revetments are designed to protect river banks against erosion and include stone rip-rap above the water level and protective mattresses laid against the bank underwater. The median diameter of stone rip-rap is defined by employing tractive force theory in reverse, with an appropriate safety factor. Thus, for a given bank slope and bankfull shear stress, the minimum stable stone size is defined. In fact, traditional methods often yield unsafe rock sizes, and it is normally necessary to base the estimation on a stability analysis including instantaneous lift and drag forces as well as the effect of flow orientation (Equation 10.8) and the effect of surface waves. Rip-rap should be graded, not uniform. The maximum rock size should be about twice the median diameter, which in turn should be twice the D_{20} diameter (Simons and Şentürk, 1977). Gradation to gravel sizes at the smaller end of the size spectrum allows interstices to be filled so that no pockets exist in which scour can be initiated. Hand-emplaced rip-rap, built like dry-stone walls, is most resistant and aesthetic but very expensive; a danger associated with bulldozing from the bank top is that avalanching segregates rock sizes down the bank. Below the water line woven willow or articulated concrete mattresses are used in large rivers. The latter, consisting of concrete blocks $0.6 \times 1.2 \times 0.2$ m linked by copper wire, is laid on Mississippi river banks by specially constructed barges (Noble, 1976).

Revetment is usually placed at susceptible locations such as concave banks, but it may affect stream behaviour by inducing erosion elsewhere. Palmer (1976) describes examples of piecemeal protection on the high-energy Molalla and Clackamas river near Portland, Oregon. Figure 10.5(d) shows a private housing development, 'Shady Dell', on the bank of the Molalla River where a revetment (A) was placed to block the river entrance to a flood bypass channel. The lack of outflow current at B caused bank erosion there, where '. . . expensive river-control structures are being built with public funds to protect private property encroachment. Shady Dell is still subject to flood damage which will probably be born by the general public through insurance' (Palmer, 1976, pp. 340–1). On the Clackamas River, revetment A (Figure 10.5e), set up to protect private property, has deflected the currents, which oscillate downstream to cause erosion at B, C, D and E. In Barton Park, public conveniences at C and a boat ramp at E have required protection, so a '. . . popular park site was chosen to take advantage of the natural river values but those same natural values are being destroyed to preserve toilets' (Palmer, 1976, p. 343). These examples of the wider spatial impact of local private action emphasize the need for the public control of river management.

Bridge crossings

Training and revetment are particularly necessary to protect the capital investment in major bridge crossings, where two river management problems concern bed scour and approach channel orientation. Scour at bridges arises because of contraction of the channel upstream and because of local turbulence and vortices generated at the bridge piers. The latter is the greater problem because it is difficult to predict and may undermine the pier if not allowed for in the bridge design. This local scour depends on the approach stream velocity and depth, the pier size, shape and orientation relative to the flow, natural variation of bed elevation as sand dunes pass, the initial difference between sediment transport rates upstream and downstream from the pier, and the protective effect of coarse gravels at or below the bed surface. Field observations are rare, but Neill (1965) noted 3–5 m of scour, followed by fill on the falling stage, during a June 1962 flood on the Beaver River, Alberta, when total floodwater depths were 8.5–9.8 m. To predict the scour depth, Inglis (1949a) suggested that bridge pier scour would increase the regime channel depth, estimated from Lacey's Equation B in Table 10.3, by two to three times. Logically, however, the scour hole depth is dependent on pier turbulence, and Shen (1971b) has successfully related it to a pier Reynolds number (Re_b) using

$$d_s = 2.2 \times 10^{-4} Re_b^{0.62} \tag{10.12}$$

in which $Re_b = v w_p / v$ and w_p is the pier width. This model overestimates scour depth in bed sediment coarser than 0.5 mm median diameter, but underestimates by up to five times the scour depth occurring when the flow approaches a pier obliquely. Correct pier design (sharp-nosed piers generate less turbulence) and orientation minimize scour, and horizontal aprons at the bed, upstream piles to disrupt the turbulent wake, and rip-rap dumps all protect against it.

Migration of the approach channel destroys correct relative flow and pier alignment and attacks bridge abutments and road or rail embankments (Figure 10.5f), until in extreme cases they may be breached and the original arch rendered redundant. It is under these circumstances that guide banks, river training works and bank protection must exploit the natural geometry, sediment transport and secondary flow processes of the river to maintain satisfactory alignment. Management must work with the river rather than attempt to stop it in its tracks.

Postscript

Insofar as man's attempts to understand his physical environment are often ultimately directed towards its management and control, this chapter provides a fitting conclusion to a review of our understanding of channel forms, sediments and processes. The study of alluvial river behaviour involves multidisciplinary co-operation, with geomorphologists concerned with morphology, geologists with sediments, engineers with management, and all three with hydraulic and sediment transport processes. It is chastening for those analysing rivers today to recall the long history of this interest. River management began with the classical riparian civilizations of Mesopotamia, Egypt, India and China, and canals excavated among the ruins of Nippur in Mesopotamia have been dated to approximately 5200 BC. Around 2250 BC China boasted a hydraulic engineer as its Emperor, although there is some suspicion that Emperor Yü's existence is as legendary as his claim to have guided nine rivers to the sea, in an early example of floodway design (Biswas, 1970)! The quantitative observation of river behaviour can be traced back to the 'Nilometers' devised from 3500 BC to measure variations in the level of the River Nile.

Truly scientific investigations began with Leonardo da Vinci (1452–1519), who would probably have been designated the father of fluvial geomorphology had he written the treatise on water whose contents he planned. By 1800 several hydraulic engineers had laid the foundation of quantitative, scientific fluvial research. Guglielmini (1697) recognized the force: resistance balance between flow and sediments as the fundamental control of fluvial morphology; Bernoulli (1738) and Chezy (1775) identified laws governing energy losses in streamflow; and Du Buat (1786) combined all this in a remarkable description of channel form and process. His comment '. . . the windings of rivers are a means used by nature to hasten the establishment of grade, in spite of the apparent excess of gradient' (Du Buat, 1786, I, 150) is an illustration of his development of ideas and terminology restated much later. In 1802 Playfair emphasized the role of rivers as erosional, transportational and depositional agents in the wider context of Huttonian geology. He introduced the appreciation of quantitative observation of sediment transport: '. . . in some instances, the water of a river in a flood contains earthy matter suspended in it, amounting to more than the 250th part of its own bulk' (Playfair, 1802, p. 106), as well as of the properties of sediments: 'the nearer that any spot is to the mountains, the larger are the gravel stones, and the less rounded is their figure' (p. 381).

In spite of this 200 year ancestry, it remains impossible to predict and explain quantitatively the three-dimensional mutual adjustment of all the degrees of freedom available to self-formed alluvial channels. Thus, the design of artificial channels remains restricted to straight reaches of uniform trapezoidal cross-section in a narrow range of stream power conditions. Before concern with aesthetics and ecology can become an automatic element of the design practice, further advances are necessary, and these will be fostered by the growing collaboration between those geomorphologists, geologists, ecologists and engineers concerned with river channel behaviour.

Appendix

Units, dimensions and variables

Physical quantities are measured by three *fundamental units* and various combinations of these which form *derived units*. The three fundamental units are mass, length and time.

(a) Mass (denoted here by the dimensional formula [M]). This is the globally constant quantity of matter in a body. The weight of a body is the force due to gravity acting on it, which varies spatially. SI units of mass, used predominantly in this book, are the kilogramme (kg) and the tonne (t; 10^3 kg). Small masses are quoted in the centimetre–gramme–second (cgs) unit, the gramme (g; 10^{-3} kg).

(b) Length [L]. The one-dimensional extent of a body, this is measured in metres (m), kilometres (km; 10^3 m) and centimetres (cm; 10^{-2} m).

(c) Time [T]. The duration of action of some process on a body is measured in seconds (s), but hours, days or years may be more appropriate time scales for the statement of sediment load transport or erosion rates.

A fourth fundamental unit [θ] which may be of significance in the analysis of river behaviour is temperature, which affects fluid viscosity and sediment transport. It is measured in degrees Centigrade (°C).

The derived units, which involve various combinations of mass, length and time, may be classified into three groups: the geometric, kinematic and dynamical units.

(a) Geometric units are powers of length, which include, with their dimensional formulae and SI units,

$$\text{length: } [L] \text{ m}$$
$$\text{area: } [L^2] \text{ m}^2$$
$$\text{volume: } [L^3] \text{ m}^3.$$

(b) Kinematic units are derived from length and time, and include

velocity: $[LT^{-1}]\,m\,s^{-1}$
acceleration: $[LT^{-2}]\,m\,s^{-2}$
volumetric discharge: $[L^3T^{-1}]\,m^3\,s^{-1}$
kinematic viscosity: $[L^2T^{-1}]\,m^2\,s^{-1}$ (or $cm^2\,s^{-1}$ in cgs units).

(c) Dynamic units incorporate mass, length and time, and include:
density (mass per unit volume): $[ML^{-3}]\,kg\,m^{-3}$ (or $g\,cm^{-3}$);
weight (mass times gravitational acceleration): $[MLT^{-2}]$ kilogramme-force (kgf);

Weight is a body force. A force is measured by the acceleration which it produces when applied to a given mass. In SI units, the newton (N) is the force which gives a 1 kg mass an acceleration of $1\,m\,s^{-2}$, while the kilogramme-force (kgf) is approximately 9.81 N. The cgs unit, the dyne (dyn; $10^{-5}\,N$), is that force giving a mass of 1 g an acceleration of $1\,cm\,s^{-2}$.

force: $[MLT^{-2}]$ newton, kgf, dyne;
specific weight (weight per unit volume): $[MLT^{-2}L^{-3}] = [ML^{-2}T^{-2}]$ kgf m^{-3};
shear stress (force per unit area): $[MLT^{-2}L^{-2}] = [ML^{-1}T^{-2}]$ kgf m^{-2}, dyn cm^{-2};
work (force times distance): $[MLT^{-2}L] = [ML^2T^{-2}]$ joule (J), erg $(10^{-7}\,J)$;

Work is the product of a force and the distance moved by the point of its application, in the direction of application. If a force of 1 N acts over 1 m, the work done is 1 joule.

energy: $[ML^2T^{-2}]$ joule (J), erg $(10^{-7}\,J)$;

Changes in energy result in work being done; thus energy and work have the same units. Potential energy $(= mgh)$ is therefore the work done in raising a body of mass m to a height h against gravity.

power (rate of doing work or expending energy): $[ML^2T^{-3}]$ watt (W = 1 joule per second, $J\,s^{-1}$), erg s^{-1} $(= 10^{-7}\,J\,s^{-1})$;

Power expenditure per unit area is $[ML^2T^{-3}L^{-2}] = [MT^{-3}]$ measured in $W\,m^{-2}$ $(J\,m^{-2}\,s^{-1})$, or erg $cm^{-2}\,s^{-1}$.

momentum (the product of mass and velocity): $[MLT^{-1}]\,kg\,m\,s^{-1}$;
dynamic viscosity: $[ML^{-1}T^{-1}]\,kg\,m^{-1}\,s^{-1}$ (or $g\,cm^{-1}\,s^{-1}$);

Units of dynamic viscosity are the poiseuille (Pl; $1\,kg\,m^{-1}\,s^{-1}$) and the poise (P; $1\,g\,cm^{-1}\,s^{-1}$).

Dimensionless numbers

Simple dimensionless variables include the width:depth ratio (F) and the slope (s), each of which is a length:length ratio. The Reynolds number is a dimensionless group – a number rendered dimensionless by the particular combination of variables, as shown using their dimensional formulae:

$$Re = \frac{\rho_w Dv}{\mu} = \frac{\text{density} \times \text{depth} \times \text{velocity}}{\text{dynamic viscosity}}$$

$$= \frac{[ML^{-3}][L][LT^{-1}]}{[ML^{-1}T^{-1}]} = \frac{[ML^{-1}T^{-1}]}{[ML^{-1}T^{-1}]} = [M^0 L^0 T^0].$$

Dimensionless ratios are numerically equivalent whichever system of units is used, as long as this system is used consistently. They are useful in scaling models; a laboratory flume requires a kinematic and dynamical *similarity* of flow conditions, as defined by the Froude and Reynolds numbers, to be representative of conditions in natural rivers.

Equations

An equation is *dimensionally balanced* if the products of the dimensional formulae of variables on the left- and right-hand sides are equal: in this case, constants will be dimensionless and will apply regardless of the units employed. Many empirical relationships have dimensioned constants, and these have been converted to SI units in the text. For example, Hack (1957) found that

$$s = 18D_{50}^{0.6} A^{-0.6},$$

where slope s is in ft per mile, D_{50} is particle diameter in millimetres and A is area in square miles. For s' to be dimensionless (metres per metre), and A' to be in square kilometres, the equation must be rewritten to include the appropriate conversion factors:

$$5280s' = 18D_{50}^{0.6} (0.386A')^{-0.6}$$

or

$$s' = 0.006D_{50}^{0.6} A'^{-0.6}.$$

In the Lacey (1930) regime equations (Table 10.3) the dimensions of the silt factor have been taken as $mm^{1/2}$ from Equation (E), and the constants in equations (A)–(D) have been adjusted to accommodate the change from fps to SI units in other variables. Blench's (1952) Equations (A), (B) and (C) are dimensionally balanced, but the bed and side factors have dimensions. Thus

$$F_b = v^2/d = [L^2T^{-2}L^{-1}] = [LT^{-2}]$$
$$F_s = v^3/w = [L^3T^{-3}L^{-1}] = [L^2T^{-3}].$$

The equations for these factors (D and E) have therefore been adjusted on the assumption that they must retain these dimensions to maintain the dimensional balance of Equations (A), (B) and (C).

References

Aastad, J. and Sognen, R. (1954) 'Discharge measurements by means of a salt solution – "the relative dilution method" ', *Publication. International Association of Scientific Hydrology*, 38, 289–92.

Abbott, J. E. and Francis, J. R. D. (1977) 'Saltation and suspension trajectories of solid grains in a water stream', *Philosophical Transactions of the Royal Society*, 284A, 225–53.

Abrahams, A. D. (1972) 'Drainage densities and sediment yields in E. Australia', *Australian Geographical Studies*, 10, 19–41.

Ackers, P. (1964) 'Experiments on small streams in alluvium', *Journal of the Hydraulics Division, American Society of Civil Engineers*, 90, 1–37.

Ackers, P. and Charlton, F. G. (1970a) 'Meander geometry arising from varying flows', *Journal of Hydrology*, 11, 230–52.

Ackers, P. and Charlton, F. G. (1970b) 'The slope and resistance of small meandering channels', *Proceedings of the Institution of Civil Engineers*, 47, Supplementary Paper 7362-S, 349–70.

Akroyd, T. N. W. (1964) *Laboratory Testing in Soil Engineering*, London, Soil Mechanics Ltd.

Al-Ansari, N. A., Al-Jabbari, M. and McManus, J. (1977) 'The effect of farming upon solid transport in the River Almond, Scotland', *Publication. International Association of Scientific Hydrology*, 122, 118–25.

Alexander, G. N. (1972) 'Effect of catchment area on flood magnitude', *Journal of Hydrology*, 16, 225–40.

Allen, J. R. L. (1965) 'A review of the origin and characteristics of recent alluvial sediments', *Sedimentology*, 5, 89–191.

Allen, J. R. L. (1968) *Current Ripples, their Relation to Patterns of Water and Sediment Motion*, Amsterdam, North-Holland.

Allen, J. R. L. (1969) 'On the geometry of current ripples in relation to stability of fluid flow', *Geografiska Annaler*, 51A, 61–96.

Allen, J. R. L. (1970a) *Physical Processes of Sedimentation*, London, George Allen & Unwin.

Allen, J. R. L. (1970b) 'A quantitative model of grain size and sedimentary structure in lateral deposits', *Geological Journal*, 7, 129–46.

Allen, J. R. L. (1976a) 'Time-lag of dunes in unsteady flows: an analysis of Nasner's data from the R. Weser, Germany', *Sedimentary Geology*, 15, 309–21.

Allen, J. R. L. (1976b) 'Computational models for dune time-lag: population structures and the effects of discharge pattern and coefficient of change', *Sedimentary Geology*, 16, 99–130.

Allen, J. R. L. (1977) 'Changeable rivers: some aspects of their mechanics and sedimentation', in Gregory, K. J. (ed.) *River Channel Changes*, Chichester, Wiley, pp. 15–46.

American Society of Civil Engineers (1963) 'Friction factors in open channels', *Journal of the Hydraulics Division, American Society of Civil Engineers*, 89, 97–143.

Amorocho, J. (1963) 'Measures of the linearity of hydrologic systems', *Journal of Geophysical Research*, 68, 2237–49.

Anderson, A. G. (1942) 'Distribution of suspended sediment in a natural stream', *Transactions of the American Geophysical Union*, 23, 678–83.

Anderson, A. G. (1967) 'On the development of stream meanders', *Proceedings of the International Association for Hydraulics Research, 12th Congress, Fort Collins*, 1, 370–8.

Anderson, D. G. (1970) 'Effects of urban development on floods in northern Virginia', *Water Supply Paper. United States Geological Survey*, 2001C.

Anderson, M. G. and Burt, T. P. (1977) 'A laboratory model to investigate the soil moisture conditions on a draining slope', *Journal of Hydrology*, 33, 383–90.

Anderson, M. G. and Burt, T. P. (1978) 'The role of topography in controlling throughflow generation', *Earth Surface Processes*, 3, 331–44.

Anderson, M. G. and Calver, A. (1977) 'On the persistence of landscape features formed by a large flood', *Transactions of the Institute of British Geographers, New Series*, 2, 243–54.

Anderson, M. G. and Richards, K. S. (1979) 'Statistical modelling of channel form and process', in Wrigley, N. (ed.) *Statistical Applications in the Spatial Sciences*, London, Pion, pp. 205–28.

Anderson, P. W. (1963) 'Variations in the chemical character of the Susquehanna River at Harrisburg, Pennsylvania', *Water Supply Paper. United States Geological Survey*, 1779B.

André, J. E. and Anderson, H. W. (1961) 'Variation of soil erodibility with geology, geographic zone, elevation and vegetation type in North California woodlands', *Journal of Geophysical Research*, 66, 3351–8.

Andrews, E. D. (1979a) 'Hydraulic adjustment of the East Fork River, Wyoming, to the supply of sediment', in Rhodes, D. D. and Williams, G. P. (ed.) *Adjustments of the Fluvial System*, Dubuque, Iowa, Kendall Hunt, pp. 69–94.

Andrews, E. D. (1979b) 'Scour and fill in a stream channel, East Fork River, Western Wyoming', *Professional Paper. United States Geological Survey*, 1117.

Andrews, E. D. (1980) 'Effective and bankfull discharges of streams in the Yampa river basin, Colorado and Wyoming', *Journal of Hydrology*, 46, 311–30.

Andrews, E. D. (1984) 'Bed-material entrainment and hydraulic geometry of gravel-bed rivers in Colorado', *Bulletin of the Geological Society of America*, 95, 371–8.

Antevs, E. (1952) 'Arroyo-cutting and filling', *Journal of Geology*, 60, 375–85.

Antropovskiy, V. I. (1972) 'Criterial relations of types of channel processes', *Soviet Hydrology*, 11, 371–81.

Armentrout, C. L. and Bissell, R. B. (1970) 'Channel slope effect on peak discharge of natural streams', *Journal of the Hydraulics Division, American Society of Civil Engineers*, 96, 307–15.

Arnett, R. R. (1978) 'Regional disparities in the denudation rate of organic sediments', *Zeitschrift für Geomorphologie*, Supplement 29, 169–79.

Bagnold, R. A. (1954) 'Experiments on a gravity free dispersion of large solid spheres in a Newtonian fluid under shear', *Proceedings of the Royal Society*, 225A, 49–63.

Bagnold, R. A. (1957) 'The flow of cohesionless grains in fluids', *Philosophical Transactions of the Royal Society*, 249A, 235–97.

Bagnold, R. A. (1960a) 'Sediment discharge and stream power: a preliminary announcement', *Circular. United States Geological Survey*, 421.

Bagnold, R. A. (1960b) 'Some aspects of the shape of river meanders', *Professional Paper. United States Geological Survey*, 282E, pp. 135–44.

Bagnold, R. A. (1966) 'An approach to the sediment transport problem from general physics', *Professional Paper. United States Geological Survey*, 422 I.

Bagnold, R. A. (1973) 'The nature of saltation and of "bed-load" transport in water', *Proceedings of the Royal Society*, 332A, 473–504.

Bagnold, R. A. (1977) 'Bed load transport by natural rivers', *Water Resources Research*, 13, 303–12.

Baker, V. R. (1977) 'Stream channel response to floods with examples from central Texas', *Bulletin of the Geological Society of America*, 88, 1057–71.

Baker, V. R. and Nummedal, D. (eds) (1978) *The Channeled Scablands*, Washington, National Aeronautics and Space Administration.

Baker, V. R. and Ritter, D. F. (1975) 'Competence of rivers to transport coarse bedload material', *Bulletin of the Geological Society of America*, 86, 975–8.

Ballantyne, C. K. (1978) 'Variations in the size of coarse clastic particles over the surface of a small sandur, Ellesmere Island, NWT, Canada', *Sedimentology*, 25, 141–7.

Barishnikov, N. B. (1967) 'Sediment transportation in river channels with flood plains', *Publication. International Association of Scientific Hydrology*, 75, 404–13.

Barnes, H. H. (1967) 'Roughness characteristics of natural channels', *Water Supply Paper. United States Geological Survey*, 1849.

Bathurst, J. C. (1979) 'Distribution of boundary shear stress in rivers', in Rhodes, D. D. and Williams, G. P. (eds) *Adjustments of the Fluvial System*, Dubuque, Iowa, Kendall Hunt, pp. 95–116.

Bathurst, J. C., Thorne, C. R. and Hey, R. D. (1979) 'Secondary flow and shear stress at river bends', *Journal of the Hydraulics Division, American Society of Civil Engineers*, 105, 1277–95.

Baulig, H. (1935) 'The changing sea level', *Publication. Institute of British Geographers*, 3.

Beaty, C. B. (1970) 'Age and estimated rate of accumulation of an alluvial fan, White Mountains. California, USA', *American Journal of Science*, 268, 50–77.

Beaty, C. B. (1974) 'Debris flows, alluvial fans and a revitalized catastrophism', *Zeitschrift für Geomorphologie*, Supplement 21, 39–51.

Beckinsale, R. P. (1969) 'Rivers as political boundaries', in Chorley, R. J. (ed.) *Water, Earth and Man*, London, Methuen, pp. 344–55.

Beckman, E. W. and Furness, L. W. (1962) 'Flow characteristics of Elkhorn River near Waterloo, Nebraska', *Water Supply Paper. United States Geological Survey*, 1498B.

Bedeus, K. and Ivicsics, L. (1963) 'Observations on the noise of bedload', *Publication. International Association of Scientific Hydrology*, 65, 384–90.

Begin, Z. B., Meyer, D. F. and Schumm, S. A. (1981) 'Development of longitudinal profiles of alluvial channels in response to base-level lowering', *Earth Surface Processes*, 6, 49–68.

Bennett, R. J. (1976) 'Adaptive adjustment of channel geometry', *Earth Surface Processes*, 1, 136–50.

Benson, M. A. (1959) 'Channel slope factor in flood frequency analysis', *Journal of the Hydraulics Division, American Society of Civil Engineers*, 85, 1–19.

Benson, M. A. (1960a) 'Characteristics of frequency curves based on a theoretical 1000-year record', in Dalrymple, T. (ed.) 'Flood frequency analyses', *Water Supply Paper. United States Geological Survey*, 1543-A.

Benson, M. A. (1960b) 'Areal flood-frequency analysis in a humid region', *Bulletin of the International Association of Scientific Hydrology*, 19, 5–15.

Benson, M. A. (1965) 'Spurious correlation in hydraulics and hydrology', *Journal of the Hydraulics Division, American Society of Civil Engineers*, 91, 35–43.

Benson, M. A. (1968) 'Uniform flood-frequency estimating methods for federal agencies', *Water Resources Research*, 4, 891–908.

Benson, M. A. and Thomas, D. M. (1966) 'A definition of dominant discharge', *Bulletin of the International Association of Scientific Hydrology*, 11, 76–80.

Bergman, D. L. and Sullivan, C. W. (1963) 'Channel changes on Sandstone Creek near Cheyenne, Oklahoma', *Professional Paper. United States Geological Survey*, 475C, 145–8.

Bernoulli, D. (1738) *Hydrodynamica, sive de viribus et motibus fluidorum commentari*, Strasbourg.

Beschta, R. L. (1978) 'Long term patterns of sediment production following road construction and logging in the Oregon Coast Range', *Water Resources Research*, 14, 1011–6.

Betson, R. P. (1964) 'What is watershed runoff?', *Journal of Geophysical Research*, 69, 1541–51.

Beven, K., Gilman, K. and Newson, M. (1979) 'Flow and flow routing in upland channel networks', *Hydrological Sciences Bulletin*, 24, 303–25.

Beven, K. and Kirkby, M. J. (1979) 'A physically based variable contributing area model of basin hydrology', *Hydrological Sciences Bulletin*, 24, 43–69.

Beverage, J. P. and Culbertson, J. K. (1964) 'Hyperconcentrations of suspended sediment', *Journal of the Hydraulics Division, American Society of Civil Engineers*, 90, 117–28.

Biswas, A. K. (1970) *History of Hydrology*, Amsterdam, North-Holland.

Blench, T. (1952) 'Regime theory for self-formed sediment-bearing channels', *Transactions of the American Society of Civil Engineers*, 117, 383–400, discussion 401–8.

Blench, T. (1969) *Mobile-bed Fluviology*, Edmonton, University of Alberta Press.

Blinco, P. H. and Simons, D. B. (1974) 'Characteristics of turbulent boundary shear stress', *Journal of the Engineering Mechanics Division, American Society of Civil Engineers*, 100, 203–20.

Bluck, B. J. (1964) 'Sedimentation of an alluvial fan in southern Nevada', *Journal of Sedimentary Petrology*, 34, 395–400.

Bluck, B. J. (1971) 'Sedimentation in the meandering R. Endrick', *Scottish Journal of Geology*, 7, 93–138.

Bluck, B. J. (1974) 'Structure and directional properties of some valley sandur deposits in southern Iceland', *Sedimentology*, 21, 533–54.

Bluck, B. J. (1976) 'Sedimentation in some Scottish rivers of low sinuosity', *Transactions of the Royal Society of Edinburgh*, 69, 425–56.

Bogardi, J. (1974) *Sediment Transport in Alluvial Streams*, Budapest, Akadémiai Kiadó.

Bogen, J. (1980) 'The hysteresis effect of sediment transport systems', *Norsk Geografisk Tidsskrift*, 34, 45–54.

Boggs, S. W. (1940) *International Boundaries*, New York, Columbia University Press.

Born, S. M. and Ritter, D. F. (1970) 'Modern terrace development near Pyramid Lake, Nevada, and its geologic implications', *Bulletin of the Geological Society of America*, 81, 1233–42.

Boyer, M. C. (1954) 'Estimating the Manning coefficient from an average bed roughness in open channels', *Transactions of the American Geophysical Union*, 35, 957–61.

Bradley, W. C. (1970) 'Effect of weathering on abrasion of granitic gravel, Colorado River, (Texas)', *Bulletin of the Geological Society of America*, 81, 61–80.

Bradley, W. C., Fahnestock, R. K. and Rowekamp, E. T. (1972) 'Coarse sediment transport by flood flows on Knik River, Alaska', *Bulletin of the Geological Society of America*, 83, 1261–84.

Bray, D. I. and Cullen, A. J. (1976) 'Study of artificial cutoffs on gravel bed rivers', *American Society of Civil Engineers, Proceedings of 3rd Annual Symposium, Waterways, Harbours and Coastal Engineering Division*, pp. 1399–417.

Bray, D. I. and Kellerhals, R. (1979) 'Some Canadian examples of the response of rivers to man-made changes', in Rhodes, D. D. and Williams, G. P. (eds) *Adjustments of the Fluvial System*, Dubuque, Iowa, Kendall Hunt, pp. 351–72.

Brayshaw, A. C. (1984) 'Characteristics and origin of cluster bedforms in coarse-grained alluvial channels', in Koster, E. H. and Steal, R. J. (eds) *Sedimentology of Gravels and Conglomerates*, Canadian Society of Petroleum Geologists, Memoir 10, 77–85.

Brice, J. C. (1960) 'Index for description of channel braiding', *Bulletin of the Geological Society of America* (Abstract), 71, 1833.

Brice, J. C. (1966) 'Erosion and deposition in the loess-mantled Great Plains, Medicine Creek drainage basin, Nebraska', *Professional Paper. United States Geological Survey*, 352-H, 255–339.

Brice, J. C. (1973) 'Meandering pattern of the White river in Indiana – an analysis', in Morisawa, M. (ed.) *Fluvial Geomorphology*, SUNY Binghamton, Publications in Geomorphology, pp. 179–200.

Bridge, J. S. (1976) 'Bed topography and grain size in open channel bends', *Sedimentology*, 23, 407–14.

Bridge, J. S. (1977) 'Flow, bed topography, grain size and sedimentary structure in open channel bends: a three dimensional model', *Earth Surface Processes*, 2, 401–16.

Bridge, J. S. and Jarvis, J. (1976) 'Flow and sedimentary processes in the meandering River South Esk, Glen Cova, Scotland', *Earth Surface Processes*, 1, 303–37.

Bridge, J. S. and Jarvis, J. (1977) 'Velocity profiles and bed shear stress over various bed configurations in a river bend', *Earth Surface Processes*, 2, 281–94.

Briggs, D. J. and Gilbertson, D. D. (1980) 'Quaternary processes and environments in the upper Thames valley', *Transactions of the Institute of British Geographers, New Series*, 5, 53–65.

Brinson, M. M. (1976) 'Organic matter losses from four watersheds in the humid tropics', *Limnology and Oceanography*, 21, 572–82.

British Standards Institution (1964a) 'Measurement of liquid flow in open channels: velocity-area methods', *BS 3680*, part 3.

British Standards Institution (1964b) 'Measurement of liquid flow in open channels: dilution methods: constant rate injection', *BS 3680*, part 2A.

British Standards Institution (1965) 'Measurement of liquid flow in open channels: thin-plate weirs and Venturi flumes', *BS 3680*, part 4A.

British Standards Institution (1971a) 'Measurement of liquid flow in open channels: the measurement of liquid level (stage)', *BS 3680*, part 7.

British Standards Institution (1971b) 'Measurement of liquid flow in open channels: specification for the installation and performance of pressure actuated liquid level measuring equipment', *BS 3680*, part 9A.

British Standards Institution (1975) 'Methods of testing soils for civil engineering purposes', *BS 1377*.

Broscoe, A. J. (1959) 'Quantitative analysis of longitudinal stream profiles of small watersheds', *Office of Naval Research, Geography Branch, Project NR-389-49, Technical Report*, 18.

Brown, D. A. (1971) 'Stream channels and flow relations', *Water Resources Research*, 7, 304–10.

Brown, E. H. (1952) 'The river Ystwyth, Cardiganshire: a geomorphological analysis', *Proceedings of the Geologists' Association*, 63, 244–68.

Brown, J. A. H. (1972) 'Hydrologic effects of a bushfire in a catchment in south-eastern New South Wales', *Journal of Hydrology*, 15, 77–96.

Brune, G. M. (1953) 'Trap efficiency of reservoirs', *Transactions of the American Geophysical Union*, 34, 407–18.

Brush, L. M. (1961) 'Drainage basins, channels and flow characteristics of selected streams in central Pennsylvania', *Professional Paper. United States Geological Survey*, 282F, pp. 145–81.

Brush, L. M., Ho, H.-W. and Singamsetti, S. R. (1962) 'A study of sediment in suspension', *Publication. International Association of Scientific Hydrology*, 59, 293–310.

Brush, L. M. and Wolman, M. G. (1960) 'Knickpoint behaviour in non-cohesive material: a laboratory study', *Bulletin of the Geological Society of America*, 71, 59–74.

Bryan, K. (1928) 'Historic evidence on changes in the channel of Rio Puerto, a tributary of the Rio Grande in New Mexico', *Journal of Geology*, 36, 265–82.

Bryan, R. B. (1977) 'Assessment of soil erodibility: new approaches and directions', in Toy, T. (ed.) *Erosion: Research Techniques, Erodibility and Sediment Delivery*, Norwich, Geobooks, pp. 57–72.

Bryant, I. D. (1983) 'Facies sequences associated with some braided river deposits of Late Pleistocene age from Southern Britain', in Collinson, J. D. and Lewin, J. (eds) *Modern and Ancient Fluvial Systems: Sedimentology and Processes*, International Association of Sedimentologists, Special Publication No. 6, 267–75.

Buchanan, J. J. and Somers, W. P. (1968) 'Stage measurement at gaging stations', *Techniques of Water-Resource Investigations of United States Geological Survey*, chapter A7, book 3.

Buchanan, J. J. and Somers, W. P. (1969) 'Discharge measurements at gaging stations', *Techniques of Water-Resource Investigations of United States Geological Survey*, chapter A8, book 3.

Buckley, A. B. (1923) 'The influence of silt on the velocity of water flowing in open channels', *Proceedings of the Institution of Civil Engineers*, 216, 183–210.

Bull, A. J., Gossling, F., Green, J. F. N., Hayward, H. A., Turner, E. A. and Wooldridge, S. W. (1934) 'The River Mole: its physiography and superficial deposits', *Proceedings of the Geologists' Association*, 45, 35–67.

Bull, W. B. (1962) 'Relation of textural (cm) patterns to depositional environment of alluvial-fan deposits', *Journal of Sedimentary Petrology*, 32, 211–6.

Bull, W. B. (1964) 'Geomorphology of segmented alluvial fans in Western Fresno county, California', *Professional Paper. United States Geological Survey, 352E*, 89–129.

Bull, W. B. (1977) 'The alluvial-fan environment', *Progress in Physical Geography*, 1, 222–70.

Bull, W. B. (1979) 'Threshold of critical power in streams', *Bulletin of the Geological Society of America*, 90, 453–64.

Buller, A. T. and McManus, J. (1972) 'Simple metric sedimentary statistics used to recognize different environments', *Sedimentology*, 18, 1–21.

Burkham, D. E. (1970) 'A method for relating infiltration rates to stream-flow rates in perched streams', *Professional Paper. United States Geological Survey*, 700D, 266–71.

Burkham, D. E. (1972) 'Channel changes of the Gila river in Safford valley, Arizona, 1846–1970', *Professional Paper. United States Geological Survey*, 655G.

Burt, T. P. (1979) 'The relationship between throughflow generation and the solute concentration of soil and streamwater', *Earth Surface Processes*, 4, 257–66.

Butcher, G. C. and Thornes, J. B. (1978) 'Spatial variability in runoff processes in an ephemeral channel', *Zeitschrift für Geomorphologie*, Supplement 29, 83–92.

Butler, P. R. (1977) 'Movement of cobbles in a gravel-bed stream during a flood season', *Bulletin of the Geological Society of America*, 88, 1072–4.

Cailleux, A. (1945) 'Distinction des galets marines et fluviatiles', *Societé Geologie*, 5ᵉ Serie, 15, 375–404.

Cailleux, A. and Tricart, J. (1963) *Initiation a l'Étude des Sables et des Galets*, Paris, Centre de Documentation Universitaire.

Calkins, D. and Dunne, T. (1970) 'A salt-tracing method for measuring channel velocities in small mountain streams', *Journal of Hydrology*, 11, 379–92.

Callander, R. A. (1969) 'Instability and river channels', *Journal of Fluid Mechanics*, 36, 465–80.

Calver, A. (1978) 'Modelling drainage headwater development', *Earth Surface Processes*, 3, 233–41.

Calver, A., Kirkby, M. J. and Weyman, D. R. (1972) 'Modelling hillslope and channel flows', in Chorley, R. J. (ed.) *Spatial Analysis in Geomorphology*, London, Methuen, pp. 197–218.

Carling, P. A. (1983) 'Threshold of coarse sediment transport in broad and narrow natural steams', *Earth Surface Processes and Landforms*, 8, 1–18.

Carlston, C. W. (1963) 'Drainage density and streamflow', *Professional Paper. United States Geological Survey*, 422C.

Carlston, C. W. (1965) 'The relation of free meander geometry to stream discharge and its geomorphic implications', *American Journal of Science*, 263, 864–85.

Carlston, C. W. (1968) 'Slope-discharge relations for eight rivers in the United States', *Professional Paper. United States Geological Survey*, 600D, pp. 45–7.

Carlston, C. W. (1969) 'Downstream variations in the hydraulic geometry of streams: special emphasis on mean velocity', *American Journal of Science*, 267, 499–509.

Carson, M. A. (1971) *The Mechanics of Erosion*, London, Pion.

Carson, M. A. and Lapointe, M. F. (1983) 'The inherent asymmetry of river meander planform', *Journal of Geology*, 91, 41–55.

Carson, M. A., Taylor, C. H. and Grey, B. J. (1973) 'Sediment production in a small Appalachian watershed during spring runoff: the Eaton Basin, 1970–1972', *Canadian Journal of Earth Sciences*, 10, 1707–34.

Carter, C. S. and Chorley, R. J. (1961) 'Early slope development in an expanding stream system', *Geological Magazine*, 98, 117–29.

Chandler, A. and Patterson, J. E. (1973) 'Digital event recorders for representative and experimental basins', in *Reports of research on representation and experimental basins, International Association of Scientific Hydrology, UNESCO, Proceedings of the Wellington Symposium*, 1, pp. 700–7.

Chang, F. F. M. (1970) 'Ripple concentration and friction factor', *Journal of the Hydraulics Division, American Society of Civil Engineers*, 96, 417–30.

Chang, H. H. (1979) 'Minimum stream power and river channel pattern', *Journal of Hydrology*, 41, 303–27.

Chang, T. P. and Toebes, G. H. (1970) 'A statistical comparison of meander planforms in the Wabash basin', *Water Resources Research*, 6, 557–78.

Charlton, F. G. (1975) 'Design of meandering channels', in Hey, R. D. and Davies, T. D. (eds) *Science, Technology and Environmental Management*, Farnborough, Saxon House, pp. 59–71.

Charlton, F. W., Brown, P. M. and Benson, R. W. (1978) 'The hydraulic geometry of some gravel rivers in Britain', *Hydraulics Research Station Report*, IT 180.

Chayes, F. (1970) 'On deciding whether trend surfaces of progressively higher order are meaningful', *Bulletin of the Geological Society of America*, 81, 1273–8.

Cheetham, G. H. (1979) 'Flow competence in relation to stream channel form and braiding', *Bulletin of the Geological Society of America*, 90, 877–86.

Chepil, W. S. (1958) 'The use of unevenly spaced hemispheres to evaluate aerodynamic forces on a soil surface', *Transactions of the American Geophysical Union*, 39, 397–404.

Cherkauer, D. S. (1972) 'Longitudinal profiles of ephemeral streams in south eastern Arizona', *Bulletin of the Geological Society of America*, 83, 353–66.

Chezy, A. de (1775) 'Memoire sur la vitesse de l'eau conduite dans une régole', MS, reprinted in *Annals des Ponts et Chaussées*, 60, 1921.

Chien, N. (1952) 'The efficiency of depth-integrating suspended-sediment sampling', *Transactions of the American Geophysical Union*, 33, 693–8.

Chien, N. (1957) 'A concept of the regime theory', *Transactions of the American Society of Civil Engineers*, 122, 785–805.

Chien, N. (1985) 'Changes in river regime after the construction of upstream reservoirs', *Earth Surface Processes and Landforms*, 10, 143–59.

Chisholm, M. (1967) 'General systems theory and geography', *Transactions of the Institute of British Geographers*, 42, 45–52.

Chitale, S. K. (1970) 'River channel patterns', *Journal of the Hydraulics Division, American Society of Civil Engineers*, 96, 201–21.

Chorley, R. J. (1958) 'Group operator variance in morphometric work with maps', *American Journal of Science*, 256, 208–18.

Chorley, R. J. (1959) 'The geomorphic significance of some Oxford soils', *American Journal of Science*, 257, 503–15.

Chorley, R. J. (1962) 'Geomorphology and general systems theory', *Professional Paper. United States Geological Survey, 500B*.

Chorley, R. J. (1964) 'Geomorphological evaluation of factors controlling shearing resistance of surface rocks in sandstone', *Journal of Geophysical Research*, 69, 1507–16.

Chorley, R. J. (1969) 'The drainage basin as the fundamental geomorphic unit', in Chorley, R. J. (ed.) *Water, Earth and Man*, London, Methuen, pp. 77–100.

Chorley, R. J. (1978) 'Bases for theory in geomorphology', in Embleton, C., Brunsden, D. and Jones, D. K. C. (eds) *Geomorphology, Present Problems and Future Prospects*, Oxford, Oxford University Press, pp. 1–13.

Chorley, R. J. and Kennedy, B. A. (1971) *Physical Geography: A Systems Approach*, London, Prentice Hall.

Chorley, R. J. and Morgan, M. A. (1962) 'Comparison of morphometric features, Unaka Mountains Tennessee and North Carolina, and Dartmoor, England', *Bulletin of the Geological Society of America*, 73, 17–34.

Chow, V. T. (1954) 'The log-probability law and its engineering application', *Proceedings of the American Society of Civil Engineers*, 80, Paper 836, 1–25.

Chow, V. T. (1959) *Open-channel Hydraulics*, Tokyo, McGraw-Hill, Kogakusha.

Christensen, B. A. (1972) 'Incipient motion on cohesionless channel banks', in Shen, H. W. (ed.) *Sedimentation*, Fort Collins, Colorado, 4.1–4.22.

Church, M. (1967) 'Observations of turbulent diffusion in a natural channel', *Canadian Journal of Earth Science*, 4, 855–72.

Church, M. (1972) 'Baffin Island sandurs: a study of arctic fluvial processes', *Bulletin of the Geological Survey of Canada*, 216.

Church, M. (1975) 'Electrochemical and fluorometric tracer techniques for streamflow measurements', *British Geomorphological Research Group Technical Bulletin*, 12, Norwich, Geo Abstracts.

Church, M. and Mark, D. M. (1980) 'On size and scale in geomorphology', *Progress in Physical Geography*, 4, 342–90.

Cleaves, E., Godfrey, A. E. and Bricker, O. P. (1970) 'Geochemical balance of a small watershed and its geomorphic implications', *Bulletin of the Geological Society of America*, 81, 3015–32.

Coffman, D. M., Keller, E. A. and Melhorn, W. N. (1972) 'New topologic relationship as an indicator of drainage network evolution', *Water Resources Research*, 8, 1497–505.

Colby, B. R. (1961) 'Effect of depth of flow on discharge of bed materials', *Water Supply Paper. United States Geological Survey*, 1498D.

Colby, B. R. (1963) 'Fluvial sediments – a summary of source, transportation, deposition and measurement of sediment discharge, *Bulletin of the United States Geological Survey*, 1181A.

Colby, B. R. (1964a) 'Scour and fill in sand-bed streams', *Professional Paper. United States Geological Survey*, 462D.

Colby, B. R. (1964b) 'Discharge of sands and mean velocity relationships in sand bed streams', *Professional Paper. United States Geological Survey*, 462A.

Colby, B. R. and Hembree, C. H. (1955) 'Computations of total sediment discharge, Niobrara River, near Cody, Nebraska', *Water Supply Paper. United States Geological Survey*, 1357, p. 187.

Cole, G. (1966) 'An application of the regional analysis of flood flows', in *River Flood Hydrology, Institution of Civil Engineers Symposium*, pp. 39–57.

Collins, D. N. (1979) 'Hydrochemistry of meltwaters draining from an alpine glacier', *Arctic and Alpine Research*, 11, 307–24.

Collinson, J. D. (1970) 'Bedforms of the Tana river, Norway', *Geografiska Annaler*, 52A, 31–55.

Conacher, A. J. (1969) 'Open systems and dynamic equilibrium in geomorphology', *Australian Geographical Studies*, 7, 153–8.

Cooke, R. U. (1974) 'The rainfall context of arroyo initiation in southern Arizona', *Zeitschrift für Geomorphologie*, Supplement 21, 63–75.

Corbett, D. N. *et al* (1943) 'Stream-gaging procedure', *Water Supply Paper. United States Geological Survey*, 888.

Costa, J. E. (1978) 'Holocene stratigraphy in flood frequency analysis', *Water Resources Research*, 14, 626–32.

Costa, J. E. (1983) 'Palaeohydraulic reconstruction of flash-flood peaks from boulder deposits in the Colorado Front Range', *Bulletin of the Geological Society of America*, 94, 986–1004.

Courtois, G. and Sauzay, G. (1966) 'Les methodes de bilan des taux de comptage de traceurs radioactifs appliquées à la mesure des débits massiques de charriage', *La Houille Blanche*, 3, 279–89.

Cowan, W. L. (1956) 'Estimating hydraulic roughness coefficients', *Agricultural Engineering*, 37, 473–5.

Crampton, C. B. (1969) 'The chronology of certain terraced river deposits in the NE Wales area', *Zeitschrift für Geomorphologie*, 13, 245–59.

Crawford, N. H. and Linsley, R. K. (1966) 'The Stanford Watershed Model Mark IV', *Technical Report*, 39, Department of Civil Engineering, Stanford University.

Crickmore, M. J. (1967) 'Measurement of sand transport in rivers with special reference to tracer methods', *Sedimentology*, 8, 175–228.

Crickmore, M. J. and Lean, G. H. (1962) 'The measurement of sand transport by means of radioactive tracers', *Proceedings of the Royal Society*, 266A, 402–21.

Croxton, F. E. (1959) *Elementary Statistics with Applications in Medicine and the Biological Sciences*, New York, Dover.

Cruff, R. W. (1965) 'Cross-channel transfer of linear momentum in smooth rectangular channels', *Water Supply Paper. United States Geological Survey*, 1592B.

Crump, E. S. (1952) 'A new method of gauging stream flow with little afflux by means of a submerged weir of triangular profile', *Proceedings of the Institution of Civil Engineers*, 1, 223–42.

Cryer, R. (1976) 'The significance and variation of atmospheric nutrient inputs in a small catchment system', *Journal of Hydrology*, 29, 121–37.

Culling, W. E. H. (1957) 'Multicyclic streams and the equilibrium theory of grade', *Journal of Geology*, 65, 259–74.

Cunnane, C. (1973) 'A particular comparison of annual maxima and partial duration series methods of flood frequency prediction', *Journal of Hydrology*, 18, 257–71.

Cunnane, C. (1979) 'A note on the Poisson assumption in partial duration series models', *Water Resources Research*, 15, 489–94.

Dal Cin, R. (1968) ' "Pebble clusters": their origin and utilization in the study of palaeocurrents', *Sedimentary Geology*, 2, 233–41.

Dalrymple, T. (1960) 'Flood frequency analysis', *Water Supply Paper. United States Geological Survey*, 1543A.

Dalrymple, T. and Benson, M. A. (1967) 'Measurement of peak discharge by the slope-area method', *Techniques of Water-Resource Investigations of United States Geological Survey*, chapter A2, book 3.

Daniels, R. B. (1960) 'Entrenchment of the Willow Drainage Ditch, Harrison County, Iowa', *American Journal of Science*, 258, 161–76.

Davies, L. H. (1974) 'Problems posed by new town development with particular reference to Milton Keynes', in *Rainfall, runoff and surface water drainage of urban catchments, Proceedings of Research Colloquium, Department of Civil Engineering, Bristol University, April 1973*, Paper 2.

Davies, T. R. and Sutherland, A. J. (1980) 'Resistance to flow past deformable boundaries', *Earth Surface Processes*, 5, 175–9.

Davis, W. M. (1899) 'The geographical cycle', *Geographical Journal*, 14, 481–504.

Davis, W. M. (1902a) 'Base-level, grade and peneplain', *Journal of Geology*, 10, 77–111.

Davis, W. M. (1902b) 'River terraces in New England', *Bulletin of the Harvard College Museum of Comparative Zoology*, 38, 281–346.

Dawdy, D. R. (1961) 'Depth-discharge relations in alluvial streams – discontinuous rating curves', *Water Supply Paper. United States Geological Survey*, 1498C.

Dawson, M. (1985) 'Environmental reconstructions of a Late Devensian terrace sequence. Some preliminary findings', *Earth Surface Processes and Landforms*, 10, 237–46.

Day, D. G. (1978) 'Drainage density changes during rainfall', *Earth Surface Processes*, 3, 319–26.

Day, T. J. (1975) 'Longitudinal dispersion in natural channels', *Water Resources Research*, 11, 909–18.

Day, T. J. (1976) 'On the precision of salt dilution gauging', *Journal of Hydrology*, 31, 293–306.

Deike, G. H. and White, W. B. (1969) 'Sinuosity in limestone solution conduits', *American Journal of Science*, 267, 230–41.

Denny, C. S. (1967) 'Fans and pediments', *American Journal of Science*, 265, 81–105.

Dietrich, W. E. and Dunne, T. (1978) 'Sediment budget for a small catchment in mountainous terrain', *Zeitschrift für Geomorphologie*, Supplement 29, 191–26.

Dietrich, W. E., Smith, J. D. and Dunne, T. (1979) 'Flow and sediment transport in a sand-bedded meander', *Journal of Geology*, 87, 305–15.

Doeglas, D. J. (1962) 'Structure of sedimentary deposits of braided rivers', *Sedimentology*, 1, 167–90.

Dooge, J. C. I. (1968) 'The hydrologic cycle as a closed system', *Bulletin. International Association of Scientific Hydrology*, 13, 58–68.

Douglas, I. (1964) 'Intensity and periodicity in denudation processes with special reference to the removal of material in solution by rivers', *Zeitschrift für Geomorphologie*, 8, 453–73.

Douglas, I. (1967) 'Man, vegetation and the sediment yield of rivers', *Nature*, 215, 925–8.

Douglas, I. (1968a) 'The effects of precipitation chemistry and catchment area lithology on the quality of river water in selected catchments in eastern Australia', *Earth Science Journal*, 2, 126–44.

Douglas, I. (1968b) 'Field methods of water hardness determination', *British Geomorphological Research Group Technical Bulletin*, 1, Norwich, Geo Abstracts.

Douglas, I. (1975) 'Flood waves and suspended sediment pulses in urbanised catchments', *Proceedings of the National Conference, Australian Institute of Civil Engineering*, pp. 61–4.

Dreimanis, A. and Vagners, U. J. (1971) 'Bimodal distribution of rock and mineral fragments in basal tills', in Goldthwait, R. P. (ed.) *Till – A Symposium*, Ohio State University Press, pp. 237–50.

du Boys, P. F. D. (1879) 'Études du régime et l'action exercée par les eaux sur un lit à fond de graviers indefinement affouilable', *Annals des Ponts et Chaussées*, Series 5, 18, 141–95.

Du Buat, L. G. (1786) *Principes d'Hydraulique, verifie par un grand nombre d'experiences, faites par ordre de Gouvernment*, 2 vols, Paris.

Dunne, T. and Black, R. D. (1970a) 'An experimental investigation of runoff production in permeable soils', *Water Resources Research*, 6, 478–90.

Dunne, T. and Black, R. D. (1970b) 'Partial area contributions to storm runoff in a small New England watershed', *Water Resources Research*, 6, 1296–311.

Dunne, T. and Leopold, L. B. (1978) *Water in Environmental Planning*, San Francisco, W. H. Freeman.

Durum, W. H., Heidel, S. G. and Tison, L. J. (1960) 'World-wide runoff of dissolved solids', *Publication. International Association of Scientific Hydrology*, 51, 618–28.

Dury, G. H. (1955) 'Bedwidth and wave-length in meandering valleys', *Nature*, 176, 31.

Dury, G. H. (1959) 'Analysis of regional flood frequency on the Nene and Great Ouse', *Geographical Journal*, 125, 223–9.

Dury, G. H. (1961) 'Bankfull discharge: an example of its statistical relationships', *Bulletin of the International Association of Scientific Hydrology*, 6(3), 48–55.

Dury, G. H. (1962) 'Results of seismic exploration of meandering valleys', *American Journal of Science*, 260, 691–706.

Dury, G. H. (1964a) 'Principles of underfit streams', *Professional Paper. United States Geological Survey*, 452A.

Dury, G. H. (1964b) 'Subsurface exploration and chronology of underfit streams', *Professional Paper. United States Geological Survey*, 452B.

Dury, G. H. (1965) 'Theoretical implications of underfit streams', *Professional Paper. United States Geological Survey*, 452C.

Dury, G. H. (1966a) 'The concept of grade', in Dury, G. H. (ed.) *Essays in Geomorphology*, London, Heinemann, pp. 211–34.

Dury, G. H. (1966b) 'Incised valley meanders on the Lower Colo River, NSW', *Australian Geographer*, 10, 17–25.

Dury, G. H. (1969) 'Relation of morphometry to runoff frequency', in Chorley, R. J. (ed.) *Water, Earth and Man*, London, Methuen, pp. 419–30.

Dury, G. H. (1970) 'A resurvey of part of the Hawkesbury River, New South Wales, after 100 years', *Australian Geographical Studies*, 8, 121–32.

Dury, G. H. (1973) 'Magnitude-frequency analysis and channel morphology', in Morisawa, M. (ed.) *Fluvial Geomorphology*, SUNY Binghamton, Publications in Geomorphology, pp. 91–121.

Dury, G. H., Hails, J. R. and Robbie, M. B. (1963) 'Bankfull discharge and the magnitude-frequency series', *Australian Journal of Science*, 26, 123–4.

Dyer, A. J. (1970) 'River discharge measurement by the rising float technique', *Journal of Hydrology*, 11, 201–12.

Dyer, K. R. and Dorey, A. P. (1974) 'Simulation of bedload transport using an acoustic pebble', *Memoires de l'institute de geologique du bassin d'Aquitaine*, 7, 377–80.

Edwards, A. M. C. (1973) 'Dissolved load and tentative solute budgets of some Norfolk rivers', *Journal of Hydrology*, 18, 201–17.

Edzwald, J. K. and O'Melia, C. R. (1975) 'Clay distributions in recent estuarine sediments', *Clays and Clay Minerals*, 23, 39–44.

Einstein, H. A. (1942) 'Formulas for the transportation of bedload', *Transactions of the American Society of Civil Engineers*, 107, 561–97.

Einstein, H. A. (1950) 'The bedload function for sediment transportation in open channel flows', *Technical Bulletin, United States Department of Agriculture*, 1026.

Einstein, H. A. and Abdel-Aal, F. M. (1972) 'Einstein bedload function at high sediment rates', *Journal of the Hydraulics Division, American Society of Civil Engineers*, 98, 137–51.

Einstein, H. A., Anderson, A. G. and Johnson, J. W. (1940) 'A distinction between bedload and suspended load in natural streams', *Transactions of the American Geophysical Union*, 21, 628–33.

Einstein, H. A. and Banks, R. B. (1950) 'Fluid resistance of composite roughness', *Transactions of the American Geophysical Union*, 31, 603–10.

Einstein, H. A. and Barbarossa, N. L. (1952) 'River channel roughness', *Transactions of the American Society of Civil Engineers*, 117, 1121–46.

Einstein, H. A. and El-Samni, E. A. (1949) 'Hydrodynamic forces on a rough wall', *Reviews of Modern Physics*, 21, 520–4.

Einstein, H. A. and Shen, H. W. (1964) 'A study of meandering in straight alluvial channels', *Journal of Geophysical Research*, 69, 5239–47.

Ellis, J. B. (1979) 'The nature and sources of urban sediments and their relation to water quality: a case study from north-west London', in Hollis, G. E. (ed.) *Man's Impact on the Hydrological Cycle in the United Kingdom*, Norwich, Geo Abstracts, pp. 199–216.

Embleton, C. and Thornes, J. (eds) (1979) *Process in Geomorphology*, London, Edward Arnold.

Embleton, C. and Whalley, B. (1979) 'Energy, forces, resistances and responses', in Embleton, C. and Thornes, J. (eds) *Process in Geomorphology*, London, Edward Arnold, pp. 11–38.

Emerson, J. W. (1971) 'Channelization: a case study', *Science*, 173, 325–6.

Emmett, W. M. (1970) 'The hydraulics of overland flow', *Professional Paper. United States Geological Survey*, 662A.

Engelund, F. (1966) 'Hydraulic resistance of alluvial streams', *Journal of the Hydraulics Division, American Society of Civil Engineers*, 92, 315–26.

Engelund, F. (1974) 'Flow and bed topography in channel bends', *Journal of the Hydraulics Division, American Society of Civil Engineers*, 100, 1631–48.

Engelund, F. and Skovgaard, O. (1973) 'On the origin of meandering and braiding in alluvial streams', *Journal of Fluid Mechanics*, 57, 289–302.

Eyles, R. J. (1966) 'Stream representation on Malayan maps', *Journal of Tropical Geography*, 22, 1–9.

Fahnestock, R. K. (1963) 'Morphology and hydrology of a glacial stream – White River, Mount Rainier, Washington', *Professional Paper. United States Geological Survey*, 422A.

Fahnestock, R. K. and Bradley, W. C. (1973) 'Knik and Matanuska rivers, Alaska: a contrast in braiding', in Morisawa, M. (ed.) *Fluvial Geomorphology*, SUNY Binghamton, Publications in Geomorphology, pp. 220–50.

Fenton, J. D. and Abbott, J. E. (1977) 'Initial movement of grains on a stream bed: the effect of relative protrusion', *Proceedings of the Royal Society*, 352A, 523–7.

Ferguson, R. I. (1973a) 'Regular meander path models', *Water Resources Research*, 9, 1079–86.

Ferguson, R. I. (1973b) 'Sinuosity of supra-glacial streams', *Bulletin of the Geological Society of America*, 84, 251–6.

Ferguson, R. I. (1973c) 'Channel pattern and sediment type', *Area*, 5, 38–41.

Ferguson, R. I. (1975) 'Meander irregularity and wavelength estimation', *Journal of Hydrology*, 26, 315–33.

Ferguson, R. I. (1976) 'Disturbed periodic model for river meanders', *Earth Surface Processes*, 1, 337–47.

Ferguson, R. I. (1977a) 'Meander sinuosity and direction variance', *Bulletin of the Geological Society of America*, 88, 212–4.

Ferguson, R. I. (1977b) 'Meander migration; equilibrium and change', in Gregory, K. J. (ed.) *River Channel Changes*, Chichester, Wiley, 235–48.

Ferguson, R. I. (1981) Personal communication.

Ferguson, R. I. (1984) 'The threshold between meandering and braiding', in Smith, K. V. A. (ed.) *Channels and Channel Control Structures, Proceedings of the First International Conference on Hydraulic Design in Water Resources Engineering*, Berlin, Springer-Verlag, 6.15–6.29.

Fergusson, J. (1863) 'On recent changes in the delta of the Ganges', *Quarterly Journal of the Geological Society of London*, 19, 321–54.

Finlayson, B. L. (1975) 'Measurement of the organic content of suspended sediments at low concentrations', *British Geomorphological Research Group Technical Bulletin*, 17, Norwich, Geo Abstracts, pp. 21–6.

Finlayson, B. L. (1978) 'Suspended solids transport in a small experimental catchment', *Zeitschrift für Geomorphologie*, 22, 192–210.

Fishman, M. J. and Downs, S. C. (1966) 'Methods for analysis of selected metals in water by atomic absorption', *Water Supply Paper. United States Geological Survey*, 1540C.

Fisk, H. N. (1939) 'Depositional terrace slopes in Louisiana', *Journal of Geomorphology*, 2, 385–410.

Fisk, H. N. (1944) 'Geological investigation of the alluvial valley of the lower Mississippi River', *US Army Corps of Engineers, Mississippi River Commission*, Vicksburg, Mississippi.

Flaxman, E. M. (1963) 'Channel stability in undisturbed cohesive soils', *Journal of the Hydraulics Division, American Society of Civil Engineers*, 89, HY2, 87–96.

Fleming, G. (1967) 'The computer as a tool in sediment transport research', *Bulletin of the International Association of Scientific Hydrology*, 12, 45–54.

Fleming, G. (1969a) 'Design curves for suspended load estimation', *Proceedings of the Institution of Civil Engineers*, 43, 1–9.

Fleming, G. (1969b) 'Suspended solids monitoring: a comparison between three instruments', *Water and Water Engineering*, 73, 377–82.

Flint, J. J. (1974) 'Stream gradient as a function of order, magnitude and discharge', *Water Resources Research*, 10, 969–73.

Folk, R. L. and Ward, W. C. (1957) 'Brazos river bar: a study in the significance of grain size parameters', *Journal of Sedimentary Petrology*, 27, 3–26.

Foster, I. D. L. (1978a) 'A multivariate model of storm-period solute behaviour', *Journal of Hydrology*, 39, 339–53.

Foster, I. D. L. (1978b) 'Seasonal solute behaviour of stormflow in a small agricultural catchment', *Catena*, 5, 151–63.

Foster, I. D. L. (1980) 'Chemical yields in runoff and denudation in a small arable catchment, East Devon, England', *Journal of Hydrology*, 47, 349–68.

Fournier, F. (1954) 'Influence des factors climatiques sur l'erosion du sol estimation des transports solides effectués en suspension par les cours d'eau', *Publication. International Association of Scientific Hydrology*, 38, 283–8.

Fournier, F. (1960) 'Debit solide des cours d'eau. Essai d'estimation de la perte en terre subie par l'ensemble du globe terrestre', *Publication. International Association of Scientific Hydrology*, 53, 19–22,

Fox, R. L. (1976) 'The urbanizing river: a case study in the Maryland Piedmont', in Coates, D. R. (ed.) *Geomorphology and Engineering*, Stroudsburg, Pennsylvania, Dowden, Hutchinson & Ross, 245–71.

Francis, J. R. D. (1973) 'Experiments on the motion of solitary grains along the bed of a water stream', *Proceedings of the Royal Society*, 332A, 443–71.

Fread, D. L. and Harbaugh, T. E. (1971) 'Open channel profiles by Newton's iteration technique', *Journal of Hydrology*, 13, 70–80.

Fredriksen, R. L. (1969) 'A battery powered proportional stream water sampler', *Water Resources Research*, 5, 1410–13.

Friedkin, J. F. (1945) 'A laboratory study of the meandering of alluvial rivers', *United States Waterways Experimental Station*, Vicksburg, Mississippi.

Friedman, G. M. (1961) 'Distinguishing between dune, beach and river sands from their textural characteristics', *Journal of Sedimentary Petrology*, 31, 514–29.

Friedman, G. M. (1962) 'On sorting, sorting coefficients and the lognormality of the grain-size distribution of sandstones', *Journal of Geology*, 70, 737–53.

Froehlich, W., Kaszowski, L. and Starkel, L. (1977) 'Studies of present-day and past river activity in the Polish Carpathians', in Gregory, K. J. (ed.) *River Channel Changes*, Chichester, Wiley, pp. 411–28.

Frogge, R. R. (1967) 'Stabilization of Frenchman River using steel jacks', *Journal of Waterways and Harbours Division, American Society of Civil Engineers*, 93, 89–108.

Frostick, L. E. and Reid, I. (1977) 'The origin of horizontal laminae in ephemeral stream channel-fill', *Sedimentology*, 24, 1–9.

Frostick, L. E. and Reid, I. (1979) 'Drainage-net controls of sedimentary parameters in sand-bed ephemeral streams', in Pitty, A. F. (ed.) *Geographical Approaches to Fluvial Processes*, Norwich, Geobooks, pp. 173–201.

Frostick, L. E. and Reid, I. (1980) 'Sorting mechanisms in coarse-grained alluvial sediments: fresh evidence from a basalt plateau gravel, Kenya', *Journal of the Geological Society of London*, 137, 431–41.

Fujiyoshi, Y. (1950) 'Theoretical treatise on the meandering of rivers', *Japan Science Review*, 1, 29–34.

Garde, R. J. and Ranga Raju, K. G. (1977) *Mechanics of Sediment Transportation and Alluvial Stream Problems*, New Delhi, Wiley.

Gardiner, V. (1978) 'Redundancy and spatial organization of drainage basin form indices: an empirical investigation of data from north-west Devon', *Transactions of the Institute of British Geographers, New Series*, 3, 416–31.

Gaunt, G. D. (1975) 'The artificial nature of the River Don north of Thorne, Yorkshire', *Yorkshire Archaeological Journal*, 47, 15–21.

Gerson, R. (1977) 'Sediment transport for desert watersheds in erodible materials', *Earth Surface Processes*, 2, 343–62.

Gessler, J. (1971) 'Aggradation and degradation', in Shen, H. W. (ed.) *River Mechanics*, vol. 1, Fort Collins, Colorado, pp. 8.1–8.23.

Geyl, W. F. (1976) 'Tidal palaeomorphs in England', *Transactions of the Institute of British Geographers, New Series*, 1, 203–24.

Gibbs, R. J. (1967) 'The geochemistry of the Amazon River System. Part 1: Factors controlling the salinity and the composition and concentration of suspended solids', *Bulletin of the Geological Society of America*, 78, 1203–32.

Gilbert, G. K. (1877) *Report on the Geology of the Henry Mountains*, United States Geological and Geographical Survey, Rocky Mountain Region, General Printing Office, Washington, DC.

Gilbert, G. K. (1917) 'Hydraulic mining debris in the Sierra Nevada', *Professional Paper. United States Geological Survey*, 105.

Glover, B. J. and Johnson, P. (1974) 'Variations in the natural chemical concentration of river water during flood flows and the lag effect', *Journal of Hydrology*, 22, 303–16.

Glover, R. E. and Florey, Q. L. (1951) 'Stable channel profiles', *United States Department of Interior Bureau of Reclamation, Hydraulic Laboratory Report*, Hyd-325.

Gorycki, M. A. (1973) 'Hydraulic drag; a meander-initiating mechanism', *Bulletin of the Geological Society of America*, 84, 175–86.

Goudie, A. S. (1970) 'Input and output considerations in estimating rates of chemical denudation', *Earth Science Journal*, 4, 59–65.

Graf, W. H. (1966) 'On the determination of the roughness coefficient in natural and artificial waterways', *Bulletin of the International Association of Scientific Hydrology*, 11, 59–68.

Graf, W. H. (1971) *Hydraulics of Sediment Transport*, New York, McGraw-Hill.

Graf, W. L. (1975) 'The impact of suburbanization on fluvial geomorphology', *Water Resources Research*, 11, 690–2.

Graf, W. L. (1977) 'The rate law in fluvial geomorphology', *American Journal of Science*, 277, 178–91.

Graf, W. L. (1979) 'The development of montane arroyos and gullies', *Earth Surface Processes*, 4, 1–14.

Gray, D. M. (1964) 'Physiographic characteristics and the runoff pattern', *National Research Council of Canada, Proceedings of Hydrology Symposium*, 4, 147–66.

Gregory, K. J. (1966) 'Dry valleys and the composition of the drainage net', *Journal of Hydrology*, 4, 327–40.

Gregory, K. J. (1974) 'Streamflow and building activity', in Gregory, K. J. and Walling, D. E. (eds) *Fluvial Processes in Instrumented Watersheds*, Institute of British Geographers Special Publication 6, 107–22.

Gregory, K. J. (1976) 'Lichens and the determination of river channel capacity', *Earth Surface Processes*, 1, 273–85.

Gregory, K. J. and Cullingford, R. A. (1974) 'Lateral variations in pebble shape in north-west Yorkshire', *Sedimentary Geology*, 12, 237–48.

Gregory, K. J. and Gardiner, V. (1975) 'Drainage density and climate', *Zeitschrift für Geomorphologie*, 19, 287–98.

Gregory, K. J. and Park, C. C. (1974) 'Adjustment of river channel capacity downstream from a reservoir', *Water Resources Research*, 10, 870–3.

Gregory, K. J. and Park, C. C. (1976) 'Stream channel morphology in north-west Yorkshire', *Revue de Géomorphologie Dynamique*, 25, 63–72.

Gregory, K. J. and Walling, D. E. (1968) 'The variation of drainage density within a catchment', *Bulletin of the International Association of Scientific Hydrology*, 13, 61–8.

Gregory, K. J. and Walling, D. E. (1973) *Drainage Basin Form and Process: A Geomorphological Approach*, London, Edward Arnold.

Grigg, N. S. (1970) 'Motion of single particles in alluvial channels', *Journal of the Hydraulics Division, American Society of Civil Engineers*, 96, 2501–18.

Gringorten, I. I. (1963) 'A plotting rule for extreme probability paper', *Journal of Geophysical Research*, 68, 813–14.

Guglielmini, D. (1697) *Della natura de'fiumi, trattato fisico-mathematico*, Bologna.

Gumbel, E. J. (1958) 'Statistical theory of floods and droughts', *Journal of the Institute of Water Engineers*, 12, 157–84.

Gupta, A. (1975) 'Streamflow characteristics in eastern Jamaica, an environment of seasonal flow and large floods', *American Journal of Science*, 275, 828–47.

Gupta, A. and Fox, H. (1974) 'Effects of high-magnitude floods on channel form: a case study in Maryland Piedmont', *Water Resources Research*, 10, 499–509.

Guy, H. P. (1964) 'An analysis of some storm period variables affecting stream sediment transport', *Professional Paper. United States Geological Survey*, 462E.

Guy, H. P., Simons, D. B. and Richardson, E. V. (1966) 'Summary of alluvial channel data from flume experiments, 1956–1961', *Professional Paper. United States Geological Survey*, 462I.

Hack, J. T. (1957) 'Studies of longitudinal stream profiles in Virginia and Maryland', *Professional Paper. United States Geological Survey*, 294B.

Hack, J. T. (1965) 'Postglacial drainage evolution and stream geometry in the Ontonagon area, Michigan', *Professional Paper. United States Geological Survey*, 504B, 1–40.

Hack, J. T. (1973a) 'Stream-profile analysis and stream-gradient index', *Journal of Research, United States Geological Survey*, 1, 421–9.

Hack, J. T. (1973b) 'Drainage adjustment in the Appalachians', in Morisawa, M. (ed.) *Fluvial Geomorphology*, SUNY Binghamton, Publications in Geomorphology, pp. 51–69.

Haible, W. W. (1980) 'Holocene profile changes along a California coastal stream', *Earth Surface Processes*, 5, 249–64.

Hales, Z. L., Shindala, A. and Denson, K. H. (1970) 'River bed degradation prediction', *Water Resources Research*, 6, 549–56.

Hammad, H. Y. (1972) 'River bed degradation after closure of dams', *Journal of the Hydraulics Division, American Society of Civil Engineers*, 98, 591–607.

Hammer, T. R. (1972) 'Stream channel enlargement due to urbanization', *Water Resources Research*, 8, 1530–40.

Happ, S. C. (1948) 'Sedimentation in the middle Rio Grande valley, New Mexico', *Bulletin of the Geological Society of America*, 59, 1191–216.

Harris, D. R. and Vita-Finzi, C. (1968) 'Kokkinopilos – a Greek badland', *Geographical Journal*, 134, 537–45.

Harvey, A. M. (1969) 'Channel capacity and the adjustment of streams to hydrologic regime', *Journal of Hydrology*, 8, 82–98.

Harvey, A. M. (1971) 'Seasonal flood behaviour in a clay catchment', *Journal of Hydrology*, 12, 129–44.

Harvey, A. M. (1975) 'Some aspects of the relations between channel characteristics and riffle spacing in meandering streams', *American Journal of Science*, 275, 470–8.

Harvey, A. M. (1977) 'Event frequency in sediment production and channel change', in Gregory, K. J. (ed.) *River Channel Changes*, Chichester, Wiley, pp. 301–15.

Harvey, A. M., Hitchcock, D. H. and Hughes, D. J. (1979) 'Event frequency and morphological adjustment of fluvial systems', in Rhodes, D. D. and Williams, G.

P. (eds) *Adjustments of the Fluvial System*, Dubuque, Iowa, Kendall Hunt, pp. 139–67.

Hayes, M. O. (1967) 'Relationship between coastal climate and bottom sediment type of the inner continental shelf', *Marine Geology*, 5, 111–32.

Heidel, S. G. (1956) 'The progressive lag of sediment concentration with flood waves', *Transactions of the American Geophysical Union*, 37, 56–66.

Helley, E. J. and Smith, W. (1971) 'Development and calibration of a pressure-difference bedload sampler', *Water Resources Division Open-File Report. United States Geological Survey*.

Helley, J. R. (1969) 'Field measurement of the initiation of large bed particle motion in Blue Creek near Klamath, California', *Professional Paper. United States Geological Survey*, 562G.

Hem, J. D. (1970) 'Study and interpretation of the chemical characteristics of natural water', *Water Supply Paper. United States Geological Survey*, 1473.

Henderson, F. M. (1961) 'Stability of alluvial channels', *Journal of the Hydraulics Division, American Society of Civil Engineers*, 87, 109–38.

Hendrickson, G. E. and Krieger, R. A. (1964) 'Geochemistry of natural waters of the Blue Grass region, Kentucky', *Water Supply Paper. United States Geological Survey*, 1700.

Herbertson, J. G. (1969) 'A critical review of conventional bedload formulae', *Journal of Hydrology*, 8, 1–26.

Herbich, J. B. and Shulits, S. (1964) 'Large scale roughness in open-channel flow', *Journal of the Hydraulics Division, American Society of Civil Engineers*, 90, 203–30.

Herschy, R. W. (1976) 'New methods of river gauging', in Rodda, J. C. (ed.) *Facets of Hydrology*, Chichester, Wiley, pp. 119–61.

Hewlett, J. D. and Hibbert, A. R. (1963) 'Moisture and energy conditions within a sloping soil mass during drainage', *Journal of Geophysical Research*, 68, 1081–7.

Hewlett, J. D. and Hibbert, A. R. (1967) 'Factors affecting the response of small watersheds to precipitation in humid areas', in Sopper, W. E. and Lull, H. W. (eds) *International Symposium on Forest Hydrology*, Oxford, pp. 275–90.

Hey, R. D. (1975) 'Design discharges for natural channels', in Hey, R. D. and Davies, J. D. (eds) *Science and Technology in Environmental Management*, Farnborough, Saxon House, pp. 73–88.

Hey, R. D. (1976) 'Impact prediction in the physical environment', in O'Riordan, T. and Hey, R. D. (eds) *Environmental Impact Assessment*, Farnborough, Saxon House, pp. 71–81.

Hey, R. D. (1978) 'Determinate hydraulic geometry of river channels', *Journal of the Hydraulics Division, American Society of Civil Engineers*, 104, 869–85.

Hey, R. D. (1979) 'Dynamic process – response model of river channel development', *Earth Surface Processes*, 4, 59–72.

Hey, R. D. and Thorne, C. R. (1975) 'Secondary flows in river channels', *Area*, 7, 191–5.

Hickin, E. J. (1969) 'A newly identified process of point bar formation in natural streams', *American Journal of Science*, 267, 999–1010.

Hickin, E. J. (1972) 'Pseudo-meanders and point dunes – a flume study', *American Journal of Science*, 272, 762–99.

Hickin, E. J. (1974) 'The development of meanders in natural river channels', *American Journal of Science*, 274, 414–42.

Hickin, E. J. and Nanson, G. C. (1975) 'The character of channel migration on the Beatton River, north-east British Columbia, Canada', *Bulletin of the Geological Society of America*, 86, 487–94.

Hill, A. R. (1973) 'Erosion of river banks composed of glacial till near Belfast, Northern Ireland', *Zeitschrift für Geomorphologie*, 17, 428–42.

Hjulström, F. (1935) 'Studies of the morphological activity of rivers as illustrated by the River Fyris', *Bulletin of the Geological Institute, University of Uppsala*, 25, 221–527.

Hjulström, F. (1949) 'Climatic changes and river patterns', *Geografiska Annaler*, 31, 83–9.

Hodges, F. (1976) 'Law relating to water', in Rodda, J. C. (ed.) *Facets of Hydrology*, Chichester, Wiley, pp. 315–30.

Holeman, J. N. (1968) 'The sediment yield of major rivers of the world', *Water Resources Research*, 4, 737–47.

Hollis, G. E. (1975) 'The effect of urbanization on floods of different recurrence Harlow, Essex', in Gregory, K. J. and Walling, D. E. (eds) *Fluvial Processes in Instrumented Watersheds*, Institute of British Geographers Special Publication 6, pp. 123–39.

Hollis, G. E. (1975) 'The effect of urbanization on floods of different recurrence interval', *Water Resources Research*, 11, 431–5.

Hollis, G. E. and Luckett, J. K. (1976) 'The response of natural river channels to urbanization: two case studies from south-east England', *Journal of Hydrology*, 30, 351–63.

Hooke, J. M. (1977) 'The distribution and nature of changes in river channel patterns: the example of Devon', in Gregory, K. J. (ed.) *River Channel Changes*, Chichester, Wiley, pp. 265–80.

Hooke, J. M. (1980) 'Magnitude and distribution of rates of river bank erosion', *Earth Surface Processes*, 5, 143–57.

Hooke, R. Le B. (1967) 'Processes on arid region alluvial fans', *Journal of Geology*, 75, 438–60.

Hooke, R. Le B. (1968) 'Steady-state relationships on arid-region alluvial fans in closed basins', *American Journal of Science*, 266, 609–29.

Hooke, R. Le B. and Rohrer, W. L. (1977) 'Relative erodibility of source-area rock types, as determined from second-order variations in alluvial-fan size', *Bulletin of the Geological Society of America*, 88, 1177–82.

Hooke, R. Le B. and Rohrer, W. L. (1979) 'Geometry of alluvial fans: effect of discharge and sediment size', *Earth Surface Processes*, 4, 147–66.

Horton, R. E. (1933) 'The role of infiltration in the hydrologic cycle', *Transactions of the American Geophysical Union*, 14, 446–60.

Horton, R. E. (1945) 'Erosional development of streams and their drainage basins: hydrophysical approach to quantitative morphology', *Bulletin of the Geological Society of America*, 56, 275–370.

Howard, A. D. (1965) 'Geomorphological systems – equilibrium and dynamics', *American Journal of Science*, 263, 303–12.

Howard, A. D. (1980) 'Thresholds in river regimes', in Coates, D. R. and Vitek, J. D. (eds) *Thresholds in Geomorphology*, London, George Allen & Unwin, pp. 227–58.

Howard, A. D., Keetch, M. E. and Vincent, C. L. (1970) 'Topological and geometrical properties of braided streams', *Water Resources Research*, 6, 1674–88.

Howe, G. M., Slaymaker, H. O. and Harding, D. M. (1967) 'Some aspects of the flood hydrology of the upper catchments of the Severn and Wye', *Transactions of the Institute of British Geographers*, 41, 33–58.

Hubbell, D. W. (1964) 'Apparatus and techniques for measuring bedload', *Water Supply Paper. United States Geological Survey*, 1748.

Hubbell, D. W. and Sayre, W. (1964) 'Sand transport studies with radioactive tracers', *Journal of the Hydraulics Division, American Society of Civil Engineers*, 90, 36–98.

Hughes, D. A. (1980) 'Floodplain inundation: processes and relationships with channel discharge', *Earth Surface Processes*, 5, 297–304.

Humpidge, H. B. and Moss, W. D. (1971) 'The development of a comprehensive computer program for the calculation of flow profiles in open channels', *Proceedings of the Institution of Civil Engineers*, 50, Paper 7406, 49–64.

Hunt, J. N. (1954) 'The turbulent transport of suspended sediment in open channels', *Proceedings of the Royal Society*, 224A, 322–35.

Hyde, G. E. and Beeman, O. (1965) 'Improvement of the navigability of the Columbia River by dredging and constriction works', *Proceedings of the Federal Inter-Agency Sedimentation Conference (1963)*, United States Department of Agriculture, Publication 970, 454–61.

Imeson, A. C. (1974) 'The origin of sediment in a moorland catchment with particular reference to the role of vegetation', in Gregory, K. J. and Walling, D. E. (eds) *Fluvial Processes in Instrumented Watersheds*, Institute of British Geographers, Special Publication 6, pp. 59–72.

Inglis, C. C. (1949a) 'The behaviour and control of rivers and canals', *Central Waterpower Irrigation and Navigation Research Station, Poona*, Research Publication 13.

Inglis, C. C. (1949b) 'The effect of variations of charge and grade on the slopes and shapes of channels', *International Association for Hydraulic Structures Research, 3rd meeting, Grenoble*, pp. II.1–II.9.

Inman, D. L. (1949) 'Sorting of sediments in the light of fluid mechanics', *Journal of Sedimentary Petrology*, 19, 51–70.

Ippen, A. T. and Drinker, P. A. (1962) 'Boundary shear stresses in curved trapezoidal channels', *Journal of the Hydraulics Division, American Society of Civil Engineers*, 88, 143–79.

Jackson, R. G. (1976) 'Depositional model of point bars in the Lower Wabash River', *Journal of Sedimentary Petrology*, 46, 579–94.

Jahns, R. H. (1947) 'Geologic features of the Connecticut valley, Massachusetts, as related to recent floods', *Professional Paper. United States Geological Survey*, 996.

Janda, R. L. (1971) 'An evaluation of procedures used in computing chemical denudation rates', *Bulletin of the Geological Society of America*, 82, 67–80.

Jarvis, R. S. and Werritty, A. (1975) 'Some comments on testing random topology stream network models', *Water Resources Research*, 11, 309–18.

Johnson, F. A. (1971) 'A note on river water sampling and testing', *Water and Water Engineering*, 75, 59–61.

Johnson, H. T., Elsewy, E. M. and Cochrane, S. R. (1980) 'A study of the infiltration characteristics of undisturbed soil under simulated rainfall', *Earth Surface Processes*, 5, 159–74.

Johnson, J. W. (1943) 'Laboratory investigation of bedload transportation and bed roughness', United States Department of Agriculture, Soil Conservation Service, Paper 50.

Johnson, J. W. (1944) 'Rectangular artificial roughness in open channels', *Transactions of the American Geophysical Union*, 25, 906–14.

Johnson, N. M., Likens, G. L., Bormann, F. H., Fisher, D. W. and Pierce, R. S. (1969) 'A working model for the variation in stream water chemistry at the Hubbard Brook Experimental Forest', *Water Resources Research*, 5, 1353–63.

Johnson, P. and Muir, T. C. (1969) 'Acoustic detection of sediment movement', *Journal of Hydraulic Research*, 7, 519–40.

Jones, A. (1971) 'Soil piping and stream channel initiation', *Water Resources Research*, 7, 602–10.

Jones, O. T. (1924) 'The longitudinal profiles of the Upper Towy drainage system', *Quarterly Journal of the Geological Society of London*, 80, 568–609.

Judson, S. and Ritter, D. F. (1964) 'Rates of regional denudation in the United States', *Journal of Geophysical Research*, 69, 3395–401.

Kalinske, A. A. (1943) 'The role of turbulence in river hydraulics', *Bulletin of the University of Iowa, Studies in Engineering*, 27, 266–79.

Kalinske, A. A. (1947) 'Movement of sediment as bedload in rivers', *Transactions of the American Geophysical Union*, 28, 615–20.

Keller, E. A. (1970) 'Bedload movement experiments: Dry Creek, California', *Journal of Sedimentary Petrology*, 40, 1339–44.

Keller, E. A. (1971) 'Areal sorting of bed load material; the hypothesis of velocity reversal', *Bulletin of the Geological Society of America*, 82, 753–6.

Keller, E. A. (1972) 'Development of alluvial stream channels: a five stage model', *Bulletin of the Geological Society of America*, 83, 1531–6.

Keller, E. A. (1975) 'Channelization: a search for a better way', *Geology*, 3, 246–8.

Keller, E. A. (1976) 'Channelization: environmental, geomorphic and engineering aspects', in Coates, D. R. (ed.), *Geomorphology and Engineering*, Stroudsburg, Pennsylvania, Dowden, Hutchinson & Ross, pp. 115–40.

Keller, E. A. and Melhorn, W. (1973) 'Bedforms and fluvial processes in alluvial stream channels: selected observations', in Morisawa, M. (ed.) *Fluvial Geomorphology*, SUNY Binghamton, Publications in Geomorphology, pp. 253–83.

Keller, E. A. and Swanson, F. J. (1979) 'Effects of large organic material on channel form and fluvial processes', *Earth Surface Processes*, 4, 361–80.

Keller, E. A. and Tally, T. (1979) 'Effects of large organic debris on channel form and fluvial processes in the coastal redwood environment', in Rhodes, D. D. and Williams, G. P. (eds) *Adjustments of the Fluvial System*, Dubuque, Iowa, Kendall Hunt, pp. 169–97.

Keller, F. J. (1962) 'Effect of urban growth on sediment discharge, northwest branch Anacostia River basin, Maryland', *Professional Paper. United States Geological Survey*, 450C, pp. 129–31.

Kellerhals, R. (1967) 'Stable channels with gravel-paved beds', *Journal of Waterways and Harbours Division, American Society of Civil Engineers*, 93, 63–84.

Kellerhals, R. and Bray, D. J. (1971) 'Sampling procedures for coarse fluvial sediments', *Journal of the Hydraulics Division, American Society of Civil Engineers*, 97, 1165–80.

Kennedy, J. F. (1963) 'The mechanics of dunes and antidunes in erodible-bed channels', *Journal of Fluid Mechanics*, 16, 521–44.

Kennedy, R. G. (1895) 'The prevention of silting in irrigation canals', *Proceedings of the Institution of Civil Engineers*, 119, 281–90.

Kennedy, V. C. (1965) 'Mineralogy and cation-exchange capacity of sediments from selected streams', *Professional Paper. United States Geological Survey*, 433D.

Kennedy, V. C. and Kouba, D. L. (1970) 'Fluorescent sand as a tracer of fluvial sediment', *Professional Paper. United States Geological Survey*, 562E.

Kesseli, J. E. (1941) 'The concept of the graded river', *Journal of Geology*, 49, 561–88.

Keulegan, G. H. (1938) 'Laws of turbulent flows in open channels', *Journal of Research, National Bureau of Standards*, 21, 707–41.

Kilpatrick, F. A. and Barnes, H. H. (1964) 'Channel geometry of piedmont streams as related to frequency of floods', *Professional Paper. United States Geological Survey*, 422E, pp. 1–10.

King, C. A. M. (1970) 'Feedback relationships in geomorphology', *Geografiska Annaler*, 52A, 147–59.

Kirkby, M. J. (1967) 'Measurement and theory of soil creep', *Journal of Geology*, 75, 359–78.

Kirkby, M. J. (1976) 'Tests of the random network model and its application to basin hydrology', *Earth Surface Processes*, 1, 197–212.

Kirkby, M. J. (1977) 'Maximum sediment efficiency as a criterion for alluvial channels', in Gregory, K. J. (ed.) *River Channel Changes*, Chichester, Wiley, pp. 429–42.

Kirkby, M. J. and Chorley, R. J. (1967) 'Throughflow, overland flow and erosion', *Bulletin of the International Association of Scientific Hydrology*, 12, 5–21.

Klimek, K. (1974) 'The retreat of alluvial river banks in the Wisloka valley (South Poland)', *Geographia Polonica*, 28, 59–75.

Knight, C. (1979) 'Urbanization and natural stream channel morphology: the case of two English new towns', in Hollis, G. E. (ed.) *Man's Impact on the Hydrological Cycle in the United Kingdom*, Norwich, Geobooks, pp. 181–98.

Knighton, A. D. (1973) 'Riverbank erosion in relation to streamflow conditions, River Bollin-Dean, Cheshire', *East Midlands Geographer*, 5, 416–26.

Knighton, A. D. (1974) 'Variation in width–discharge relation and some implications for hydraulic geometry', *Bulletin of the Geological Society of America*, 85, 1059–76.

Knighton, A. D. (1975a) 'Variations in at-a-station hydraulic geometry', *American Journal of Science*, 275, 186–218.

Knighton, A. D. (1975b) 'Channel gradient in relation to discharge and bed material characteristics', *Catena*, 2, 263–74.

Knighton, A. D. (1979) 'Comments on log-quadratic relations in hydraulic geometry', *Earth Surface Processes*, 4, 205–9.

Knighton, A. D. (1980) 'Longitudinal changes in size and sorting of stream-bed material in four English rivers', *Bulletin of the Geological Society of America*, 91, 55–62.

Knox, J. C. (1972) 'Valley alluviation in south-western Wisconsin', *Annals of the Association of American Geographers*, 62, 401–10.

Knox, J. C. (1975) 'Concept of the graded stream', in Melhorn, W. N. and Flemal, R. C. (eds) *Theories of Landform Development*, SUNY Binghamton, Publications in Geomorphology, pp. 169–98.

Kohler, M. A. and Linsley, R. K. (1951) 'Predicting the runoff from storm rainfall', *United States Weather Bureau*, Research Paper 34.

Kolosseus, H. J. and Davidian, J. (1961) 'Flow in an artificially roughened channel', *Professional Paper. United States Geological Survey*, 424B, pp. 25–6.

Komura, S. and Simons, D. B. (1967) 'River bed degradation below dams', *Journal of the Hydraulics Division, American Society of Civil Engineers*, 93, 1–14.

Konditerova, E. A. and Popov, I. V. (1966) 'Relation between changes in the horizontal and vertical characteristics of river channels', *Soviet Hydrology*, 5, 515–27.

Kondrat'yev, N. Ye. (1968) 'Hydromorphological principles of computations of free meandering. 1. Signs and indexes of free meandering', *Soviet Hydrology*, 7, 309–35.

Kondrat'yev, N. Ye. and Popov, I. V. (1967) 'Methodological prerequisites for conducting network observations of the channel process', *Soviet Hydrology*, 6, 273–97.

Krigström, A. (1962) 'Geomorphological studies of sandur plains and their braided rivers in Iceland', *Geografiska Annaler*, 44, 328–46.

Krinsley, D. H. and Doornkamp, J. C. (1973) *Atlas of Quartz Sand Surface Textures*, Cambridge, Cambridge University Press.

Krumbein, W. C. (1937) 'Sediments and exponential curves', *Journal of Geology*, 45, 577–601.

Krumbein, W. C. (1941a) 'Measurement and geological significance of shape and roundness of sedimentary particles', *Journal of Sedimentary Petrology*, 11, 64–72.

Krumbein, W. C. (1941b) 'The effects of abrasion on the size and shape of rock fragments', *Journal of Geology*, 49, 482–520.

Krumbein, W. C. (1942) 'Settling velocity and flume behaviour of non-spherical particles', *Transactions of the American Geophysical Union*, 23, 621–32.

Krumbein, W. C. and Orme, A. R. (1972) 'Field mapping and computer simulation of braided stream networks', *Bulletin of the Geological Society of America*, 83, 3369–80.

Kuenen, P. H. (1956) 'Experimental abrasion of pebbles. 2. Rolling by current', *Journal of Geology*, 64, 336–68.

Kuenen, P. H. (1959) 'Experimental abrasion. 3. Fluviatile action on sand', *American Journal of Science*, 257, 172–90.

Kunkle, G. R. (1965) 'Computation of ground-water discharge to streams during floods, or to individual reaches during baseflow, by use of specific conductance', *Professional Paper. United States Geological Survey*, 525D, pp. 207–10.

Kunkle, S. H. and Comer, G. H. (1971) 'Estimating suspended sediment concentrations in streams by turbidity measurements', *Journal of Soil and Water Conservation*, 26, 18–20.

Lacey, G. (1930) 'Stable channels in alluvium', *Proceedings of the Institution of Civil Engineers*, 229, 259–92.

Lacey, G. and Pemberton, W. (1972) 'A general formula for uniform flow in self-formed alluvial channels', *Proceedings of the Institution of Civil Engineers*, 53, 373–81.

Lane, E. W. (1937) 'Stable channels in erodible materials', *Transactions of the American Society of Civil Engineers*, 102, 123–94.

Lane, E. W. (1952) 'Progress report on results of studies on design of stable channels', United States Department of Interior, Bureau of Reclamation, *Hydraulic Laboratory Report Hyd-352*.

Lane, E. W. (1955a) 'The importance of fluvial morphology in hydraulic engineering', *Proceedings of the American Society of Civil Engineers*, 81, 1–17.

Lane, E. W. (1955b) 'Design of stable channels', *Transactions of the American Society of Civil Engineers*, 120, 1234–79.

Lane, E. W. (1957) 'A study of the shape of channels formed by natural streams flowing in erodible material', *MRD Sediment Series 9*, US Army Engineer Division, Missouri River.

Lane, E. W. and Carlson, E. J. (1954) 'Some observations on the effect of particle shape on the movement of coarse sediments', *Transactions of the American Geophysical Union*, 35, 453–62.

Lane, E. W., Carlson, E. J. and Hanson, O. S. (1949) 'Low temperature increases sediment transportation in Colorado river', *Civil Engineering*, 19, 619–20.

Lane, E. W. and Kalinske, A. A. (1939) 'The relation of suspended to bed material in rivers', *Transactions of the American Geophysical Union*, 20, 637–41.

Lane, E. W. and Lei, K. (1950) 'Streamflow variability', *Transactions of the American Society of Civil Engineers*, 115, 1084–134.

Lane, E. W., Lin, P. N. and Liu, H. K. (1959) 'The most efficient stable channel for comparatively clear water in non-cohesive material', *Report CER59HKL5*, Colorado State University.

Langbein, W. B. (1949) 'Annual floods and the partial duration flood series', *Transactions of the American Geophysical Union*, 30, 879–81.

Langbein, W. B. (1962) 'Hydraulics of river channels as related to navigability', *Water Supply Paper. United States Geological Survey*, W 1539W.

Langbein, W. B. (1964a) 'Geometry of river channels', *Journal of the Hydraulics Division, American Society of Civil Engineers*, 90, 301–12.

Langbein, W. B. (1964b) 'Profiles of rivers of uniform discharge', *Professional Paper. United States Geological Survey*, 501B, pp. 119–24.

Langbein, W. B. (1965) 'Geometry of river channels: closure of discussion', *Journal of the Hydraulics Division, American Society of Civil Engineers*, 91, 297–313.

Langbein, W. B. (1966) 'A random walk model of hydraulic friction', *Bulletin of the International Association of Scientific Hydrology*, 11, 5–9.

Langbein, W. B. and Dawdy, D. R. (1964) 'Occurrence of dissolved solids in surface waters in the United States', *Professional Paper. United States Geological Survey*, 501D, pp. 115–17.

Langbein, W. B. and Leopold, L. B. (1964) 'Quasi-equilibrium states in channel morphology', *American Journal of Science*, 262, 782–94.

Langbein, W. B. and Leopold, L. B. (1966) 'River meanders – theory of minimum variance', *Professional Paper. United States Geological Survey*, 422H.

Langbein, W. B. and Leopold, L. B. (1968) 'River channel bars and dunes – theory of kinematic waves', *Professional Paper. United States Geological Survey*, 422L.

Langbein, W. B. and Schumm, S. A. (1958) 'Yield of sediment in relation to mean annual precipitation', *Transactions of the American Geophysical Union*, 39, 1076–84.

Langford, K. J. (1976) 'Change in yield of water following a bushfire in a forest of *Eucalyptus Regnans*', *Journal of Hydrology*, 29, 87–114.

Laronne, J. B. and Carson, M. A. (1976) 'Interrelationships between bed morphology and bed material transport for a small gravel bed channel', *Sedimentology*, 23, 67–85.

Laury, R. L. (1971) 'Stream bank failure and rotational slumping; preservation and significance in the geological record', *Bulletin of the Geological Society of America*, 82, 1251–66.

Le Ba Hong and Davies, T. R. H. (1979) 'A study of stream braiding: summary', *Bulletin of the Geological Society of America*, 90, 1094–5.

Leeder, M. R. (1979) ' "Bedload" dynamics: grain–grain interactions in water flows', *Earth Surface Processes*, 4, 229–40.

Leliavsky, S. (1955) *An Introduction to Fluvial Hydraulics*, London, Constable.

Lemmens, M. and Roger, M. (1978) 'Influence of ion exchange on dissolved load of alpine meltwaters', *Earth Surface Processes*, 3, 179–87.

Leopold, L. B. (1953) 'Downstream change of velocity in rivers', *American Journal of Science*, 251, 606–24.

Leopold, L. B. (1968) 'Hydrology for urban land planning – a guidebook on the hydrologic effects of urban land use', *Circular. United States Geological Survey*, 554.

Leopold, L. B. (1969) 'Quantitative comparison of some aesthetic factors among rivers', *Circular. United States Geological Survey*, 620.

Leopold, L. B. (1970) 'An improved method for size distribution of stream bed gravel', *Water Resources Research*, 6, 1357–66.

Leopold, L. B. (1973a) 'River channel change with time: an example', *Bulletin of the Geological Society of America*, 84, 1845–60.

Leopold, L. B. (1973b) 'Hydrologic research on instrumented watersheds', in *Results of Research on Representative and Experimental Basins, IASH/UNESCO, Proceedings of the Wellington Symposium, 1970*, 2, pp. 135–50.

Leopold, L. B. (1978) 'El Asunto del Arroyo', in Embleton, C., Brunsden, D. and Jones, D. K. C. (eds) *Geomorphology: Present Problems and Future Prospects*, Oxford, Oxford University Press, pp. 25–39.

Leopold, L. B., Bagnold, R. A., Wolman, M. G. and Brush, L. M. (1960) 'Flow resistance in sinuous or irregular channels', *Professional Paper. United States Geological Survey*, 282D, pp. 111–35.

Leopold, L. B. and Bull, W. B. (1979) 'Base level, aggradation and grade', *Proceedings of the American Philosophical Society*, 123, 168–202.

Leopold, L. B. and Emmett, W. W. (1976) 'Bedload measurements, East Fork River, Wyoming', *Proceedings of the National Academy of Sciences*, 73, 1000–4.

Leopold, L. B. and Langbein, W. B. (1962) 'The concept of entropy in landscape evolution', *Professional Paper. United States Geological Survey*, 500A.

Leopold, L. B. and Langbein, W. B. (1963) 'Association and indeterminacy in geomorphology', in Albritton, C. C. (ed.) *The Fabric of Geology*, Stanford, California, Freeman, Cooper & Co., pp. 184–92.

Leopold, L. B. and Maddock, T. Jr (1953a) 'The hydraulic geometry of stream channels and some physiographic implications', *Professional Paper. United States Geological Survey*, 252, pp. 1–57.

Leopold, L. B. and Maddock, T. (1953b) 'Relation of suspended-sediment concentration to channel scour and fill', *Proceedings of the 5th Hydraulic Conference, University of Iowa, Bulletin* 34, 159–78.

Leopold, L. B. and Maddock, T. (1954) *The Flood Control Controversy*, New York, Ronald Press.

Leopold, L. B. and Marchand, M. O'B. (1968) 'On the quantitative inventory of the riverscape', *Water Resources Research*, 4, 709–17.

Leopold, L. B. and Miller, J. P. (1954) 'A postglacial chronology for some alluvial valleys in Wyoming', *Water Supply Paper. United States Geological Survey*, 1261, pp. 1–90.

Leopold, L. B. and Miller, J. P. (1956) 'Ephemeral streams – hydraulic factors and their relation to the drainage net', *Professional Paper. United States Geological Survey*, 282A.

Leopold, L. B. and Wolman, M. G. (1956) 'Floods in relation to the river channel', *Publication. International Association of Scientific Hydrology*, 42, 3, 85–98.

Leopold, L. B. and Wolman, M. G. (1957) 'River channel patterns – braided, meandering and straight', *Professional Paper. United States Geological Survey*, 282B.

Leopold, L. B. and Wolman, M. G. (1960) 'River meanders', *Bulletin of the Geological Society of America*, 71, 769–94.

Leopold, L. B., Wolman, M. G. and Miller, J. P. (1964) *Fluvial Processes in Geomorphology*, San Francisco, W. H. Freeman.

Lewin, J. (1976) 'Initiation of bedforms and meanders in coarse grained sediment', *Bulletin of the Geological Society of America*, 87, 281–5.

Lewin, J., Cryer, R. and Harrison, D. I. (1974) 'Sources for sediments and solutes in mid-Wales', in Gregory, K. J. and Walling, D. E. (eds) *Fluvial Processes in Instrumented Watersheds*, London, Institute of British Geographers Special Publication 6, pp. 73–85.

Lewin, J., Hughes, D. and Blacknell, C. (1977) 'Incidence of river erosion', *Area*, 9, 177–80.

Lewin, J. and Manton, M. M. (1975) 'Welsh floodplain studies: the nature of floodplain geometry', *Journal of Hydrology*, 25, 37–50.

Lewis, L. A. (1966) 'The adjustment of some hydraulic variables at discharges less than 1 cfs', *Professional Geographer*, 18, 230–4.

Lewis, L. A. (1969) 'Some fluvial geomorphic characteristics of the Manati Basin, Puerto Rico', *Annals of the Association of American Geographers*, 59, 280–93.

Lewis, W. V. (1945) 'Nick points and the curve of water erosion', *Geological Magazine*, 82, 256–66.

Li, R.-M., Simons, D. B. and Stevens, M. A. (1976) 'Morphology of cobble streams in small watersheds', *Journal of the Hydraulics Division, American Society of Civil Engineers*, 102, 1101–17.

Lighthill, M. J. and Whitham, G. B. (1955) 'On kinematic waves. II. A theory of traffic flow on long crowded roads', *Proceedings of the Royal Society*, 229A, 317–45.

Likens, G. E., Bormann, F. H., Johnson, N. M. and Pierce, R. S. (1967) 'The calcium, magnesium, potassium and sodium budgets for a small forested ecosystem', *Ecology*, 48, 772–85.

Limerinos, J. T. (1969) 'Relation of the Manning coefficient to measured bed roughness in stable natural channels', *Professional Paper. United States Geological Survey*, 650D, pp. 215–21.

Lindley, E. S. (1919) 'Regime channels', *Proceedings of the Punjab Engineering Congress*, 7, 63–74.

Linsley, R. K., Kohler, M. A. and Paulhus, J. L. H. (1949) *Applied Hydrology*, New York, McGraw-Hill.

Lisle, T. (1979) 'A sorting mechanism for a riffle pool sequence', *Bulletin of the Geological Society of America*, 90, 1142–57.

Liu, H. K. (1957) 'Mechanics of sediment-ripple formation', *Journal of the Hydraulics Division, American Society of Civil Engineers*, 83, Paper 1197.

Livesey, R. H. (1965) 'Channel armoring below Fort Randall dam', *Proceedings of the Federal Inter-Agency Sedimentation Conference, 1963*, United States Department of Agriculture, Publication 970, pp. 461–70.

Livingstone, D. A. (1963) 'Chemical composition of rivers and lakes', *United States Geological Survey, Professional Paper* 440-G.

Loughran, R. J. (1976) 'The calculation of suspended-sediment transport from concentration v. discharge curves', *Catena*, 3, 45–61.

Lovera, F. and Kennedy, J. F. (1969) 'Friction factors for flat-bed flows in sand channels', *Journal of the Hydraulics Division, American Society of Civil Engineers*, 95, 1227–34.

Lusby, G. C. (1970) 'Hydrologic and biotic effects of grazing versus non-grazing near Grand-Junction, Colorado', *Professional Paper. United States Geological Survey*, 700B, pp. 232–6.

Lyles, L. and Woodruff, N. P. (1972) 'Boundary-layer flow structure: effects on detachment of noncohesive particles', in Shen, H. W. (ed.) *Sedimentation*, Fort Collins, Colorado, pp. 2.1–2.16.

McCarthy, G. T. (1938) 'The unit hydrograph and flood routing', unpublished manuscript, presented at conference of North Atlantic Division, US Army Corps of Engineers, 24 June 1938.

McGilchrist, C. A. and Woodyer, K. D. (1968) 'Statistical tests for common bankfull frequency in rivers', *Water Resources Research*, 4, 331–4.

McGilchrist, C. A., Woodyer, K. D. and Chapman, T. G. (1968) 'Recurrence intervals between exceedances of selected river levels. 1. Introduction and a Markov model', *Water Resources Research*, 4, 183–9.

McGowen, J. H. and Garner, L. E. (1970) 'Physiographic features and stratification types of coarse-grained point bars: modern and ancient examples', *Sedimentology*, 14, 77–111.

McHenry, J. R., Coleman, N. L., Willis, J. C., Gill, A. C., Sansom, O. W. and Carroll, B. R. (1970) 'Effect of concentration gradients on the performance of a nuclear sediment concentration gauge', *Water Resources Research*, 6, 538–48.

Mackin, J. H. (1948) 'Concept of the graded river', *Bulletin of the Geological Society of America*, 59, 463–512.

Mackin, J. H. (1963) 'Rational and empirical methods of investigation in geology', in Albritton, C. C. Jr (ed.) *The Fabric of Geology*, Reading, Massachusetts, Addison–Wesley, pp. 135–63.

McManus, J. (1979) 'The evolution of fluviatile sediments as demonstrated by Q Da–Md analysis', *Earth Surface Processes*, 4, 141–6.

McPherson, H. J. and Rannie, W. F. (1970) 'Geomorphic effects of the May 1967 flood in Graburn watershed, Cypress Hills, Alberta, Canada', *Journal of Hydrology*, 9, 307–21.

McQueen, I. S. (1961) 'Some factors influencing streambank erodibility', *Professional Paper. United States Geological Survey*, 424B, pp. 28–9.

Madden, E. B. (1965) 'Channel design for modified sediment regime conditions on the Arkansas River', *Proceedings of the Federal Inter-Agency Sedimentation Conference, 1963*, United States Department of Agriculture Publication 970, pp. 335–53.

Maddock, T. (1969) 'The behaviour of straight open channels with movable beds', *Professional Paper. United States Geological Survey*, 622A.

Maddock, T. (1970) 'Indeterminate hydraulics of alluvial channels', *Journal of the Hydraulics Division, American Society of Civil Engineers*, 96, 2309–23.

Mahmood, K. and Shen, H. W. (1971) 'Regime concept of sediment-transporting canals and rivers', in Shen, H. W. (ed.) *River Mechanics*, vol. 2, Fort Collins, Colorado, pp. 30.1–30.39.

Maizels, J. K. (1979) 'Proglacial aggradation and changes in braided channel patterns during a period of glacier advance: an Alpine example', *Geografiska Annaler*, 61A, 87–101.

Maizels, J. K. (1983) 'Palaeovelocity and palaeodischarge determination for coarse gravel deposits', in Gregory, K. J. (ed.) *Background to Palaeohydrology*, Chichester, Wiley, 101–39.

Manning, R. (1891) 'On the flow of water in open channels and pipes', *Transactions of the Institution of Civil Engineers of Ireland*, 20, 161–207.

Mao, S. W. and Flook, L. R. Jr (1971) 'Link canal design practices in West Pakistan', in Shen, H. W. (ed.) *River Mechanics*, vol. 2, Fort Collins, Colorado, Appendix A.

Martinec, J. (1972) 'Comment on the paper "On river meanders" by Ch. T. Yang', *Journal of Hydrology*, 15, 249–51.

Masch, F. D. (chairman) (1968) 'Erosion of cohesive sediments; Task Committee on erosion of cohesive materials', *Journal of the Hydraulics Division, American Society of Civil Engineers*, 94, 1017–49.

Matalas, N. C. and Benson, M. A. (1968) 'Note on the standard error of the coefficient of skewness', *Water Resources Research*, 4, 204–5.

Meade, R. H. (1969) 'Errors in using modern stream-load data to estimate natural rates of denudation', *Bulletin of the Geological Society of America*, 80, 1265–74.

Mein, R. G., Laurenson, E. M. and McMahon, T. A. (1974) 'Simple nonlinear model for flood estimation, *Journal of the Hydraulics Division, American Society of Civil Engineers*, 100, 1507–18.

Melhorn, W. N., Keller, E. A. and McBane, R. A. (1975) 'Landscape aesthetics numerically defined (LAND system): application to fluvial environments', *Technical Report*, 37, *Studies in Fluvial Geomorphology No. 1*, Purdue University Water Resources Research Centre, West Lafayette, Indiana.

Melton, M. A. (1957) 'An analysis of the relations among elements of climate, surface properties and geomorphology', *Office of Naval Research, Geography Branch, Project NR 389–042: Technical Report* 11, Columbia University.

Melton, M. A. (1958) 'Correlation structure of morphometric properties of drainage systems and their controlling agents', *Journal of Geology*, 66, 442–60.

Melton, M. A. (1961) 'Discussion: the effect of sediment type on the shape and stratification of some modern fluvial deposits', *American Journal of Science*, 259, 231–3.

Menard, H. W. (1950) 'Sediment movement in relation to current velocity', *Journal of Sedimentary Petrology*, 20, 148–60.

Meyer, L. D. (1971) 'Soil erosion by water on upland areas', in Shen, H. W. (ed.) *River Mechanics*, vol. 2, Fort Collins, Colorado, pp. 27.1–27.25.

Meyer-Peter, E. and Müller, R. (1948) 'Formulas for bed-load transport', *Proceedings of the International Association of Hydraulic Research, 3rd Annual Conference, Stockholm*, pp. 39–64.

Miall, A. D. (1973) 'Markov chain analysis applied to an ancient alluvial plain succession', *Sedimentology*, 20, 347–64.

Miall, A. D. (1977) 'A review of the braided-river depositional environment', *Earth Science Reviews*, 13, 1–62.

Middleton, G. V. (1976) 'Hydraulic interpretation of sand size distributions', *Journal of Geology*, 84, 405–26.

Miller, A. A. (1939) 'Attainable standards of accuracy in the determination of preglacial sea levels by physiographic methods; *Journal of Geomorphology*, 2, 95–115.

Miller, C. R. and Borland, W. M. (1963) 'Stabilization of Fivemile and Muddy Creeks', *Journal of the Hydraulics Division, American Society of Civil Engineers*, 89, HY1, 67–98.

Miller, J. P. (1958) 'High mountain streams; effects of geology on channel characteristics and bed material', *New Mexico State Bureau of Mines and Mineral Resources*, Memoir 4.

Miller, J. P. and Leopold, L. B. (1963) 'Simple measurements of morphological changes in river channels and hill slopes', in *Changes of Climate*, UNESCO Arid Zonè Research Series, 20, pp. 421–7.

Miller, T. K. and Onesti, L. J. (1979) 'The relationship between channel shape and sediment characteristics in the channel perimeter', *Bulletin of the Geological Society of America*, 80, 301–4.

Milne, J. A. (1979) 'The morphological relationships of bends in confined stream channels in upland Britain', in Pitty, A. F. (ed.) *Geographical Approaches to Fluvial Processes*, Norwich, Geobooks, pp. 241–60.

Milton, L. E. (1966) 'The geomorphic irrelevance of some drainage net laws', *Australian Geographical Studies*, 4, 89–95.

Mirajgaoker, A. G. and Charlu, K. L. N. (1963) 'Natural roughness effects in rigid open channels', *Journal of the Hydraulics Division, American Society of Civil Engineers*, 89, 29–44.

Mollard, J. D.(Undated) *Landforms and Surface Materials of Canada: A Stereoscopic Airphoto Atlas and Glossary* (loose-leaf).

Morgan, R. P. C. (1980) 'Soil erosion and conservation', *Progress in Physical Geography*, 4, 24–47.

Morisawa, M. (1962) 'Quantitative geomorphology of some watersheds in the Appalachian plateau', *Bulletin of the Geological Society of America*, 73, 1025–46.

Morisawa, M. (1968) *Streams: Their Dynamics and Morphology*, New York, McGraw-Hill.

Morisawa, M. (1971) 'Evaluating riverscapes', in Coates, D. R. (ed.) *Environmental Geomorphology*, SUNY Binghamton, Publications in Geomorphology, pp. 91–106.

Morisawa, M. and Laflure, E. (1979) 'Hydraulic geometry, stream equilibrium and urbanization', in Rhodes, D. D. and Williams, G. P. (eds) *Adjustments of the Fluvial System*, Dubuque, Iowa, Kendall-Hunt, pp. 333–50.

Mosley, M. P. (1975a) 'Channel changes on the River Bollin, Cheshire, 1872–1973', *East Midlands Geographer*, 6, 185–99.

Mosley, M. P. (1975b) 'Meander cutoffs on the River Bollin, Cheshire, in July 1973', *Revue de Géomorphologie Dynamique*, 24, 21–31.

Mosley, M. P. and Zimpfer, G. L. (1978) 'Hardware models in geography', *Progress in Physical Geography*, 2, 438–61.

Moss, A. J. (1972) 'Initial fluviatile fragmentation of granitic quartz', *Journal of Sedimentary Petrology*, 42, 905–16.

Mueller, J. E. (1968) 'Introduction to hydraulic and topographic sinuosity indexes', *Annals of the Association of American Geographers*, 58, 371–85.

Mueller, J. E. (1972) 'Re-evaluation of the relationship of master streams to drainage basins', *Bulletin of the Geological Society of America*, 83, 3471–4.

Muir, T. C. (1970) 'Bed load discharge of River Tyne, England', *Bulletin of the International Association of Scientific Hydrology*, 15, 35–9.

Mycielska-Dowgiałło, E. (1977) 'Channel pattern changes during the last glaciation and Holocene, in the northern part of the Sandomierz basin and the middle part of the Vistula valley, Poland', in Gregory, K. J. (ed.) *River Channel Changes*, Chichester, Wiley, pp. 75–87.

Naden, P. S. (1981) 'Gravel bedforms: deductions from sediment movement', *University of Leeds, Department of Geography, Working Paper 314*.

Nagabushanaiah, H. S. (1967) 'Meandering of rivers', *Bulletin of the International Association of Scientific Hydrology*, 12, 28–43.

Nakamura, R. (1971) 'Runoff analysis by electrical conductance of water', *Journal of Hydrology*, 14, 197–212.

Nanson, G. C. (1974) 'Bedload and suspended-load transport in a small, steep, mountain stream', *American Journal of Science*, 274, 471–86.

Nash, J. E. (1959) 'Discussion of "A study of bankfull discharge" by M. Nixon', *Proceedings of the Institution of Civil Engineers*, 14, 403–6.

Nash, J. E. (1960) 'A unit hydrograph study with particular reference to British catchments', *Proceedings of the Institution of Civil Engineers*, 17, 249–82.

Nash, J. E. and Shaw, B. L. (1966) 'Flood frequency as a function of catchment characteristics', *Institution of Civil Engineers, River Flood Hydrology Symposium*, pp. 115–36.

Natural Environment Research Council (NERC) (1975) *Flood Study Report*, Natural Environment Research Council, UK.

Neff, E. L. (1967) 'Discharge frequency compared to long term sediment yields, *Publication. International Association of Scientific Hydrology*, 75, 236–42.

Negev, M. (1967) *A sediment model on a digital computer*, Report 76, Stanford University, California.

Neill, C. R. (1965) 'Measurements of bridge scour and bed changes in a flooding sand-bed river', *Proceedings of the Institution of Civil Engineers*, 30, 415–35.

Neu, H. A. (1967) 'Transverse flow in a river due to earth's rotation', *Journal of the Hydraulics Division, American Society of Civil Engineers*, 93, 149–65.

Nevin, C. (1946) 'Competency of moving water to transport debris', *Bulletin of the Geological Society of America*, 57, 651–74.

Newson, M. (1980) 'The geomorphological effectiveness of floods – a contribution stimulated by two recent events in mid-Wales', *Earth Surface Processes*, 5, 1–16.

Nixon, M. (1959) 'A study of the bankfull discharges of rivers in England and Wales', *Proceedings of the Institution of Civil Engineers*, 12, 157–75.

Noble, C. A. and Palmquist, R. C. (1968) 'Meander growth in artificially straightened streams', *Proceedings of the Iowa Academy of Science*, 75, 234–42.

Noble, C. C. (1976) 'The Mississippi River flood of 1973', in Coates, D. R. (ed.) *Geomorphology and Engineering*, Stroudsburg, Pennsylvania, Dowden, Hutchinson & Ross, pp. 79–98.

Nordin, C. F. (1971) 'Statistical properties of dune profiles', *Professional Paper. United States Geological Survey*, 562F.

Nordin, C. F. and Algert, J. H. (1966) 'Spectral analysis of sand waves', *Journal of the Hydraulics Division, American Society of Civil Engineers*, 92, 95–114.

Nordin, C. F. and Beverage, J. P. (1965) 'Sediment transport in the Rio Grande, New Mexico', *Professional Paper. United States Geological Survey*, 462F.

Nordin, C. F. and Dempster, G. R. (1963) 'Vertical distribution of velocity and suspended sediment, Middle Rio Grande, New Mexico', *Professional Paper. United States Geological Survey*, 462B.

Novak, I. (1973) 'Predicting coarse sediment transport: the Hjulström curve revisited', in Morisawa, M. (ed.) *Fluvial Geomorphology*, SUNY Binghamton, Publications in Geomorphology, pp. 13–25.

Novak, P. (1957) 'Bedload meters – development of a new type and determination of their efficiency with the aid of scale models', *Proceedings of the International Association of Hydraulic Research, 7th Annual Conference, Lisbon*, 1, A9.1–A9.11.

Nunally, N. R. (1967) 'Definition and identification of channel and overbank deposits and their respective roles in floodplain formation', *Professional Geographer*, 19, 1–4.

O'Brien, M. P. (1933) 'Review of the theory of turbulent flow and its relation to sediment transportation', *Transactions of the American Geophysical Union*, 14, 487–91.

Olsen, O. J. and Florey, Q. L. (1952) 'Sedimentation studies in open channels – boundary shear and velocity distribution by membrane analogy, analytical and finite difference methods', *United States Department of Interior, Bureau of Reclamation, Hydrualic Laboratory Report* SP 34.

Ongley, E. D. (1968) 'An analysis of the meandering tendency of Serpentine Cave, New South Wales', *Journal of Hydrology*, 6, 15–32.

O'Riordan, T. and More, R. J. (1969) 'Choice in water use', in Chorley, R. J. (ed.) *Water, Earth and Man*, London, Methuen, pp. 547–73.

Østrem, G. (1964) 'A method of measuring water discharge in turbulent streams', *Geographical Bulletin*, 21, 21–43.

Ouma, J. P. B. M. (1967) 'Fluviatile morphogenesis of roundness: the Hacking River, New South Wales, Australia', *Publication. International Association of Scientific Hydrology*, 75, 319–43.

Oxley, N. C. (1974) 'Suspended sediment delivery rates and the solute concentration of stream discharge in two Welsh catchments', in Gregory, K. J. and Walling, D. E. (eds) *Fluvial Processes in Instrumented Watersheds*, London, Institute of British Geographers Special Publication 6, pp. 141–54.

Page, K. and Nanson, G. (1982) 'Concave-bank benches and associated floodplain formation', *Earth Surface Processes and Landforms*, 7, 529–43.

Paintal, A. S. (1971) 'Concept of critical shear stress in loose boundary open channels', *Journal of Hydraulic Research*, 9, 91–113.

Palmer, L. (1976) 'River management criteria for Oregon and Washington', in Coates, D. R. (ed.) *Geomorphology and Engineering*, Stroudsburg, Pennsylvania, Dowden, Hutchinson & Ross, pp. 329–46.

Palmquist, R. C. (1975) 'Preferred position model and subsurface symmetry of valleys', *Bulletin of the Geological Society of America*, 86, 1392–8.

Park, C. C. (1976) 'The relationship of slope and stream channel form in the River Dart, Devon', *Journal of Hydrology*, 29, 139–47.

Park, C. C. (1977a) 'World wide variations in hydraulic geometry exponents of stream channels: an analysis and some observations', *Journal of Hydrology*, 33, 133–46.

Park, C. C. (1977b) 'Man-induced changes in stream channel capacity', in Gregory, K. J. (ed.) *River Channel Changes*, Chichester, Wiley, pp. 121–44.

Park, C. C. (1978) 'Channel bank material and cone penetrometer studies: an empirical evaluation', *Area*, 10, 227–30.

Park, C. C. (1979) 'Tin streaming and channel changes – some preliminary observations from Dartmoor, England', *Catena*, 6, 235–44.

Park, C. C. (1981) 'Man, river systems and environmental impacts', *Progress in Physical Geography*, 5, 1–31.

Parker, G. (1976) 'On the cause and characteristic scale of meandering and braiding in rivers', *Journal of Fluid Mechanics*, 76, 459–80.

Parker, G. (1978) 'Self-formed straight rivers with equilibrium banks and mobile bed, Part 1; the sand-silt river', *Journal of Fluid Mechanics*, 89, 109–25.

Parker, G. (1979) 'Hydraulic geometry of active gravel rivers', *Journal of the Hydraulics Division, American Society of Civil Engineers*, 105, 1185–201.

Parker, G. and Andres, D. (1976) 'Detrimental effects of river channelization', *Waterways, Harbors and Coastal Engineering Division, American Society of Civil Engineers, Proceedings of the 3rd Annual Symposium, Fort Collins*, pp. 1248–66.

Parsons, D. A. (1965) 'Vegetative control of streambank erosion', *Proceedings of the Federal Inter-Agency Sedimentation Conference, 1963*, United States Department of Agriculture Publication 970, pp. 130–6.

Partheniades, E. and Paaswell, R. E. (1970) 'Erodibility of channels with cohesive boundary', *Journal of the Hydraulics Division, American Society of Civil Engineers*, 96, 755–71.

Passega, R. (1964) 'Grain size representation by *C-M* patterns as a geological tool', *Journal of Sedimentary Petrology*, 34, 830–47.

Patton, P. C., Baker, V. R. and Kochel, R. C. (1979) 'Slack-water deposits: a geomorphic technique for the interpretation of fluvial palaeohydrology', in Rhodes, D. D. and Williams, G. P. (eds) *Adjustments of the Fluvial System*, Dubuque, Iowa, Kendall Hunt, pp. 225–53.

Pearce, T. H. (1971) 'Short distance fluvial rounding of volcanic detritus', *Journal of Sedimentary Petrology*, 41, 1069–72.

Penman, H. L. (1950) 'The water balance of the Stour catchment area', *Journal of the Institution of Water Engineers*, 4, 457–69.

Penning-Rowsell, E. and Townshend, J. R. G. (1978) 'The influence of scale on the factors affecting stream channel slope', *Transactions of the Institute of British Geographers, New Series*, 3, 395–415.

Peterson, D. F. and Mohanty, P. K. (1960) 'Flume studies of flow in steep, rough channels', *Journal of the Hydraulics Division, American Society of Civil Engineers*, 86, 55–76.

Petts, G. E. (1977) 'Channel response to flow regulation: the case of the River Derwent, Derbyshire', in Gregory, K. J. (ed.) *River Channel Changes*, Chichester, Wiley, pp. 145–64.

Petts, G. E. (1979) 'Complex response of river channel morphology subsequent to reservoir construction', *Progress in Physical Geography*, 3, 329–62.

Petts, G. E. and Lewin, J. (1979) 'Physical effects of reservoirs on river systems', in Hollis, G. E. (ed.) *Man's Impact on the Hydrological Cycle in the United Kingdom*, Norwich, Geobooks, pp. 79–91.

Picard, M. D. and High, L. R. (1973) *Sedimentary Structures of Ephemeral Streams*, Amsterdam, Elsevier.

Pickup, G. (1975) 'Downstream variations in morphology, flow conditions and sediment transport in an eroding channel', *Zeitschrift für Geomorphologie*, 19, 443–59.

Pickup, G. (1976) 'Adjustment of stream-channel shape to hydrologic regime', *Journal of Hydrology*, 30, 365–73.

Pickup, G. (1977) 'Simulation modelling of river channel erosion', in Gregory, K. J. (ed.) *River Channel Changes*, Chichester, Wiley, pp. 47–60.

Pickup, G. and Rieger, W. A. (1979) 'A conceptual model of the relationship between channel characteristics and discharge', *Earth Surface Processes*, 4, 37–42.

Pickup, G. and Warner, R. F. (1976) 'Effects of hydrologic regime on magnitude and frequency of dominant discharge', *Journal of Hydrology*, 29, 51–75.

Pierce, R. S., Hornbeck, J. W., Likens, G. E. and Bormann, F. H. (1973) 'Effect of elimination of vegetation on stream water quantity and quality', in *Results of research on representative and experimental basins, IASH/UNESCO, Proceedings of the Wellington Symposium, 1970*, 1, pp. 311–28.

Pinder, G. F. and Jones, J. F. (1969) 'Determination of the groundwater component of peak discharge from the chemistry of total runoff', *Water Resources Research*, 5, 438–45.

Pittman, E. D. and Ovenshine, A. T. (1968) 'Pebble morphology in the Merced River (California)', *Sedimentary Geology*, 2, 125–40.

Pitty, A. F. (1968) 'The scale and significance of solutional loss from the limestone tract of the southern Pennines', *Proceedings of the Geologists' Association*, 79, 153–77.

Pizzuto, J. E. (1984) 'Equilibrium bank geometry and the width of shallow streams', *Earth Surface Processes and Landforms*, 9, 199–207.

Playfair, J. (1802) *Illustration of the Huttonian Theory of the Earth*, Edinburgh.

Plumley, W. J. (1948) 'Black Hills terrace gravels: a study in sediment transport', *Journal of Geology*, 56, 526–78.

Poston, T. and Stewart, I. (1978) *Catastrophe Theory and its Applications*, London, Pitman.

Potter, P. E. (1955) 'The petrology and origin of the Lafayette gravel. Part 1: mineralogy and petrology', *Journal of Geology*, 63, 1–38.

Potter, W. D. (1958) 'Upper and lower frequency curves for peak rates of runoff', *Transactions of the American Geophysical Union*, 39, 100–5.

Potter, W. D., Stovicek, F. K. and Woo, D. C. (1968) 'Flood frequency and channel cross-section of a small natural stream', *Bulletin of the International Association of Scientific Hydrology*, 13, 66–76.

Powell, K. E. C. (1978) 'Weed growth – a factor of channel roughness', in Herschy, R. W. (ed.) *Hydrometry*, Chichester, Wiley, pp. 327–52.

Prandtl, L. (1952) *Essentials of Fluid Dynamics*, London, Blackie & Sons.

Prasad, R. (1970) 'Numerical methods of calculating flow profiles', *Journal of the Hydraulics Division, American Society of Civil Engineers*, 96, 75–86.

Price, R. K. (1973) 'Flood routing methods for British rivers', *Proceedings of the Institution of Civil Engineers*, 55, 913–30.

Prus-Chacinski, T. (1954) 'Patterns of motion in open channel bends', *Publication. International Association of Scientific Hydrology*, 38, 3, 311–18.

Putnam, W. C. (1942) 'Geomorphology of the Ventura region, California', *Bulletin of the Geological Society of America*, 53, 691–754.

Rainwater, F. H. and Guy, H. P. (1961) 'Some observations on the hydrochemistry and sedimentation of the Chamberlin Glacier area, Alaska', *Professional Paper. United States Geological Survey*, 414C.

Rainwater, F. H. and Thatcher, L. L. (1960) 'Methods for collection and analysis of water samples', *Water Supply Paper. United States Geological Survey*, 1454.

Rana, S. A., Simons, D. B. and Mahmood, K. (1973) 'Analysis of sediment sorting in alluvial channels', *Journal of the Hydraulics Division, American Society of Civil Engineers*, 99, 1967–80.

Rango, A. (1970) 'Possible effects of precipitation modification on stream channel geometry and sediment yield', *Water Resources Research*, 6, 1765–70.

Rathbun, R. E. and Nordin, C. F. (1971) 'Tracer studies of sediment transport processes', *Journal of the Hydraulics Division, American Society of Civil Engineers*, 97, 1305–16.

Raudkivi, A. J. (1963) 'Study of sediment ripple formation', *Journal of the Hydraulics Division, American Society of Civil Engineers*, 89, 15–33.

Raudkivi, A. J. (1967) 'Analysis of resistance in fluvial channels', *Journal of the Hydraulics Division, American Society of Civil Engineers*, 93, 73–84.

Reich, B. M. (1962) 'Soil conservation service design hydrographs', *The Civil Engineer in South Africa*, 4, 77–87.

Reich, B. M. (1970) 'Flood series compared to rainfall extremes', *Water Resources Research*, 6, 1655–67.

Reid, I., Frostick, L. E. and Layman, J. T. (1985) 'The incidence and nature of bedload transport during flood flows in coarse-grained alluvial channels', *Earth Surface Processes and Landforms*, 10, 33–44.

Reid, I., Layman, J. T. and Frostick, L. E. (1980) 'The continuous measurement of bedload discharge', *Journal of Hydraulic Research*, 18, 243–9.

Rendon-Herrero, O. (1974) 'Estimation of washload produced on certain small watersheds', *Journal of the Hydraulics Division, American Society of Civil Engineers*, 100, 835–48.

Reynolds, O. (1883) 'An experimental investigation of the circumstances which determine whether the motion of water shall be direct or sinuous, and of the laws of resistance in parallel channels', *Philosophical Transactions of the Royal Society*, 174, 935–82.

Reynolds, R. C. (1971) 'Analysis of Alpine waters by ion electrode methods', *Water Resources Research*, 7, 1333–7.

Rhodes, D. D. (1977) 'The *b–f–m* diagram: graphical representation and interpretation of at-a-station hydraulic geometry', *American Journal of Science*, 277, 73–96.

Richards, K. S. (1972) 'Meanders and valley slope', *Area*, 4, 288–90.

Richards, K. S. (1973) 'Hydraulic geometry and channel roughness – a non-linear system', *American Journal of Science*, 273, 877–96.

Richards, K. S. (1976a) 'Complex width–discharge relations in natural river sections', *Bulletin of the Geological Society of America*, 87, 199–206.

Richards, K. S. (1976b) 'The morphology of riffle–pool sequences', *Earth Surface Processes*, 1, 71–88.

Richards, K. S. (1977a) 'Slope form and basal stream relationships: some further comments', *Earth Surface Processes*, 2, 87–95.

Richards, K. S. (1977b) 'Channel and flow geometry: a geomorphological perspective', *Progress in Physical Geography*, 1, 65–102.

Richards, K. S. (1978a) 'Simulation of flow geometry in a riffle–pool stream', *Earth Surface Processes*, 3, 345–54.

Richards, K. S. (1978b) 'Channel geometry in the riffle–pool sequence', *Geografiska Annaler*, 60A, 23–7.

Richards, K. S. (1979) 'Channel adjustment to sediment pollution by the china clay industry in Cornwall, England', in Rhodes, D. D. and Williams, G. P. (eds) *Adjustments of the Fluvial System*, Dubuque, Iowa, Kendall-Hunt, pp. 309–31.

Richards, K. S. (1980) 'A note on changes in channel geometry at tributary junctions', *Water Resources Research*, 16, 241–4.

Richards, K. S. (1981) 'Evidence of Flandrian valley alluviation in Staindale, North York Moors', *Earth Surface Processes and Landforms*, 6, 183–6.

Richards, K. S. (1984) 'Some observations on suspended sediment dynamics in Storbregrova, Jotunheimen', *Earth Surface Processes and Landforms*, 9, 101–12.

Richards, K. S. and Milne, L. (1979) 'Problems in the calibration of an acoustic device for the observation of bedload transport', *Earth Surface Processes*, 4, 335–46.

Richardson, E. V., Simons, D. B. and Haushild, W. L. (1962) 'Boundary form and resistance to flow in alluvial channels', *Bulletin of the International Association of Scientific Hydrology*, 7, 48–52.

Richardson, E. V., Simons, D. B. and Posakony, G. J. (1961) 'Sonic depth sounder for laboratory and field use', *Circular. United States Geological Survey*, 450.

Riggs, H. C. (1973) 'Regional analysis of streamflow characteristics', *Techniques of Water-Resource Investigations of the United States Geological Survey*, chap. B3, book 4.

Riley, S. J. (1969) 'A simplified levelling instrument: the A-frame', *Earth Science Journal*, 3, 51–2.

Riley, S. J. (1972) 'A comparison of morphometric measures of bankfull', *Journal of Hydrology*, 17, 23–31.

Riley, S. J. and Taylor, G. (1978) 'The geomorphology of the upper Darling River system with special reference to the present fluvial system', *Proceedings of the Royal Society of Victoria*, 90, 89–102.

Robinson, A. M. (1976) 'The effects of urbanization on stream channel morphology', *Proceedings of the National Symposium on Urban Hydrology, Hydraulics and Sediment Control*, University of Kentucky, pp. 115–27.

Robinson, A. R. and Albertson, M. L. (1952) 'Artificial roughness standard for open channels', *Transactions of the American Geophysical Union*, 33, 881–8.

Rockwell, T. K., Johnson, D. L., Keller, E. A. and Dembroff, G. R. (1985) 'A late Pleistocene-Holocene soil chronosequence in the central Ventura basin, Southern California, USA', in Richards, K. S., Arnett, R. R. and Ellis, S. (eds) *Geomorphology and Soils*, London, Allen & Unwin, 309–27.

Rodda, J. C. (1969) 'The flood hydrograph', in Chorley, R. J. (ed.) *Water, Earth and Man*, London, Methuen, pp. 405–18.

Rodda, J. C. (1976) 'Basin studies', in Rodda, J. C. (ed.) *Facets of Hydrology*, Chichester, Wiley, pp. 257–97.

Roehl, J. W. (1962) 'Sediment source areas, delivery ratios and influencing morphological factors', *Publication. International Association of Scientific Hydrology*, 59, 202–13.

Rogers, W. F. (1972) 'New concept in hydrograph analysis', *Water Resources Research*, 8, 973–81.

Rotnicki, K. and Borówka, R. K. (1985) 'Definition of subfossil meandering palaeochannels', *Earth Surface Processes and Landforms*, 10, 215–25.

Royse, C. F. (1968) 'Recognition of fluvial environments by particle-size characteristics', *Journal of Sedimentary Petrology*, 38, 1171–8.

Rozovskii, I. L. (1961) *Flow of Water in Bends of Open Channels*, Jerusalem, Israel Program for Scientific Translations.

Rubey, W. W. (1933a) 'Settling velocities of gravel, sand and silt particles', *American Journal of Science*, 225, 325–38.

Rubey, W. W. (1933b) 'Equilibrium conditions in debris-laden streams', *14th Annual Meeting, Transactions of the American Geophysical Union*, pp. 479–505.

Rubey, W. W. (1938) 'The force required to move particles on a stream bed', *Professional Paper. United States Geological Survey*, 189E, pp. 121–41.

Ruhe, R. V. (1952) 'Topographic discontinuities of the Des Moines lake', *American Journal of Science*, 250, 46–56.

Ruhe, R. V. (1971) 'Stream regimen and man's manipulation', in Coates, D. R. (ed.) *Environmental Geomorphology*, SUNY Binghamton, Publications in Geomorphology, pp. 9–23.

Russell, R. D. and Taylor, R. E. (1937) 'Roundness and shape of Mississippi river sands', *Journal of Geology*, 45, 225–67.

Rust, B. (1972) 'Pebble orientation in fluvial sediments', *Journal of Sedimentary Petrology*, 42, 384–8.

Salisbury, N. E. (1980) 'Thresholds and valley widths in the South River basin, Iowa', in Coates, D. R. and Vitek, J. D. (eds) *Thresholds in Geomorphology*, London, George Allen & Unwin, pp. 103–29.

Sangal, B. P. and Biswas, A. K. (1970) 'The 3-parameter log-normal distribution and its applications in hydrology', *Water Resources Research*, 6, 505–15.

Sartz, R. S. (1973) 'Effect of land use on the hydrology of small watersheds in south-western Wisconsin', in *Results of research on representative and experimental basins, IASH/UNESCO, Proceedings of the Wellington Symposium, 1970*, 1, pp. 286–95.

Sayre, W. W. and Chang, F. M. (1968) 'A laboratory investigation of open-channel dispersion processes for dissolved, suspended and floating dispersants', *Professional Paper. United States Geological Survey*, 433E.

Schaffernak, F. (1950) *Flussmorphologie und Flussbau*, Vienna, Springer-Verlag.

Scheidegger, A. E. (1966) 'Stochastic branching processes and the law of stream orders', *Water Resources Research*, 2, 199–203.

Schick, A. P. (1974) 'Formation and obliteration of desert stream terraces – a conceptual analysis', *Zeitschrift für Geomorphologie*, Supplement 21, 88–105.

Schofield, A. N. and Wroth, P. (1968) *Critical State Soil Mechanics*, London, McGraw-Hill.

Schoklitsch, A. (1933) 'Uber die Verkleinerung der Geschiebe der Flusslaufen', *Sitzungsberichte der Akademie der Wissenschaften in Wien*, 142, 343–66.

Schumm, S. A. (1954) 'The relation of drainage basin relief to sediment loss', *Publication. International Association of Scientific Hydrology*, 36, 216–19.

Schumm, S. A. (1956) 'The evolution of drainage systems and slopes in badlands at Perth Amboy, New Jersey', *Bulletin of the Geological Society of America*, 67, 597–646.

Schumm, S. A. (1960a) 'The shape of alluvial channels in relation to sediment type', *Professional Paper. United States Geological Survey*, 352B, pp. 17–30.

Schumm, S. A. (1960b) 'The effect of sediment type on the shape and stratification of some modern fluvial deposits', *American Journal of Science*, 258, 177–84.

Schumm, S. A. (1961a) 'The effect of sediment type on the shape and stratification of some modern fluvial deposits – a reply', *American Journal of Science*, 259, 234–9.

Schumm, S. A. (1961b) 'Dimensions of some stable alluvial channels', *Professional Paper. United States Geological Survey*, 424B, pp. 26–7.

Schumm, S. A. (1961c) 'Effect of sediment characteristics on erosion and deposition in ephemeral stream channels', *Professional Paper. United States Geological Survey*, 352C, pp. 31–70.

Schumm, S. A. (1963) 'Sinuosity of alluvial rivers on the Great Plains', *Bulletin of the Geological Society of America*, 74, 1089–100.

Schumm, S. A. (1968a) 'River adjustment to altered hydrologic regimen – Murrumbidgee River and palaeochannels, Australia, *Professional Paper. United States Geological Survey*, 598.

Schumm, S. A. (1968b) 'Speculations concerning palaeohydrologic controls of terrestrial sedimentation', *Bulletin of the Geological Society of America*, 79, 1573–88.

Schumm, S. A. (1969a) 'Geomorphic implications of climatic changes', in Chorley, R. J. (ed.) *Water, Earth and Man*, London, Methuen, pp. 525–34.

Schumm, S. A. (1969b) 'River metamorphosis', *Journal of the Hydraulics Division, American Society of Civil Engineers*, 95, 255–73.

Schumm, S. A. (1971) 'Fluvial geomorphology; channel adjustments and river metamorphosis', in Shen, H. W. (ed.) *River Mechanics*, vol. 1, Fort Collins, Colorado, pp. 5.1–5.22.

Schumm, S. A. (1973) 'Geomorphic thresholds and complex response of drainage systems', in Morisawa, M. (ed.) *Fluvial Geomorphology*, SUNY Binghamton, Publications in Geomorphology, pp. 299–309.

Schumm, S. A. (1975) 'Episodic erosion: A modification of the geomorphic cycle', in Melhorn, W. N. and Flemal, R. C. (eds) *Theories of Landform Development*, SUNY Binghamton, Publications in Geomorphology, pp. 69–85.

Schumm, S. A. (1977) *The Fluvial System*, New York, Wiley-Interscience.

Schumm, S. A. (1979) 'Geomorphic thresholds: the concept and its applications', *Transactions of the Institute of British Geographers, New Series*, 4, 485–515.

Schumm, S. A. and Hadley, R. F. (1957) 'Arroyos and the semi-arid cycle of erosion', *American Journal of Science*, 255, 161–74.

Schumm, S. A. and Khan, H. R. (1972) 'Experimental study of channel patterns', *Bulletin of the Geological Society of America*, 83, 1755–70.

Schumm, S. A. and Lichty, R. W. (1963) 'Channel widening and flood plain construction along Cimarron river in south-western Kansas', *Professional Paper. United States Geological Survey*, 352D, pp. 71–88.

Schumm, S. A. and Lichty, R. W. (1965) 'Time, space and causality in geomorphology', *American Journal of Science*, 263, 110–19.

Schumm, S. A. and Stevens, M. A. (1973) 'Abrasion in place: a mechanism for rounding and size reduction of coarse sediments in rivers', *Geology*, 1, 37–40.

Scobey, F. C. (1939) 'Flow of water in irrigation and similar canals', *Technical Bulletin. United States Department of Agriculture*, 652.

Scott, C. H. and Culbertson, J. K. (1971) 'Resistance to flow in flat-bed sand channels', *Professional Paper. United States Geological Survey*, 750B, pp. 254–8.

Shea, J. H. (1974) 'Deficiencies of clastic particles of certain sizes', *Journal of Sedimentary Petrology*, 44, 985–1003.

Shen, H. W. (1971a) 'Stability of alluvial channels', in Shen, H. W. (ed.) *River Mechanics*, vol. 1, Fort Collins, Colorado, pp. 16.1–16.33.

Shen, H. W. (1971b) 'Scour near piers', in Shen, H. W. (ed.) *River Mechanics*, vol. 2, Fort Collins, Colorado, pp. 23.1–23.25.

Shen, H. W. and Hung, C. S. (1972) 'An engineering approach to total bed-material load by regression analysis', in Shen, H. W. (ed.) *Sedimentation*, Fort Collins, Colorado, pp. 14.1–14.17.

Shen, H. W. and Komura, S. (1968) 'Meandering tendencies in straight alluvial channels', *Journal of the Hydraulics Division, American Society of Civil Engineers*, 94, 997–1046.

Shepherd, R. G. (1979) 'River channel and sediment responses to bedrock lithology and stream capture, Sandy Creek drainage, central Texas', in Rhodes, D. D. and Williams, G. P. (eds) *Adjustments of the Fluvial System*, Dubuque, Iowa, Kendall-Hunt, pp. 255–75.

Sherman, L. K. (1932) 'Streamflow from rainfall by the unit-graph method', *Engineering News Record*, 108, 501–5.

Shields, A. (1936) 'Anwendung der Aehnlichkeitsmechanik und der turbulenzforschung auf die geschiebebewegung', *Mitteilung der Preussischen versuchsanstalt fuer Wasserbau und Schiffbau*, Heft 26, Berlin.

Shoobert, J. (1968) 'Australian landform examples number 12. Underfit stream of the Osage type: head of the Port Hacking River', *Australian Geographer*, 10, 523–4.

Shotton, F. W. (1953) 'The Pleistocene deposits of the area between Coventry, Rugby and Leamington and their bearing upon the topographical development of the Midlands', *Philosophical Transactions of the Royal Society*, 237B, 209–60.

Shreve, R. L. (1966) 'Statistical law of stream numbers', *Journal of Geology*, 74, 17–37.

Shreve, R. L. (1967) 'Infinite topologically random channel networks', *Journal of Geology*, 75, 178–86.

Shreve, R. L. (1974) 'Variation of mainstream length with basin area in river networks', *Water Resources Research*, 10, 1167–77.

Shulits, S. (1936) 'Fluvial morphology in terms of slope, abrasion and bedload', *Transactions of the American Geophysical Union*, 17, 440–4.

Shulits, S. (1941) 'Rational equation of riverbed profile', *Transactions of the American Geophysical Union*, 22, 622–30.

Shulits, S. (1955) 'Graphical analysis of trend profile of a shortened section of river', *Transactions of the American Geophysical Union*, 36, 649–54.

Sigafoos, R. S. (1964) 'Botanical evidence of floods and floodplain deposition', *Professional Paper. United States Geological Survey*, 485A.

Simons, D. B. and Albertson, M. L. (1960) 'Uniform water-conveyance channels in alluvial material', *Journal of the Hydraulics Division, American Society of Civil Engineers*, 86, 33–71.

Simons, D. B. and Li, R. M. (1979) 'Analysis of watersheds and river systems', in Shen, H. W. (ed.) *Modeling of Rivers*, New York, Wiley, pp. 11.1–11.58.

Simons, D. B. and Richardson, E. V. (1962) 'The effect of bed roughness on depth-discharge relations in alluvial channels', *Water Supply Paper. United States Geological Survey*, 1498E.

Simons, D. B. and Richardson, E. V. (1966) 'Resistance to flow in alluvial channels', *Professional Paper. United States Geological Survey*, 422J.

Simons, D. B., Richardson, E. V. and Albertson, M. L. (1961) 'Flume studies using medium sand (0.45 mm), *Water Supply Paper. United States Geological Survey*, 1498A.

Simons, D. B., Richardson, E. V. and Haushild, W. L. (1963) 'Some effects of fine sediment on flow phenomena', *Professional Paper. United States Geological Survey*, 1498G.

Simons, D. B., Richardson, E. V. and Nordin, C. F. (1965) 'Bedload equation for ripples and dunes', *Professional Paper. United States Geological Survey*, 462H.

Simons, D. B. and Şentürk, F. (1977) *Sediment Transport Technology*, Fort Collins, Colorado, Water Resources Publications.

Simons, D. B., Stevens, M. A. and Duke, J. H., Jr. (1973) 'Predicting stages on sand-bed rivers', *Journal of Waterways, Harbours and Coastal Engineering Division, American Society of Civil Engineers*, 99, 231–44.

Singh, K. P. and Sinclair, R. A. (1972) 'Two-distribution method for flood-frequency analysis', *Journal of the Hydraulics Division, American Society of Civil Engineers*, 98, 29–44.

Sissons, J. B. (1979) 'The later lakes and associated fluvial terraces of Glen Roy, Glen Spean and vicinity', *Transactions of the Institute of British Geographers, New Series*, 4, 12–29.

Sissons, J. B. and Smith, D. E. (1965) 'Raised shorelines associated with the Perth readvance in the Forth valley and their relation to glacial isostasy', *Transactions of the Royal Society of Edinburgh*, 66, 143–68.

Skempton, A. W. and Bishop, A. W. (1950) 'The measurement of shear strength of soils', *Geotechnique*, 2, 90–118.

Skibinski, J. (1967) 'Bedload transport at flood time', *Publication. International Association of Scientific Hydrology*, 75, 41–7.

Slaymaker, H. O. (1972) 'Patterns of present sub-aerial erosion and landforms in mid-Wales', *Transactions of the Institute of British Geographers*, 55, 47–67.

Small, R. J. (1973) 'Braiding terraces in the Val d'Herens, Switzerland', *Geography*, 58, 129–35.

Smalley, I. J. and Vita-Finzi, C. (1969) 'The concept of "system" in the earth sciences', *Bulletin of the Geological Society of America*, 80, 1591–4.

Smart, J. S. (1978) 'The analysis of drainage network composition', *Earth Surface Processes*, 3, 129–70.

Smart, J. S. and Werner, C. (1976) 'Applications of the random model of drainage basin composition', *Earth Surface Processes*, 1, 219–33.

Smith, D. D. and Wischmeier, W. H. (1957) 'Factors affecting sheet and rill erosion', *Transactions of the American Geophysical Union*, 39, 285–91.

Smith, D. G. (1973) 'Aggradation of the Alexandra–North Saskatchewan River, Banff Park, Alberta', in Morisawa, M. (ed.) *Fluvial Geomorphology*, SUNY Binghamton, Publications in Geomorphology, pp. 201–19.

Smith, D. G. (1976) 'Effect of vegetation on lateral migration of anastomosed channels of a glacier meltwater river', *Bulletin of the Geological Society of America*, 87, 857–60.

Smith, N. D. (1970) 'The braided stream depositional environment: comparison of the Platte River with some Silurian clastic rocks, N. Central Appalachians', *Bulletin of the Geological Society of America*, 81, 2993–3014.

Smith, N. D. (1974) 'Sedimentology and bar formation in the Upper Kicking Horse River, a braided outwash stream', *Journal of Geology*, 82, 205–23.

Smith, T. R. (1974) 'A derivation of the hydraulic geometry of steady-state channels from conservation principles and sediment transport laws', *Journal of Geology*, 82, 98–104.

Smith, T. R. and Dunne, T. (1977) 'Watershed geochemistry: the control of aqueous solutions by soil materials in a small watershed', *Earth Surface Processes*, 2, 421–6.

Sneed, E. D. and Folk, R. L. (1958) 'Pebbles in the Lower Colorado River, Texas: a study in particle morphogenesis', *Journal of Geology*, 66, 114–50.

Snyder, F. F. (1938) 'Synthetic unit graphs', *Transactions of the American Geophysical Union*, 19, 447–54.

Soni, J. P., Garde, R. J. and Ranga Raju, K. G. (1980) 'Aggradation in streams due to overloading', *Journal of the Hydraulics Division, American Society of Civil Engineers*, 106, 117–32.

Sopper, W. E. and Lynch, J. A. (1973) 'Changes in water yield following partial forest cover removal on an experimental watershed', in *Results of research on representative and experimental basins, IASH/UNESCO, Proceedings of the Wellington Symposium, 1970*, 1, pp. 369–89.

Speight, J. G. (1965a) 'Meander spectra of the Angabunga River', *Journal of Hydrology*, 3, 1–15.

Speight, J. G. (1965b) 'Flow and channel characteristics of the Angabunga River, Papua', *Journal of Hydrology*, 3, 16–36.

Spencer, D. W. (1963) 'The interpretation of grain size distribution curves of clastic sediments', *Journal of Sedimentary Petrology*, 33, 180–90.

Starkel, L. (1966) 'Post-glacial climate and the moulding of European relief', in *World Climate from 8000 to 0 BC*, London, Royal Meteorological Society, pp. 15–33.

Starkel, L. and Thornes, J. B. (eds) (1981) *Palaeohydrology of river basins. Guide to subproject A of IGCP project No. 158*, British Geomorphology Research Group Technical Bulletin 28, 107.

Sternberg, H. (1875) 'Untersuchungen über Längen- und Querprofil Geschiebeführende Flüsse', *Zeitschrift für Bauwesen*, 25, 483–506.

Stevens, M. A. and Simons, D. B. (1971) 'Stability analysis for coarse granular material on slopes', in Shen, H. W. (ed.) *River Mechanics*, vol. 1, Fort Collins, Colorado, pp. 17.1–17.27.

Stevens, M. A., Simons, D. B. and Richardson, E. V. (1975) 'Non-equilibrium river form', *Journal of the Hydraulics Division, American Society of Civil Engineers*, 101, 557–66.

Stewart, J. H. and Lamarche, V. C. (1967) 'Erosion and deposition produced by the floods of December 1964 on Coffee Creek, Trinity County, California', *Professional Paper. United States Geological Survey*, 422K.

Stoddart, D. R. (1978) 'Geomorphology in China', *Progress in Physical Geography*, 2, 187–236.

Stokes, G. G. (1851) 'On the effect of the internal friction of fluids on the motion of pendulums', *Transactions of the Cambridge Philosophical Society*, 9, 51–2.

Strahler, A. N. (1950) 'Equilibrium theory of erosional slopes approached by frequency distribution analysis', *American Journal of Science*, 248, 673–96, 800–14.

Strahler, A. N. (1952) 'Hypsometric (area-altitude) analysis of erosional topography', *Bulletin of the Geological Society of America*, 63, 1117–42.

Strahler, A. N. (1956) 'The nature of induced erosion and aggradation', in Thomas, W. L. (ed.) *Man's Role in changing the Face of the Earth*, Chicago, University of Chicago Press, pp. 621–38.

Strand, R. I. (1971) *Turkey Creek channel stability studies, Farwell Unit, middle Loup division, Pick–Sloan Missouri basin program*, Bureau of Reclamation, Denver, Colorado, Engineering and Research Centre, REC-ERC-71-9.

Strickler, A. (1923) 'Beiträge zur Frage der Geschwindigheitsformel und der Rauhigkeitszahlen für Strome, Kanale und Geschlossene Leitungen', *Mitteilungen des Eidgenössischer Amtes für Wasserwirtschaft*, Bern, Switzerland, 16g.

Stringham, G. E., Simons, D. B. and Guy, H. P. (1969) 'The behavior of large particles falling in quiescent liquids', *Professional Paper. United States Geological Survey*, 562C.

Stuart, T. A. (1953) *Spawning migration, reproduction, and young stages of loch trout*, Freshwater and Salmon Fisheries Research Station, Edinburgh, HMSO.

Sundborg, A. (1956) 'The river Klarälven, a study of fluvial processes', *Geografiska Annaler*, 38, 127–316.

Sundborg, A. (1967) 'Some aspects of fluvial sediments and fluvial morphology', *Geografiska Annaler*, 49A, 333–43.

Surkan, A. J. (1974) 'Simulation of storm velocity effects on flow from distributed channel networks', *Water Resources Research*, 10, 1149–60.

Tanner, W. F. (1971) 'The river profile', *Journal of Geology*, 79, 482–92.

Teisseyre, A. K. (1977) 'Pebble clusters as a directional structure in fluvial gravels: modern and ancient examples', *Geologia Sudetica*, 12, 79–89.

Thakur, T. R. and Scheidegger, A. E. (1970) 'Chain model of river meanders', *Journal of Hydrology*, 12, 25–47.

Thorne, C. R. and Lewin, J. (1979) 'Bank processes, bed material movement and planform development in a meandering river', in Rhodes, D. D. and Williams, G. P. (eds) *Adjustments of the Fluvial System*, Dubuque, Iowa, Kendall-Hunt, pp. 117–37.

Thornes, J. B. (1970) 'The hydraulic geometry of stream channels in the Xingu–Araguaia headwaters', *Geographical Journal*, 136, 376–82.

Thornes, J. B. (1977) 'Channel changes in ephemeral streams: observations, problems and models', in Gregory, K. J. (ed.) *River Channel Changes*, Chichester, Wiley, pp. 317–35.

Tiefenbrun, A. J. (1965) 'Bank stabilization of Mississippi River between the Ohio and Missouri Rivers', *Proceedings of the Federal Inter-Agency Sedimentation Conference, 1963*, United States Department of Agriculture Publication 970, pp. 387–99.

Tinkler, K. J. (1970) 'Pools, riffles and meanders', *Bulletin of the Geological Society of America*, 81, 2.

Tinkler, K. J. (1971) 'Active valley meanders in south-central Texas and their wider implications', *Bulletin of the Geological Society of America*, 82, 1873–99.

Toler, L. G. (1965) 'Relation between chemical quality and water discharge in Spring Creek, South western Georgia', *Professional Paper. United States Geological Survey*, 525C, pp. 209–13.

Tovey, N. K. (1978) 'A scanning electron microscope study of the liquefaction potential of sand grains', in Whalley, W. B. (ed.) *Scanning Electron Microscopy in the Study of Sediments*, Norwich, Geo Abstracts, pp. 83–93.

Tracy, H. J. and Lester, C. M. (1961) 'Resistance coefficients and velocity distribution; smooth rectangular channel', *Water Supply Paper. United States Geological Survey*, 1592A.

Tricart, J. and Schaeffer, R. (1950) 'L'indice d'émoussée des galets', *Revue de Geomorphologie Dynamique*, 1, 151–79.

Tricart, J. and Vogt, H. (1967) 'Quelques aspects du transport des alluvions grossieres et du façonnement des lits fluviaux', *Geografiska Annaler*, 49A, 351–66.

Trimble, S. W. (1975) 'Denudation studies: Can we assume stream steady state?', *Science*, 188, 1207–8.

Trimble, S. W. (1977) 'The fallacy of stream equilibrium in contemporary denudation studies', *American Journal of Science*, 277, 876–87.

Troeh, F. R. (1965) 'Landform equations fitted to contour maps', *American Journal of Science*, 263, 616–27.

Truhlar, J. F. (1978) 'Determining suspended sediment loads from turbidity records', *Hydrological Sciences Bulletin*, 23, 409–17.

Twidale, C. R. (1964) 'Erosion of an alluvial bank at Birdwood, South Australia', *Zeitschrift für Geomorphologie*, 8, 189–211.

Tywoniuk, N. and Warnock, R. G. (1973) 'Acoustic detection of bedload transport', *Proceedings of the Hydrological Symposium No. 9, National Research Council of Canada*, pp. 728–35.

Ursic, S. J. and Dendy, F. E. (1965) 'Sediment yields from small watersheds under various land uses and forest covers', *Proceedings of the Federal Inter-Agency Sedimentation Conference, 1963*, United States Department of Agriculture, Publication 970, pp. 47–52.

Van den Burg, B. (1975) 'A three-dimensional law of the wall for turbulent shear flows', *Journal of Fluid Mechanics*, 70, 149–60.

Van Denburgh, A. S. and Feth, J. H. (1965) 'Solute erosion and chloride balance in selected river basins of the western conterminous United States', *Water Resources Research*, 1, 537–41.

Vanoni, V. A. (1946) 'Transportation of suspended sediment by water', *Transactions of the American Society of Civil Engineers*, 3, 67–133.

Vanoni, V. A., Brooks, N. H. and Kennedy, J. F. (1961) 'Lecture notes on sediment transportation and channel stability', *W. M. Keck Laboratory of Hydraulics and Water Resources, Caltech*, Report No. KH-R-1.

Vanoni, V. A. and Hwang, L. S. (1967) 'Relation between bedforms and friction in streams', *Journal of the Hydraulics Division, American Society of Civil Engineers*, 93, 121–44.

Vanoni, V. A. and Nomicos, G. N. (1960) 'Resistance properties of sediment laden streams', *Transactions of the American Society of Civil Engineers*, 125, 1140–67.

Visher, G. S. (1969) 'Grain size distributions and depositional processes', *Journal of Sedimentary Petrology*, 39, 1074–106.

Walker, P. H., Woodyer, K. D. and Hutka, J. (1974) 'Particle-size measurements by Coulter Counter of very small deposits and low suspended sediment concentrations in streams', *Journal of Sedimentary Petrology*, 44, 673–9.

Walling, D. E. (1971) 'Streamflow from instrumented catchments in south-east Devon', in Gregory, K. J. and Ravenhill, W. L. D. (eds) *Exeter Essays in Geography*, Exeter University, pp. 55–81.

Walling, D. E. (1974) 'Suspended sediment and solute yields from a small catchment prior to urbanization', in Gregory, K. J. and Walling, D. E. (eds) *Fluvial Processes in Instrumented Watersheds*, Institute of British Geographers Special Publication 6, pp. 169–92.

Walling, D. E. (1975) 'Solute variations in small catchment streams: some comments', *Transactions of the Institute of British Geographers*, 64, 141–7.

Walling, D. E. (1977) 'Limitations of the rating curve technique for estimating suspended sediment loads, with particular reference to British rivers', *Publication. International Association for Scientific Hydrology*, 122, 34–48.

Walling, D. E. (1979) 'The hydrological impact of building activity: a study near Exeter', in Hollis, G. E. (ed.) *Man's Impact on the Hydrological Cycle in the United Kingdom*, Norwich, Geobooks, pp. 135–51.

Walling, D. E. and Foster, I. D. L. (1975) 'Variations in the natural chemical concentration of river water during flood flows, and the lag effect: some further comments', *Journal of Hydrology*, 26, 237–44.

Walling, D. E. and Teed, A. (1971) 'A simple pumping sampler for research into suspended sediment transport in small catchments', *Journal of Hydrology*, 13, 325–37.

Walling, D. E. and Webb, B. W. (1975) 'Spatial variation of river water quality: a survey of the River Exe', *Transactions of the Institute of British Geographers*, 65, 155–71.

Ward, P. R. B. and Wurzel, F. (1968) 'The measurement of river flow with radioactive isotopes', *Bulletin of the International Association of Scientific Hydrology*, 13, 40–9.

Watson, R. A. (1969) 'Explanation and prediction in geology', *Journal of Geology*, 77, 488–94.

Wentworth, C. K. (1919) 'A laboratory and field study of cobble abrasion', *Journal of Geology*, 27, 507–21.

Werner, C. and Smart, J. S. (1973) 'Some new methods of topologic classification of channel networks', *Geographical Analysis*, 5, 271–95.

Werritty, A. (1972) 'Accuracy of stream link lengths derived from maps', *Water Resources Research*, 8, 1255–64.

Werritty, A. and Ferguson, R. I. (1980) 'Pattern changes in a Scottish braided river over 1, 30 and 200 years', in Cullingford, R. A., Davidson, D. A. and Lewin, J. (eds) *Timescales in Geomorphology*, Chichester, Wiley, pp. 53–68.

Weyman, D. R. (1970) 'Throughflow on slopes and its relation to the stream hydrograph', *Bulletin of the International Association of Scientific Hydrology*, 15, 25–33.

Weyman, D. R. (1973) 'Measurement of downslope flow of water in a soil', *Journal of Hydrology*, 20, 267–88.

Weyman, D. R. (1975) *Runoff Processes and Streamflow Modelling*, Oxford, Oxford University Press.

Whalley, W. B. (1976) *Properties of Materials and Geomorphological Explanation*, Oxford, Oxford University Press.

Wheeler, D. A. (1979) 'The overall shape of longitudinal profiles of streams', in Pitty, A. F. (ed.) *Geographical Approaches to Fluvial Processes*, Norwich, Geobooks, pp. 241–60.

Whipkey, R. Z. (1965) 'Subsurface stormflow from forested slopes', *Bulletin of the International Association of Scientific Hydrology*, 10, 74–85.

White, C. M. (1940) 'The equilibrium of grains on the bed of a stream', *Proceedings of the Royal Society*, 174A, 322–38.

White, S. J. (1970) 'Plane bed thresholds of fine-grained sediments', *Nature*, 228, 152–3.

Wilcock, D. N. (1967) 'Coarse bedload as a factor determining bed slope', *Publication. International Association of Scientific Hydrology*, 75, 143–50.

Wilcock, D. N. (1971) 'Investigation into the relations between bedload transport and channel shape', *Bulletin of the Geological Society of America*, 82, 2159–76.

Williams, G. P. (1978a) 'Bankfull discharge of rivers', *Water Resources Research*, 14, 1141–54.

Williams, G. P. (1978b) 'Hydraulic geometry of river cross sections – theory of minimum variance', *Professional Paper. United States Geological Survey*, 1029.

Williams, G. P. (1984) 'Palaeohydrological methods and some examples from Swedish fluvial environments. II – River meanders', *Geografiska Annaler*, 66A, 89–102.

Williams, P. B. and Kemp, P. H. (1971) 'Initiation of ripples on flat sediment beds', *Journal of the Hydraulics Division, American Society of Civil Engineers*, 97, 505–22.

Williams, P. F. and Rust, B. R. (1969) 'The sedimentology of a braided river', *Journal of Sedimentary Petrology*, 39, 649–79.

Williams, R. B. G. and Robinson, D. A. (1981) 'Weathering of sandstone by the combined action of frost and salt', *Earth Surface Processes*, 6, 1–9.

Wilson, L. (1973) 'Variations in mean annual sediment yield as a function of mean annual precipitation', *American Journal of Science*, 273, 335–49.

Winkley, B. R. (1971) 'Practical aspects of river regulation and control', in Shen, H. W. (ed.) *River Mechanics*, vol. 1, Fort Collins, Colorado, pp. 19.1–19.79.

Woldenberg, M. J. (1971) 'A structural taxonomy of spatial hierarchies', in Chisholm, M., Frey, A. and Haggett, P. (eds) *Regional Forecasting*, London, Butterworths, pp. 147–75.

Wolfenden, P. J. and Lewin, J. (1977) 'Distribution of metal pollutants in floodplain sediments', *Catena*, 4, 309–17.

Wolfenden, P. J. and Lewin, J. (1978) 'Distribution of metal pollutants in active stream sediments', *Catena*, 5, 67–78.

Wolman, M. G. (1954) 'A method of sampling coarse bed material', *Transactions of the American Geophysical Union*, 35, 951–6.

Wolman, M. G. (1955) 'The natural channel of Brandywine Creek, Pennsylvania', *Professional Paper. United States Geological Survey*, 271.

Wolman, M. G. (1959) 'Factors influencing erosion of a cohesive river bank', *American Journal of Science*, 257, 204–16.

Wolman, M. G. (1967) 'A cycle of sedimentation and erosion in urban river channels', *Geografiska Annaler*, 49A, 385–95.

Wolman, M. G. and Brush, L. M. (1961) 'Factors controlling the size and shape of stream channels in coarse, non-cohesive sands', *Professional Paper. United States Geological Survey*, 282G, pp. 183–210.

Wolman, M. G. and Eiler, J. P. (1958) 'Reconaissance study of erosion and deposition produced by the flood of August 1955 in Connecticut', *Transactions of the American Geophysical Union*, 39, 1–14.

Wolman, M. G. and Gerson, R. (1978) 'Relative scales of time and effectiveness of climate in watershed geomorphology', *Earth Surface Processes*, 3, 189–208.

Wolman, M. G. and Leopold, L. B. (1957) 'River flood plains: some observations on their formation', *Professional Paper. United States Geological Survey*, 282C, pp. 87–107.

Wolman, M. G. and Miller, J. P. (1960) 'Magnitude and frequency of forces in geomorphic processes', *Journal of Geology*, 68, 54–74.

Wolman, M. G. and Schick, A. P. (1967) 'Effects of construction on fluvial sediment: urban and suburban areas of Maryland', *Water Resources Research*, 3, 451–62.

Womack, W. R. and Schumm, S. A. (1977) 'Terraces of Douglas Creek, north-western Colorado: an example of episodic erosion', *Geology*, 5, 72–6.

Wong, S. T. (1963) 'A multivariate statistical model for predicting mean annual flood in New England', *Annals of the Association of American Geographers*, 53, 298–311.

Wood, P. A. (1978) 'Fine sediment mineralogy of source rocks and suspended sediment, Rother catchment, West Sussex', *Earth Surface Processes*, 3, 255–63.

Woodford, A. O. (1951) 'Stream gradients and Monterey sea valley', *Bulletin of the Geological Society of America*, 62, 799–852.

Woodyer, K. D. (1968) 'Bankfull frequency in rivers', *Journal of Hydrology*, 6, 114–42.

Yalin, M. S. (1963) 'An expression for bed-load transportation', *Journal of the Hydraulics Division, American Society of Civil Engineers*, 89, 221–51.

Yalin, M. S. (1971) 'On the formation of dunes and meanders', *Proceedings of the 14th International Congress of the Hydraulic Research Association, Paris*, 3, C 13, pp. 1–8.

Yalin, M. S. (1977) *Mechanics of Sediment Transport*, Oxford, Pergamon.

Yalin, M. S. and Finlayson, G. D. (1972) 'On the velocity distribution of the flow carrying sediment in suspension', in Shen, H. W. (ed.) *Sedimentation*, Fort Collins, Colorado, pp. 8.1–8.18.

Yang, C. T. (1971a) 'Potential energy and stream morphology', *Water Resources Research*, 7, 311–22.

Yang, C. T. (1971b) 'Formation of riffles and pools', *Water Resources Research*, 7, 1567–74.

Yang, C. T. (1971c) 'On river meanders', *Journal of Hydrology*, 13, 231–53.

Yang, C. T. (1972) 'Unit stream power and sediment transport', *Journal of the Hydraulics Division, American Society of Civil Engineers*, 98, 1805–26.

Yang, C. T. (1973) 'Incipient motion and sediment transport', *Journal of the Hydraulics Division, American Society of Civil Engineers*, 99, 1679–704.

Yang, C. T. and Stall, J. B. (1976) 'Applicability of the unit stream power equation', *Journal of the Hydraulics Division, American Society of Civil Engineers*, 102, 559–68.

Yatsu, E. (1955) 'On the longitudinal profile of the graded river', *Transactions of the American Geophysical Union*, 36, 655–63.

Yorke, T. H. and Davis, W. J. (1971) 'Effects of urbanization on sediment transport in Bel Pre Creek basin, Maryland', *Professional Paper. United States Geological Survey*, 750B, pp. 218–23.

Zeller, J. (1967) 'Meandering channels in Switzerland', *Publication. International Association of Scientific Hydrology*, 75, 174–86.

Zimmerman, R. C., Goodlett, J. C. and Comer, G. H. (1967) 'The influence of vegetation on channel form of small streams', *Publication. International Association of Scientific Hydrology*, 75, pp. 255–75.

Zingg, T. (1935) 'Beitrag zur Schotteranalyse', *Schweizerische Mineralogische und Petrographische Mitteilungen*, 15, 39–140.

Subject index

Index of rivers and streams

Printed and bound by CPI Group (UK) Ltd, Croydon, CR0 4YY

17/10/2024

01775656-0016